Backflow Prevention Theory and Practice

Third Edition

Robin L. Ritland

Les O'Brien

University of Florida

Office of Professional and Workforce Development

Center for Training, Research and Education for Environmental Occupations (TREEO)

Cover photos courtesy of the Center for Training, Research and Education
for Environmental Occupations, University of Florida

Copyright © 1990, 2004, 2019 by the Center for Training, Research and Education
for Environmental Occupations Office of Professional and Workforce Development
University of Florida

ISBN 978-1-5249-9258-3

Kendall/Hunt Publishing Company has the exclusive rights to reproduce this work,
to prepare derivative works from this work, to publicly distribute this work,
to publicly perform this work and to publicly display this work.

All rights reserved. No part of this publication may be reproduced,
stored in a retrieval system, or transmitted, in any form or by any
means, electronic, mechanical, photocopying, recording, or otherwise,
without the prior written permission of Kendall/Hunt Publishing Company.

Published in the United States of America

TABLE OF CONTENTS

CHAPTER ONE: INTRODUCTION TO BACKFLOW PREVENTION 1

Introduction.. 1
    Backflow/Cross-Connections... 1
    Non-potable Source/Contaminant/Pollutant 2
    Backpressure/Backsiphonage.. 3
    Public Health Significance.. 4
    Backflow Prevention... 6

Purpose and Scope 8
Summary... 8
Chapter One Review... 9

CHAPTER TWO: THE HAZARDS OF BACKFLOW 11

    Biological Pollution/Contamination 11
    Backflows Short-circuit the Water Treatment Process.................. 13
    Public Health Significance.. 14
    Monetary Costs of Biological Contamination................................. 14
    Biological Hazards: Case Histories 15
    Factors That Affect the Magnitude of a Biological Backflow Incident 16
    Chemical Pollution/Contamination... 18
    Public Health Significance... 21
    Monetary Cost of Chemical Contaminants 21
    Chemical Hazards: Case Histories 22
    Factors That Affect the Magnitude of a Chemical Backflow Incident..... 26

Summary... 27
Chapter Two Review... 29

CHAPTER THREE: LAWS AND RESPONSIBILITY 31

Laws.. 31
    Federal Regulations 31
    State Regulations 33
    Regulations in Other States.. 35
    Plumbing Codes 35

Responsibilities and Liabilities.. 36
    Liabilities and Damages 36

TABLE OF CONTENTS

Individual Responsibilities 39
- Water Purveyors 39
- Consumer 43
- Regulatory Officials 44
- Plumbing Inspector 45
- Consulting Engineers 45
- Contractors, Plumbers, and Testers 45
- Code of Conduct 46
- Others with Some Level of Responsibility 46

Summary 47
Chapter Three Review 48

CHAPTER FOUR: FUNDAMENTALS OF BACKFLOW 51

Cross-Connections 51
- Types of Cross-Connections 51
- Why Cross-Connections Are Created 53

Understanding Backflow 59
- Pressure Principles 59
- Water: Under Static Conditions 60
- Atmospheric Pressure 64
- Water Movement 66
- Effects of the Weight of Water 66
- Effects of Atmospheric Pressure 68
- Effects of Water Temperature 70
- Effects of Water Velocity 71
- Mechanical Devices 73
- Backpressure 74
- Backsiphonage 76

Summary 78
Chapter Four Review 79

CHAPTER FIVE: METHODS AND MECHANISMS FOR PREVENTING BACKFLOW 81

Devices and Approved Assemblies 81
Mechanical Backflow Preventers 84
- Atmospheric Vacuum Breaker 84
- Hose Bibb Vacuum Breaker 88
- Pressure Vacuum Breaker 88
- Backflow Protection for Irrigation Systems 90
- Spill-resistant Vacuum Breaker Assembly 90
- Double Check Valve Assembly 94

TABLE OF CONTENTS

    Reduced Pressure Principle Assembly 97
    Air Gap: Approved Non-mechanical Backflow Prevention 106
    Summary of Approved Backflow Prevention Methods 109
Other Backflow Preventers **109**
    Barometric Loop 110
    Single Check Valve 111
    Dual Check Valve 112
    Auxiliary Methods for Preventing Backflow 113
    Detector Check 114
    Double Check Detector Assembly 114
    Reduced Pressure Detector Assembly 115
Commercial Fire Sprinkler Systems **115**
    Wet Pipe Fire Sprinkler System 116
    Deluge Sprinkler System 117
    Combined Dry Pipe-Preaction Sprinkler System 117
    Dry Pipe Sprinkler System 118
    Preaction Sprinkler System 118
    Residential Fire Sprinkler System (Single Family) 118
Assembly Installation **118**
    Thermal Expansion 119
    Temperature and Pressure Valve 120
    Freeze Protection 120
Summary **121**
Chapter Five Review **123**

CHAPTER SIX: FIELD TESTING **127**

Test Gauges **127**
    Differential Pressure Gauge 127
    Sight Tube 131
Testing **132**
    Testing the Reduced Pressure Principle Assembly 133
    Step-by-step Testing Procedure for the Reduced Pressure Principle
        Assembly 135
    Testing the Reduced Pressure Principle Assembly with a Leaking
        Outlet Shut-off Valve 143
    Testing the Reduced Pressure Detector Assembly 143
    Testing the Double Check Valve Assembly 144
    Step-by-step Testing Procedures for the DCVA, Differential Pressure
        Gauge Single-hose Method 144
    Testing the DCVA with a Leaking Inlet Shut-off Valve, Differential
        Pressure Gauge Method 146
    Testing the DCVA with a Leaking Outlet Shut-off Valve, Differential
        Pressure Gauge Method 147
    Sight Tube Method 149

TABLE OF CONTENTS

    Step-by-step Testing Procedures for the DCVA, Sight Tube Method 149
    Testing the DCVA with a Leaking Inlet Shut-off Valve, Sight Tube Method 151
    Testing the Double Check Detector Assembly. 153
    Testing the Pressure Vacuum Breaker 153
    Step-by-step Testing Procedures for the Pressure Vacuum Breaker 154
    Testing the Pressure Vacuum Breaker with a Leaking Inlet Shut-off Valve 157
    Step-by-step Testing Procedures for the Spill-resistant Vacuum Breaker 158
    Testing the Spill-resistant Vacuum Breaker with a Leaking Inlet
       Shut-off Valve. 159
    Testing Follow-up. 159
Summary **162**
Chapter Six Review **163**

CHAPTER SEVEN: TROUBLESHOOTING, MAINTENANCE, AND REPAIR. 169

    Troubleshooting Reduced Pressure Principle Assemblies 169
    Testing the Reduced Pressure Principle Assembly with a Leaking Outlet
       Shut-off Valve. 169
Maintenance and Repair **178**
    Safety 180
    Repair 180
Items to Check When the Backflow Preventer Fails **181**
    Reduced Pressure Principle Assembly and Reduced Pressure Detector
       Assembly. 181
    Double Check Valve Assembly and Double Check Detector Assembly 182
    Pressure Vacuum Breaker and Spill-Resistant Vacuum Breaker Assembly 183
    Atmospheric Vacuum Breaker Device. 183
    Residential Dual Check Device 184
    General Items to Check When the Backflow Preventer Fails 184
    Corrosion. 185
    Water Quality. 186
Summary **186**
Chapter Seven Review **187**

CHAPTER EIGHT: DEVELOPING A CROSS-CONNECTION
CONTROL PROGRAM 189

Getting Started **189**
Administration of the Program **191**
    An Ordinance. 192
    Service Contracts 192
    Policies and Rules. 193
    Standard Operating Procedures 194
    Administrative Authority. 195

TABLE OF CONTENTS

Seven Elements of a Cross-Connection Control Program. **195**
    Establishing Legal Authority 196
    Plan Review of New Construction 197
    Using Standards and Specifications to Define "Approved" Assemblies 198
    Testing and Maintenance.. 199
    Record Keeping.. 201
    Program for Surveying and Retrofitting Existing Facilities............. 203
    Training and Education 206

Developing a Program for Dealing with Emergencies **209**

Program Manual **212**

Summary **212**

SUGGESTED SUPPLEMENTARY READINGS **215**

GLOSSARY **219**

INDEX. **225**

APPENDIX **227**

Appendix A	AWWA Policy Statement	229
Appendix B	Florida Department of Environmental Protection Rule 62-555.360	231
Appendix C	Abbreviations	245
Appendix D	Field Test Procedures: RP, DCVA, PVB, SVB	247
Appendix E	Test and Maintenance Report Form........................	253
Appendix F	Troubleshooting Guide..................................	257
Appendix G	Test Kits: Suppliers and Repair Locations	261
Appendix H	Repair Parts Suppliers	265
Appendix I	Florida Building Code – Plumbing	269
Appendix J	FCCC & HR Approval Process	271
Appendix K	Testing the RP Chart	273
Appendix L	A.S.S.E. Numbers: Approved Assemblies and Devices	275
Appendix M	Building a Model Backflow Prevention Program	277
Appendix N	Elements of Program Ordinance	279
Appendix O	Form Letters: Required Annual Testing, Follow-up, Final, and Repair	283
Appendix P	List of High Hazard Facilities	289
Appendix Q	CCC Questionnaire	293
Appendix R	Survey Inspection Forms	295
Appendix S	Sample Inspection Letters: Initial Letter Follow-up Letter	299
Appendix T	List of Common Cross-Connection Locations	303
Appendix U	Incident Report Form	307
Appendix V	Selecting the Proper Backflow Prevention Assembly	311
Appendix W	Nomenclature Chart: Valves, Shut-Off, Check, Needle and Test Cocks	315
Appendix X	Lawn Irrigation Systems	317

TABLE OF CONTENTS

Appendix Y   Article on Reclaimed Water (Re-use) 319
Appendix Z   Enclosures for Freeze Protection 333

TABLE OF FIGURES

Figure 2-1	Transmission Route of Giardia Lamblia.	13
Figure 2-2	Parts per Billion Analogy	22
Figure 4-1	Tank Truck Creates Cross-Connection	52
Figure 4-2	Submerged Bath Tub Inlet Creates a Cross-Connection.	53
Figure 4-3	Jet Truck Creates a Cross-Connection	54
Figure 4-4	Cross-Connection Created at Swimming Pool	55
Figure 4-5	"Plumber's Helper" Creates a Cross-Connection	56
Figure 4-6	Siphon Chemical Mixer	56
Figure 4-7	A "Bath-to-Shower" Adaptor Creates a Cross-Connection	57
Figure 4-8	A Hose Creates a Cross-Connection at a Wastewater Treatment Plant	57
Figure 4-9	Common Kitchen Spray Nozzle Creates a Cross-Connection	58
Figure 4-10	Comparison between the Weight of Water and the Weight of Air	59
Figure 4-11	Comparison of Pressures between Two Blocks of the Same Dimensions	60
Figure 4-12	The Weight of a 1-foot Column of Water	61
Figure 4-13	The Weight of a 2-foot Column of Water = 0.866 psi.	62
Figure 4-14	A Comparison of Water Pressures in Two Different Styles of Water Towers	62
Figure 4-15	Water Pressure Is Dependent on Depth	63
Figure 4-16	Determining the Height of a Water Column.	63
Figure 4-17	The Weight of Atmospheric Pressure on the Earth's Surface	64
Figure 4-18	Atmospheric Pressure: Sea Level versus the Mountains	65
Figure 4-19	Absolute versus Gauge Pressure	65
Figure 4-20	Determining the Direction of Water Movement in a Pipe by Comparing Pressure Gauge Readings	66
Figure 4-21	Water Movement in a U-tube.	67
Figure 4-22	Water Movement through a Siphon.	67
Figure 4-23	Drinking from a Straw Illustrates the Effects of Negative Pressure.	68
Figure 4-24	The Creation of a Total Vacuum within a System Theoretically Causes Water to Rise 33.9 Feet	69
Figure 4-25	A Graph of the Relationship between Temperature and Pressure at a Constant Volume	70
Figure 4-26	How the Effects of the Ideal Gas Law Could Cause Backflow from a Boiler.	71
Figure 4-27	A Graph of the Relationship between Velocity and Pressure	71
Figure 4-28a	An Illustration of How a Venturi Can Cause Backflow.	72
Figure 4-28b	An Illustration of How the Water Pressure Falls as the System Pressure Drops	72

TABLE OF FIGURES

Figure 4-29	An Illustration of How a Pump Can Cause Backpressure-Backflow.	73
Figure 4-30	An Illustration of How a Pump Can Cause Backsiphonage-Backflow.	74
Figure 4-31	A Cross-Connection to an Irrigation Well.	75
Figure 4-32	Backsiphonage at a Restaurant.	77
Figure 4-33	Backflow in a Laboratory Caused by an Aspirator.	78
Figure 5-1	Shut-off Valves.	82
Figure 5-2	Test Cock "Blow-out Proof" Stem.	83
Figure 5-3	Atmospheric Vacuum Breaker Normal Flow	85
Figure 5-4	Atmospheric Vacuum Breaker During Backsiphonage	86
Figure 5-5	Atmospheric Vacuum Breaker In-line on an Irrigation System	87
Figure 5-6	Atmospheric Vacuum Breaker Installed on a Laboratory Sink.	87
Figure 5-7	Hose Bibb Vacuum Breaker.	88
Figure 5-8	Pressure Vacuum Breaker.	89
Figure 5-9	Shut-off Valves Are Allowed Downstream on Pressure Vacuum Breakers.	90
Figure 5-10	Pop-up Irrigation Heads.	91
Figure 5-11	Elevated Irrigation Heads	91
Figure 5-12	Spill-resistant Vacuum Breaker With No Flow.	92
Figure 5-13	Spill-resistant Vacuum Breaker during Normal Flow Conditions	92
Figure 5-14	Spill-resistant Vacuum Breaker	93
Figure 5-15	Spill-resistant Vacuum Breaker during Normal Flow	94
Figure 5-16	Spill-resistant Vacuum Breaker	94
Figure 5-17	Double Check Valve Assembly	95
Figure 5-18	Double Check Valve Assemblies in Parallel	96
Figure 5-19	Double Check Valve Assembly Installed in a Vault	96
Figure 5-20	Reduced Pressure Principle Assembly	97
Figure 5-21	RP: Normal Flow.	98
Figure 5-22	RP during Backpressure Conditions	99
Figure 5-23	RP Before Backsiphonage Conditions.	100
Figure 5-24	RP during Backsiphonage Conditions	100
Figure 5-25	RP: Failing Check Valve #2 under Backpressure and Backsiphonage Conditions.	102
Figure 5-26	Backpressure with a Leaking Check Valve #2.	102
Figure 5-27	Drain Line Attached to an RP	103
Figure 5-28	Spitting: The Relief Valve Opens When the Supply Pressure Drops.	104
Figure 5-29	The Relief Valve Closes When Supply Pressure Increases	104
Figure 5-30	Dumping.	105
Figure 5-31	The RP Will Adjust to the Drop in Supply Pressure.	106
Figure 5-32	Air Gap at a Booster Pump	107
Figure 5-33	Absolute Minimum Air Gap Separation	107
Figure 5-34	Barometric Loop	110
Figure 5-35	Single Check Valve	111

TABLE OF FIGURES

Figure 5-36	Dual Check Valve	112
Figure 5-37	Double Check Detector Assembly	115
Figure 5-38	Typical Piping System for a Customer	119
Figure 5-39	Freeze Protection for Backflow Preventers	121
Figure 6-1	Differential Gauge: Diaphragm	128
Figure 6-2	Differential Test Kit	129
Figure 6-3	Differential Test Kit: Without Separate Bleed Valves	129
Figure 6-4	Differential Test Kit: With Pressure Differences Illustrated	130
Figure 6-5	Digital Gauge	131
Figure 6-6	Determining the Pressure Differential Across Check Valve #1 on an RP	136
Figure 6-7	Determining the Opening Point of the Relief Valve on an RP	137
Figure 6-8	Testing an RP with a Leaking Outlet Shut-off Valve	138
Figure 6-9	Determining That Check Valve #2 of an RP Will Hold Tight Against Backpressure	139
Figure 6-10	Confirm That the Outlet Shut-off Valve Is Not Leaking	141
Figure 6-11	Testing the Differential Pressure Across Check Valve #2	142
Figure 6-12	Testing Check Valve #1 on a DCVA	145
Figure 6-13	Testing Check Valve #2 on a DCVA	146
Figure 6-14	Testing Check Valve #2 with Water Flowing to Customer through a Leaking Outlet Check Valve	148
Figure 6-15	Testing Check Valve #1 of the DCVA with Sight Tube	149
Figure 6-16	Testing Check Valve #2 of the DCVA with Sight Tube	150
Figure 6-17	Testing the Opening Point of the Air Inlet Valve in a PVB with a Five-valve Differential Pressure Gauge	154
Figure 6-18	Testing the Opening Point of the Air Inlet Valve in a PVB with a Three-valve Differential Pressure Gauge	156
Figure 6-19	Verifying That the Single Check of a PVB Will Hold Back 1.0 psi in the Direction of Flow	157
Figure 6-20	Testing the Check Valve in a PVB with a Leaking Inlet Shut-off Valve	158
Figure 6-21	Testing a SVB With a Leaking Inlet Shut-off Valve	159
Figure 6-22	Approval Tag	160
Figure 6-23	Red Tag	160
Figure 7-1	Failing Check Valve #1	170
Figure 7-2	Failing Check Valve #1	170
Figure 7-3	Testing an RP with a Leaking Outlet Shut-off Valve	171
Figure 7-4	RP: Failing Check Valve #2 during Backpressure Conditions	173
Figure 7-5	RP: Failing Check Valve #2 under Backpressure and Backsiphonage Conditions	174
Figure 7-6	Backpressure with a Leaking Check Valve #2	175
Figure 7-7	RP: Clogged Sensing Line under Backpressure Conditions	176
Figure 7-8	Clogged Sensing Line with 0 psi in the Sensing Line	177
Figure 7-9	Wear on the Plastic Check Valve Guide	179
Figure 8-1	Flowchart of Steps to Have Plumbing Plans Approved	194
Figure 8-2	Suggested Emergency Response Flowchart	211

TABLE OF TABLES

Table 2-1	Common Waterborne Diseases	12
Table 2-2	Channels of Transmission of Infection	16
Table 2-3	Secondary Drinking Water Standards	19
Table 2-4	Maximum Contaminant Levels ("Primary Standards") for Inorganic Compounds	20
Table 2-5	Maximum Contaminant Levels for Volatile Organic Contaminants	23
Table 2-6	Maximum Contaminant Levels for Synthetic Organic Contaminants	24
Table 5-1	Summary of Requirements for an Approved Assembly	84
Table 5-2	Organizations That Establish Standards for Backflow Prevention Assemblies	84
Table 5-3	Summary of Backflow Prevention Methods	109
Table 6-1	Testing the DCVA – Single Hose Differential Pressure Gauge Method	148
Table 7-1	Testing an RP with a Leaking Outlet Shut-off Valve	172
Table 7-2	Troubleshooting the RP	178
Table 7-3	Troubleshooting the DCVA	178

ACKNOWLEDGMENTS

Author
    Robin Ritland

Managing Editor
    Allison Keating, University of Florida TREEO Center, Gainesville, FL

Editing
    Jim Clifton, City of Manchester, CT
    Kate Ziemak, University of Florida TREEO Center, Gainesville, FL

Graphics and Illustrations
    Office of Institutional Research, University of Florida, Gainesville, FL
    Kyle Latner, University of Florida TREEO Center, Gainesville, FL
    Les O'Brien, University of Florida TREEO Center, Gainesville, FL
    Candace Hollinger, Illustrator

Technical Review
    Gerald Buhr, Buhr and Associates, PA, Lutz, FL
    Fred DeJong, Baltimore, MD
    Charles Hettel (deceased)
    Paul Johnson, Wiginton Fire Protection Engineering, Inc., Orlando, FL
    Mark Inman, American Pump and Supply, Inc., Tallahassee, FL
    Les O'Brien, University of Florida TREEO Center, Gainesville, FL
    Mark Palmer, Florida Keys Aqueduct Authority, Key West, FL
    Jack Poole, Poole Consulting, Olathe, KS
    Jim Purzycki, BAVCO, Long Beach, CA

CHAPTER ONE

INTRODUCTION TO BACKFLOW PREVENTION

Introduction

Two Florida backflow incidents will be discussed in this chapter to introduce the topics of backflow and cross-connection. The public health significance of backflow prevention in the past and present will be emphasized. Later chapters will more thoroughly discuss how cross-connections are created, how backflow occurs, and the impact that backflows have on public health.

On May 1, 1988 the *Orlando Sentinel* reported that 5,700 Edgewater, Florida residents were without water for 24 hours. Propylene glycol, a chemical used in paint manufacturing, backflowed into the city water supply when a valve malfunctioned inside the Coronado Paint Co. There was a cross-connection between the potable water system and the pipe or vat containing propylene glycol. Because the pressure in the pipeline at the paint factory was greater than the city's water-line pressure, the contaminant backflowed into municipal lines.[1]

In May 1988, a worker for the City of Belle Glade, Florida died the day after drinking from a faucet that was used to dilute pesticides in crop-dusting planes. While officials cannot verify how the poison entered the water, the pesticide probably backflowed into the potable water system at some point when the pesticide in the plane's holding tank was being diluted. A hose was likely connected to the faucet to extend the potable water line to the holding tank of the crop-dusting plane. The hose could have easily been submerged in the pesticide, creating a cross-connection between the potable water supply and the toxic pesticide. If the faucet was not protected by a backflow preventer, the pesticide could backflow through the cross-connection into the potable water line. This could occur if, for instance, the potable water supply pressure was reduced below atmospheric pressure. According to the *Gainesville Sun*, the worker died from "complications due to insecticide intoxication and chronic alcoholism."[2] Intoxication is a medical term that means poisoning by a drug or toxic substance.

Backflow/Cross-Connections

Understanding the events and conditions that cause backflow is necessary to comprehend how the previously mentioned incidents occurred. Technically, **backflow** is a reversal of the normal flow of a liquid or gas. A potentially hazardous backflow can occur whenever a cross-connection exists. A **cross-connection** is a link between a potable water system and a

Backflow is a reversal of the normal flow of a liquid or gas.

A cross-connection is a link between a potable water system and a non-potable system.

INTRODUCTION TO BACKFLOW PREVENTION

non-potable system. Potable water is both safe for consumption and is aesthetically pleasing.

In the Edgewater backflow incident, a direct, **permanent cross-connection** provided a physical link between the drinking water supply and the non-potable source. In this case, the pipes containing the pollutant (propylene glycol) and the pipes containing drinking water were directly connected to one another; only a valve prevented the free flow of the pollutant into the potable drinking water. When this valve failed, propylene glycol backflowed into the potable water system.

In Belle Glade, the cross-connection was probably created by a hose. A hose connected to a faucet becomes a temporary extension of the potable water system. When the tip of the hose was submerged in the pesticide, a temporary, indirect cross-connection between the potable drinking water system and the tank containing the insecticide was created. This physical link between the potable system and non-potable source provides a pathway that allows movement in either direction. Normally the pressure maintained in the potable water system will cause water to flow from the potable water system into the non-potable source. However, any event that greatly reduces the potable line pressure can create conditions where the non-potable source will backflow into the potable system.

Non-potable Source/Contaminant/Pollutant

A **non-potable water** source or system can be loosely defined as any liquid, gas, or other substance that can be diluted, dissolved, suspended or mixed in water and that adversely affects the quality of the water. The non-potable substances can be classified as either "low hazards" or "high hazards."

A substance is a **high hazard** (health hazard) if it can adversely affect human health and safety. The introduction of a **contaminant** into the potable water system creates a health hazard. Obviously, both the insecticide and the chemical propylene glycol could be considered contaminants. Most insecticides are toxic and many are deadly if ingested in sufficient quantities. Propylene glycol irritates the eyes and skin on contact, and can cause heart and urological damage in large doses.

Obviously, **low hazards** (non-health hazards) do not affect public health, but they do affect the aesthetic quality of the drinking water. The introduction of a **pollutant** such as iron into the potable water system creates a non-health hazard. Iron is an essential vitamin that many people do not consume in sufficient quantities; however, it can become a nuisance if elevated levels are found in potable water. Iron can stain laundry and plumbing fixtures. Sulfur is another example of a pollutant. At low levels, sulfur is non-toxic even though it creates a rotten-egg odor. However, sulfur can also be a contaminant, and at very high levels it is toxic. Propylene glycol can also be classified as either a pollutant or a contaminant, depending

---

A permanent cross-connection is a link between the potable water supply and any other non-potable system designed to remain in place.

Non-potable water is water that is contaminated by any liquid, gas or solid that can be diluted, dissolved, suspended or mixed with water that adversely affects the quality of the water.

A substance is a high hazard (health hazard) if it can adversely affect human health and safety.

Low hazards (non-health hazards) do not affect public health.

A pollutant is a substance that deteriorates the aesthetic quality of water or other materials but is not harmful to health.

on the concentration. At low concentrations, propylene glycol is relatively non-toxic, and it is routinely added to a variety of products such as prescription drugs, potato chips or other foods as a preservative. In high dosage it becomes a contaminant and a health hazard.

Backpressure/Backsiphonage

The two types of backflow are backpressure and backsiphonage. **Backpressure** occurs when the non-potable system's pressure exceeds the potable water system pressure. This can occur through a rise in the non-potable pressure, a drop in the potable pressure or a combination of both.

Increases in non-potable pressure above potable water-line pressure can be created, for instance, by booster pumps or temperature increases (e.g., in a boiler). A reduction in the potable water supply pressure can occur whenever water use (demand) exceeds the amount being supplied by the water purveyor. Examples of heavy water use include water-line flushing, firefighting or breaks in water mains. Smaller water demands, such as small water-line breaks, can reduce the supply line pressure sufficiently to create a backflow.

The Edgewater backflow incident apparently occurred because the pressure in the paint factory pipelines was greater than the water pressure supplied by the city water lines, so a backpressure-backflow occurred.

The other type of backflow is **backsiphonage**. Backsiphonage occurs when the supply line pressure falls below atmospheric pressure, which is 14.7 psi at sea level. In this situation, atmospheric pressure creates a higher pressure on the non-potable side of the cross-connection. The higher atmospheric pressure produces a backflow from the non-potable system into the potable system. For a backflow incident to be termed backsiphonage, the potable system pressure must fall below atmospheric pressure. Thus, events that cause backpressure by reducing potable side pressure (e.g., firefighting), can also cause backsiphonage. Normally, atmospheric pressure on the potable side would balance the effect of atmospheric pressure on the non-potable side. However, a large water demand such as a fire actually creates a vacuum in the potable water lines thus, the pressure falls below 14.7 psi and backsiphonage can occur.

In the Belle Glade incident, experts speculate that a reduction in the city's water supply pressure below atmospheric pressure caused the cross-connection. Atmospheric pressure pushed the toxic mixture of pesticide and water into the hose and potable water line.

In both incidents, an undesirable pressure differential occurred. That is, the pressure on the potable-water side was lower than the pressure on the non-potable source side. Backflow can occur whenever this type of pressure differential exists and a cross-connection provides the link between the two sources.

> **The two types of backflow are backpressure and backsiphonage.**
>
> **Backpressure occurs when the customer's pressure exceeds the supply pressure.**
>
> **Backsiphonage occurs when the supply line pressure falls below atmospheric pressure.**

INTRODUCTION TO BACKFLOW PREVENTION

Public Health Significance

The backflow examples given above illustrate serious consequences. In Belle Glade a man died. While no one was killed in Edgewater, 5,700 persons were inconvenienced for 24 hours by a single backflow incident. Residents of Edgewater were told not to use the water for drinking or bathing until the city had flushed the contaminant from the piping system. Further, because of the adverse health effects propylene glycol can have if consumed in large quantities, this incident could have had much more serious consequences. These two examples illustrate that the hazards associated with backflow can have serious implications for public health and safety.

Backflows have been occurring since the development of public water systems, often with catastrophic results. Introductions of poisonous chemicals or pathogenic organisms into the potable water supply have resulted in death and disease of epidemic proportions. The fact that backflows still occur today seems even more unacceptable considering that water potability (water quality) was scrutinized as far back as 2000 B.C. and water distribution systems were established in Alexandria, Egypt prior to 47 B.C.[3] However, little attention was given to the potential hazards of backflow until the last two centuries.

In London in 1854, Dr. John Snow was the first to link the spread of the disease cholera with the drinking water system. Snow believed the disease to be spread by contamination of the potable water supply. He discovered that the residents' sewage line had been leaking into the well where the community obtained their drinking water. In 17 weeks, the cholera epidemic claimed 700 lives.[4] However, Snow's hypothesis that water was spreading the disease was not widely accepted until many years later when the causal link between microorganisms and disease was finally established by Louis Pasteur's experiments and Koch's postulate on the identification of disease-causing organisms. It was not until 1872 that cross-connections (loosely defined) were recognized as possible sources of contamination, according to a citation appearing in the American Water Works Association (AWWA) proceedings of 1897.[5] However, little was done about preventing the spread of disease through cross-connections until well into the 20th century. The epidemic at the 1933 Chicago World's Fair emphasized the risk of ignoring this potential hazard. At the World's Fair, antiquated plumbing and fixtures allowed a backflow that resulted in 98 deaths and 1,409 cases of amoebic dysentery. This incident spurred a change in the plumbing laws to provide better protection against the potential threat of cross-connections.[6]

The quality of the drinking water in the United States today is significantly better than it was in 1854 or 1933. This is due in large part to laws, such as the Safe Drinking Water Act, which require that community water systems provide potable water. However, the quality of the water can be degraded very quickly when a cross-connection exists. It is estimated that approximately 100,000 cross-connections are created every day of the year in this country.[7] While many of these are probably **temporary cross-connections**, it has been shown through previous examples that a temporary cross-connection can pose just as deadly a threat as a permanent cross-connection. Moreover, a higher potential for cross-connections and hazardous backflow incidents exists today, since drinking water is used for a variety of other purposes such as air heating/cooling systems, food processing, waste disposal, recreation, irrigation, and fire control. Case histories follow that illustrate each of these problems.[8]

These examples demonstrate how easily backflow incidents can occur, but do not show the total impact that backflow incidents have had, and will continue to have, on public health. In the period from 1903 to 1984, 245 backflow incidents related to biological contaminants were documented; it is conservatively estimated that these incidents resulted in well over 60,000 cases of dysentery, diarrhea, and typhoid, including 272 deaths from typhoid.[9] However, the reported cases may be only the tip of the iceberg. Recall that an estimated 100,000 cross-connections are created every day of the year.[7] Further, the number of backflow incidents will potentially increase as our water distribution systems continue to expand in size. The more plumbing connections that exist; the greater the chances for cross-connections.

CASE HISTORY 1: Chromates from an Air Conditioning System [7]

In New York City in September of 1974, 20 city employees became ill after drinking water that contained chromates. Apparently, a cross-connection existed between the potable water system and the air conditioning system. The air conditioning system for the building, except for the fifth and sixth floors, was located on the top floor of a 32-story building. When the air conditioning system on the fifth and sixth floors broke down, this small system was interconnected with the air conditioning system for the rest of the building. After repairs were made to the air conditioning system on the 5th and 6th floor, but before the two air conditioning systems were disconnected, a service repairman connected a hose from the potable water supply to the air conditioning system. Because the water supply pressure on the fifth and sixth floors was lower than the pressure created by the building's air conditioning system, chromates contained in the system backpressure-backflowed into the potable water supply.

CASE HISTORY 2: Water into Wine[8]

In December of 1970 in Cincinnati, Ohio, a water supply valve to a wine-distilling tank was accidentally left open in a local winery. This is an example of a **direct cross-connection**. As the wine fermented, a greater pressure was created in the vat than the city's water line. As a result, wine entered the drinking water. By backpressure-backflow, wine was supplied to

---

A temporary cross-connection is a link between the potable water supply and any other non-potable system created with removable sections, swivel or change-over devices, garden hoses, and other non-permanent methods.

A direct cross-connection is a link between the potable water supply and any other non-potable system that is subject to both back-siphonage and backpressure.

An indirect cross-connection is a temporary link between the potable water supply and any other non-potable system that is subject to backsiphonage only.

some Cincinnati residents. Since wine is not a threat to public health and safety (if consumed in moderation), it is considered a pollutant or low hazard. In fact, some individuals probably did not view it as a hazard at all.

CASE HISTORY 3: Wastewater into Public Mains[6]

One of two water mains in the town of Newton, Kansas was temporarily taken out of service. During this time, someone opened a fire hydrant to obtain water and left it open when no water flowed from the hydrant. When water service was restored, the large water demand created by the open valve lowered the water pressure in the system enough for wastewater from the toilets of 10 families to backsiphonage-backflow into the potable water supply for two days. Because the lines were not flushed or cleaned, 2,500 people became ill with digestive disorders.

CASE HISTORY 4: A Lost Football Season - 1969[6]

A backflow caused the cancellation of a football season for Holy Cross College when the players and coaches drank from a fountain that was contaminated with infectious hepatitis. Apparently some children playing around the irrigation heads were carriers of the virus. Urine had collected in the sprinkler boxes where the children had played. The contamination back-siphon-backflowed when a heavy fire demand in the neighboring town of Worchester, MA created a negative pressure in the water line. Because the drinking fountain and the irrigation system for the field were on the same line and the irrigation system lacked a backflow preventer, the contaminated water was dispensed at the drinking fountain.

Backflow Prevention

Backflow must be prevented because of the major impact that backflow can have on public health. The **Safe Drinking Water Act (SDWA)** requires that sanitary surveys be conducted at water treatment plants to assure that they are producing and distributing water that is safe to drink. Backflow prevention programs are essential to ensure that the drinking water supply remains safe. To that end, each state's rules and regulations clearly require municipalities to establish cross-connection control programs. "Community water systems and all public water systems shall establish and implement a routine cross-connection control program to detect and control cross-connections and prevent backflow of contaminants into the water system."[10] This is a major step toward eliminating cross-connections and the hazards associated with backflow.

Under the **SDWA**, the federal government granted a number of states primacy. **Primacy** means that states have primary enforcement responsibility. In order to be granted primacy, each state must adopt regulations at least equal to the federal regulations for protecting public health. However, this

The Safe Drinking Water Act (SDWA) was established in 1974.

Primacy means that the states have primary enforcement responsibility.

still allows states great flexibility, resulting in many different plumbing regulations being developed. In addition, a number of different organizations have developed their own guidelines and standards. For instance, the American Society of Sanitary Engineering (**ASSE**), the American Water Works Association (**AWWA**), the Foundation for Cross-Connection Control and Hydraulic Research (**FCCC & HR**), the American Society for Testing and Materials (**ASTM**), the American Society of Mechanical Engineers (**ASME**), the International Association of Plumbing and Mechanical Officials (**IAPMO**), the Underwriters Laboratory (**UL**), the Factory Mutual (**FM**), the American National Standards Institute (**ANSI**), and the National Sanitation Foundation (**NSF**) are involved in testing or setting standards for backflow preventers. Therefore, engineers, plumbing contractors and anyone else involved in water distribution not only have to be aware of the different guidelines, but they must also be familiar with the plumbing regulations adopted by each municipality, county, or state in their operating area. The tremendous variation in water regulations related to backflow prevention has complicated and slowed the enforcement of backflow prevention. For example, each of the many plumbing codes treats the problem of cross-connections and backflow prevention slightly differently.

The single most significant problem hampering the elimination of cross-connections and prevention of backflow is that little has been done to educate the water consumer about the problems associated with backflow. While many water purveyors recognize that they could be found negligent if they do not take action to protect the public water supply from a backflow incident, the 1975 National Interim Primary Regulations state: "contaminants added to the water under circumstances controlled by the user . . . are excluded" from the definition of maximum contaminant.[11] This has been interpreted (by legal precedent) to mean that consumers are legally liable for their own actions. However, few consumers are aware of their responsibilities, and most cross-connections are inadvertently created by individuals who are unaware of the hazards of backflow.

In fact, many states have done little to educate the people involved: water purveyors, water distribution personnel, plumbers, plumbing inspectors (building inspectors), engineers, plumbing contractors, irrigation contractors, heating/cooling contractors, wastewater operators, home owners, handymen, or virtually any other individual in a position to tap into or alter a potable water line for any purpose. An alteration can be as simple as connecting a hose to a faucet.

Even if individuals are familiar with the inherent dangers of cross-connections, they may accidentally create one if they do not fully understand the hydraulic conditions that cause backflow or if they have not had adequate training in the proper installation and maintenance of backflow prevention assemblies. Many individuals intentionally install cross-connections because of convenience. For instance, those who are not aware of the significant advances in backflow prevention assemblies might install an inadequate device instead of the appropriate form of protection because they fear that the appropriate backflow preventer would significantly reduce line pressure. The problem is further compounded if plumbing installations are not inspected or if plumbing inspectors are not adequately trained in backflow prevention. Education is essential to eliminate cross-connections and prevent backflow.

Purpose and Scope

The purpose of this manual is to provide necessary training and education to those who design, alter, install or inspect potable and non-potable water supply systems. Two basic approaches to preventing backflow hazards are presented: backflow prevention and cross-connection control. The first approach covers hydraulic conditions that allow backflow events to occur, the methods available to eliminate these physical conditions, and mechanisms that can be used to prevent backflow. The second approach involves inspecting and surveying programs designed to identify and eliminate cross-connections. Eliminating the physical link reduces the potential for a hazardous backflow incident.

An effective backflow prevention program must include seven essential elements:

- legal authority
- plan review of new construction
- standards and specifications to define "approved" backflow prevention assemblies
- testing and maintenance program
- record keeping
- program for surveying and retrofitting existing facilities
- training utility personnel and educating consumers and others

Because backflow prevention assemblies have a limited life span and because plumbing is installed and altered frequently, a backflow prevention program will not be effective if a "fix-it-and-forget-it" philosophy is adopted. An ongoing program that includes testing, monitoring, and continuing education is essential to adequately deal with the problem of backflow. An education program must include the practical, applied aspects of backflow prevention as well as theoretical considerations.

Summary

This textbook augments the backflow tester training course and the program manager's series by providing more detailed terminology, and principles and hazards associated with backflow. Further, it covers the essentials of a backflow prevention program and it provides the information necessary to develop an effective cross-connection control program. The textbook itself is designed to serve as a reference source for future implementation and maintenance of a program.

Chapter One Review

(Please use answers below for questions 1-1 through 1-7.)

BACKFLOW	BACKPRESSURE
BACKSIPHONAGE	CONSUMPTION
HEALTH	NON-HEALTH
LINK	PERMANENT
SEVEN	INDIRECT

1-1 _____ is a reversal of the normal direction of flow. *Backflow*

1-2 A cross-connection is a _____ between the potable water system and some other non-potable system or source. *Link*

1-3 Potable water is safe for _____. *Consumption*

1-4 Another term for a direct cross-connection is a _____ cross-connection. *Permanent*

1-5 A _____ hazard is a high hazard, not a low hazard. *Health*

1-6 Two types of backflow are _____ and _____. *Backsiphonage & Backpressure*

1-7 There are _____ essential components of an effective cross-connection control program. *Seven*

REFERENCES

1. "Edgewater Lifts Ban on Water", *Orlando Sentinel*, May 1, 1988. Orlando, FL.

2. "Examiner: City Worker Killed by Pesticides", *Gainesville Sun*, June 26, 1988. Gainesville, FL.

3. *Water & Man's Health (Technical Series No. 5)*, 1962. Department of State, Agency for International Development, Washington, DC.

4. *Water Borne Disease Control*, 1982. (Manual #1 - Water and Human Health Home Study Course 3014-G), Centers for Disease Control-Center for Professional Development and Training, U.S. Department of Health and Human Services, Washington, DC.

5. *Cross-Connection and Backflow Prevention (2nd edition)*, 1974. American Water Works Assn., New York, NY.

6. *Cross-Connection Control Manual*, 1975. (EPA 430-9-73-002). U.S. Environmental Protection Agency, Water Supply Division.

7. *Cross-Connection Control Training Package*, 1985. American Water Works Assn., New York, NY.

8. *50 Cross-connection Questions, Answers and Illustrations Relating to Backflow Prevention Devices and Protection of Potable Water Supply*, (F-50-1), Watts Regulator Company.

9. *Manual of Cross-Connection Control (9th Edition)*, 1994. Foundation for Cross-Connection Control and Hydraulic Research, University of Southern California, University Park, CA.

10. "Permitting, Construction, Operation, and Maintenance of Public Water Systems," *Florida Administrative Code*. 1997, amended August 28, 2003. (62-555.360, F.A.C.). Florida Department of Environmental Protection, Tallahassee, FL.

11. *National Interim Primary Drinking Water Regulations*. 1976. (EPA 570/9-76-003). U.S. Environmental Protection Agency, Office of Water Supply, Washington, DC.

CHAPTER TWO

THE HAZARDS OF BACKFLOW

In Chapter 1, two backflow incidents in Florida helped define backflow terms such as cross-connection, backsiphonage, backpressure, and potable water. Case histories emphasized the public health significance of backflows and illustrated the need for backflow prevention programs.

In this chapter, the hazards created by backflows will be discussed more thoroughly. Not only will the effects on public health be addressed, but the financial consequences of backflows will also be reviewed. This chapter also provides a detailed look at the variables that determine who is affected by a backflow incident.

Biological Pollution/Contamination

Biological agents can be classified as either pollutants or contaminants. **Biological pollutants** (low or non-health hazards) are agents that do not cause disease but that do decrease the aesthetic quality of the water. For instance, *Lactobacillus* is a type of bacteria that is used in the production of yogurt; therefore, it is considered a beneficial bacterium. However, if large numbers of these bacteria pollute the potable water supply, they can create unwanted odors or taste or otherwise affect the appearance of the water. Maintaining the aesthetic quality of the water is important from a public health viewpoint, because if the water is not aesthetically pleasing, the water users may switch to an alternate water source that may not be safe. In addition, biological pollutants can interfere with the water treatment process.

Biological contaminants (high or health hazards) are agents that can cause disease. There are a number of different biological agents (bacteria, viruses, protozoa, algae, etc.) that can cause disease. Any of these biological agents that are primarily spread through water are termed waterborne diseases.

While some of the diseases listed in Table 2-1 are not very serious, others can be fatal. All of these organisms can be spread through a backflow incident. For instance, *Giardia lamblia* is not normally found in well water, but it is found in surface waters. While modern water treatment processes such as flocculation and filtration are used to ensure that this organism is removed from potable water systems that use surface water sources, a cross-connection between raw surface water and the potable water system could allow this organism to backflow into the potable water supply (Figure 2-1). An improperly developed private well that is contaminated with surface water could also be the source of the organism. Since one form of the organism, the o-cyst, is resistant to chlorine, a backflow of this biological agent into the water could result in an outbreak of Giardiasis.[1]

> Biological pollutants (low hazards) are agents that do not cause disease, but that do decrease the aesthetic quality of water.
>
> Biological contaminants (high hazards) are agents that can cause disease.

THE HAZARDS OF BACKFLOW

Table 2-1   Common Waterborne Diseases

DISEASE	CAUSATIVE ORGANISM	SYMPTOM
Gastroenteritis	Salmonella	Acute diarrhea and vomiting
Typhoid	Salmonella typhi	Inflamed intestine, enlarged spleen, high temperature *Fatal*
Dysentery	Shigella	Diarrhea
Cholera	Vibrio cholerae	Vomiting, severe diarrhea, rapid dehydration, mineral loss *Highly fatal*
Infectious hepatitis	Virus	Yellowed skin, enlarged liver, abdominal pain, lasts up to 4 months *Low mortality*
Amoebic dysentery	Entamoeba	Mild diarrhea, chronic dysentery *Sometimes fatal*
Giardiasis	Giardia lamblia	Diarrhea, cramps, nausea and general weakness Lasts 1 week to 7 months
Cryptosporidiosis	Cryptosporidium parvum	Severe intestinal distress Lasts 2 days to 4 weeks *May be fatal*

Legionella, the bacterium that causes Legionnaires' disease, can also be spread through the potable water system. The bacterium seems to survive in chlorinated water, and it is commonly found in devices that permit the stagnation of water and/or the accumulation of solids, such as cooling towers, evaporation condensers, and hot water tanks.[1] Therefore, if a cross-connection exists between any of these and the potable water system, the organism can contaminate the potable water system. It is essential that backflow preventers be installed on these devices to prevent the spread of this bacterium. Legionnaires' Disease is a very serious disease. It is fatal to 15–20% of its victims, if left untreated.[2] The disease causes a severe form of pneumonia. This bacterium does not appear to be spread by ingestion of the water; it is spread by aerosol (fine water droplets). Thus, individuals can be exposed by breathing infected water either from air heating, cooling systems, or potentially from exposure to the organism during showers. The bacterium has also been isolated from shower heads.[2]

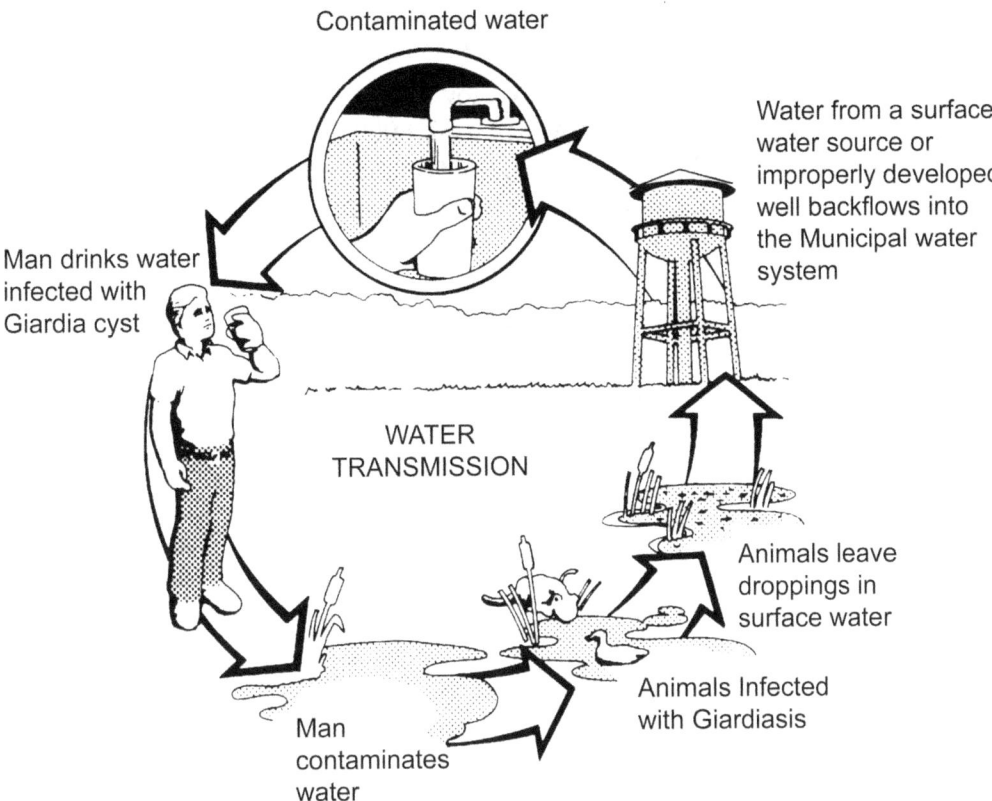

Figure 2-1 **Transmission Route of Giardia Lamblia**
The cyst form of Giardia lamblia is resistant to chlorine; therefore, a backflow that contaminates the potable water supply with this organism has overcome the treatment process.

Backflows Short-circuit the Water Treatment Process

Modern treatment processes can effectively kill or remove most of the organisms listed in Table 2-1. However, when contaminants backflow into the distribution system, treatment becomes somewhat irrelevant since the treatment process has been bypassed. Contaminants or pollutants (biological or chemical) introduced during a backflow will quickly react with the low level of disinfectant maintained throughout the distribution system. This leaves a disinfectant-free environment in which biological agents can survive or remain viable. Bacteria and viruses are considered viable if they can cause disease.

Many water treatment systems now use chloramines for disinfection instead of free chlorine to prevent the formation of trihalomethanes, which are known to cause cancer. Trihalomethanes are formed when free chlorine reacts with organics naturally found in water. **Chloramines** are created when ammonia is used in combination with chlorine. Chloramines are somewhat less effective as disinfectants than free chlorine; therefore, systems that use chloramines can be less effective in dealing with contaminants that are added to the potable water system during backflow.

Chloramines are created when ammonia is used in combination with chlorine.

THE HAZARDS OF BACKFLOW

Public Health Significance

The impact on public health can best be shown by showing the number of people that have been affected by backflows. Well over 60,000 cases of dysentery, diarrhea, and typhoid and 272 deaths from typhoid were caused by backflows from 1903 to 1984.[3] While modern water treatment processes have greatly improved the quality of drinking water, data from the Centers for Disease Control for the years 1971–1984 reveal that 108,000 people were made ill by waterborne pathogens from public or private water supplies.[4] Undoubtedly, many cases were caused by the backflow of contaminants into the potable water supply. Distribution system deficiencies (cross-connections) caused more waterborne disease outbreaks in the United States between 1946 and 1970 than any other reported factor.[5] From 1971 to 1974 backflows in municipal water systems caused 32% of confirmed waterborne disease outbreaks,[6] and backflows caused 47% of the disease outbreaks in municipal water systems from 1975 to 1976.[7] Clearly, the backflow of biological contaminants into potable water supplies can have serious impacts on public health.

Monetary Costs of Biological Contamination

Much evidence exists to highlight the problems that biological contamination creates in public health; however, monetary loss can also be a repercussion of a backflow incident, usually the result of lawsuits. Victims of a backflow incident, or their family and friends, generally seek reimbursement for lost wages and medical care, as well as compensation for loss of life. Also, punitive damages may be sought and assessed by the court where negligence is found.

Biological contamination can also generate financial costs through destruction of property. Significant property destruction might result in a loss of business and a loss of wages for the employees of that business. For instance, if a backflow of biological contaminants occurred at a restaurant and patrons became ill, the restaurant would likely lose a significant number of customers. In fact, when a restaurant is the source of a publicized disease outbreak, its loss of business will frequently be so great that it will be forced to close.

Removing or killing the biological contaminant and restoring the distribution system result in additional costs. This may require that the water lines be cleaned and flushed and that the chlorine levels be elevated within the line. When certain "persistent" chemicals contaminate the water pipe, these pipes must be removed and replaced at great expense. Also, the water customers must be provided with an alternative safe water source until the water quality in the potable water system is restored. This might take a few hours, a few days, or longer.

Biological Hazards: Case Histories

The following case histories demonstrate the impact on public health and the financial consequences of a backflow incident. These actual backflow incidents also emphasize just how easily backflows can and do occur.[8,9]

These three case histories illustrate just a few of the different biological agents that can produce disease. Table 2-1 lists eight waterborne diseases common in the United States, their causative organism, and the public health significance of each disease.

CASE HISTORY 5: Contaminated Meat - 1979[8]

A backflow at the Swift and Company packing plant in Marshalltown, Iowa resulted in the contamination of $2,000,000 worth of meat. The plant shut down for an extended period while an inspection was conducted to determine the cause of the contamination, monitor decontamination and sterilization procedures at the plant. As a result, 200 employees were laid off from the packing plant. The investigation determined that a cross-connection existed between the potable water supply and wastewater from the kill floor. The meat was contaminated when an employee sprayed the contaminated water on hog carcasses and cuttings during the normal cleaning process. The cause of the backflow is unclear, but an improper hydraulic gradient was somehow created "because the wrong pipe was hooked up to a newly drilled well." The total cost of this backflow incident was estimated at $3,000,000 to $5,000,000.

CASE HISTORY 6: River Water from a Drinking Fountain - 1967[9]

A backflow of river water into the potable water system occurred when an industrial plant worker cross-connected the fire system line to a bubbler. The fire system used a non-potable source river water. Apparently, the worker mistook the fire system for the potable system. Seven people developed infectious hepatitis, and over a hundred people became ill with gastroenteritis (a bacterial infection that causes nausea, vomiting, and diarrhea).

CASE HISTORY 7: Gastroenteritis at College Cafeteria - 1967[9]

A backflow occurred at a small private college in Pennsylvania after a water line broke in the school cafeteria. A cross-connection existed between the potable water system and the wastewater system. The broken water line lowered the water pressure in the potable water line, creating a hydraulic gradient that allowed the backflow to occur. Apparently, the wastewater contained sufficient quantities of the bacterium *Shigelli sonnei*

THE HAZARDS OF BACKFLOW

to completely exhaust the chlorine residual in the water supply. As a result, 700 students and faculty contracted gastroenteritis from drinking the water or eating food that had been prepared using the contaminated water.

Factors That Affect the Magnitude of a Biological Backflow Incident

A single backflow incident can potentially affect hundreds or even thousands of people. The larger the water distribution system; the greater the potential for a single backflow incident to affect a large number of people. Other factors that determine how many people will be affected by a biological backflow incident include: how quickly it is determined that a backflow has occurred (detection time), the onset time of the disease, the pathogenicity and virulence of the organism, the susceptibility of the exposed individual, and the survival time of the biological agent.

Detection Time: Because most biological contaminants are not detectable through odor, taste, or appearance, a disease outbreak is often the first indication that a backflow of biological contaminants has occurred. Most waterborne diseases have more than one mode of transmission, (Table 2-2), which means a disease outbreak may not clearly indicate that a backflow has occurred.

In other words, the disease can be spread by methods other than water. For instance, bacillary dysentery can be passed from one individual to another by soiled hands. Therefore, not only will those individuals who ingested the contaminated water become ill, but many other people who come into contact with (shake hands with) these infected individuals will also become ill. A person can be a carrier of disease long before and long after they show any symptoms of that disease. Since the disease is not limited to the specific geographic area where the backflow occurred, it

Table 2-2   Channels of Transmission of Infection

DISEASE	FROM	THROUGH	TO
Amoebiasis	man	food, flies, contact, water	man
Bacillary dysentery	man	soiled hands, food, milk, flies	man
Echinococcosis	animals	food, water, soiled objects, contact	man
Infectious hepatitis	man	food, milk, shellfish, water	man
Leptospirosis	man	food, contact, water	man
Paratyphoid fever	man	contact, food, milk, shellfish, soiled hands, water	man
Typhoid fever	man	food, milk, shellfish, contact, water	man

becomes less obvious that the distribution system is spreading the disease. This can also make it more difficult to determine the location of the cross-connection and may delay its elimination and lengthen the exposure time of individuals to the backflow hazard.

Even though drinking water regulations require frequent monitoring of the biological quality of the water, monitoring is usually not frequent enough to detect biological contamination due to backflow before it has caused disease. Monitoring frequencies vary from daily to monthly based on the population served. (Monitoring is required more frequently when the distribution system serves a larger population.) Obviously, in larger systems, more people could potentially be affected by contamination; therefore, the objective is to detect problems more quickly in large systems.

Water is not monitored for every waterborne hazard. This would simply be too costly and time-consuming to justify. The water is monitored for a minimum free chlorine residual of 0.2 parts per million (ppm) or minimum combined chlorine residual of 0.6 ppm where chloramines are used. A chlorine residual less than this might indicate that some sort of contamination has occurred. However, testing water for coliform bacteria is a better method for determining if the potable water system has been contaminated. Coliforms are a group of bacteria normally found in the intestines of humans and other warm-blooded animals. They are considered good indicator organisms since they signal the possible presence of other pathogenic organisms. Because they survive longer in water than most pathogenic bacteria, it is assumed that the absence of coliforms assures absence of pathogenic bacteria. Testing water for "Total Coliforms" is the accepted method for determining the potability of water in regard to biological quality.

Onset Time is the amount of time that passes between the ingestion of (or contact with) the biological agent and the first disease symptoms. The onset time for waterborne diseases is quite long, ranging from a few hours for some organisms to a few days for others. A long onset time means that more individuals can ingest the biological agent before the contamination is discovered.

Pathogenicity is the ability of an organism to make an individual ill. **Virulence** is how ill the organism will make an individual. For instance, *Vibrio cholerae*, the bacterium that causes cholera, is very virulent. It can kill infected individuals. On the other hand, *Salmonella*, the bacterium that causes gastroenteritis, usually does not kill people, but it will make infected individuals very ill. Also, a greater number of pathogenic organisms present in the water increases the chance that a person will become ill. A certain number of organisms must be present for a person to become ill. However, this minimum number can vary considerably depending on the type of organism and the susceptibility of the exposed person. Individuals who are already ill or weak, such as older people or newborns, are more susceptible than young, healthy individuals. Generally, the immune systems of the very old and the very young are not as effective in fighting diseases.

THE HAZARDS OF BACKFLOW

Survival Time: As mentioned earlier, while modern water treatment processes can remove pre-existing biological contaminants, they do little to protect against biological agents that are introduced during backflow incidents. After water leaves the treatment plant, normally low levels of chloramines (or chlorine) are maintained in the water to prevent biological growth. However, biological contaminants and pollutants (as well as many chemicals) that are introduced through backflow quickly react with the low levels of chloramine. Consequently, the chloramines are depleted, leaving a disinfectant-free environment in which biological agents can now survive. Survival time for different biological agents in disinfectant-free water varies, ranging from a few hours or days to several months. Viruses remain viable for very long periods of time. In fact, some have been known to remain viable for months or even years.

When the survival time is very short and the onset time is very long, the organisms may die before a backflow of biological contaminants is detected. In other words, the organisms may no longer be present in the water by the time people start becoming ill. This makes it difficult to determine the cause of the illness. Thus, backflows of biological contaminants likely occur more frequently than realized.

The number of people affected by a backflow incident is also dependent on the presence of backflow prevention assemblies/devices. **Isolation**, or internal protection, is the term used when backflow preventers are installed at each possible point of contamination. **Containment**, or service protection, is the term used when backflow prevention assemblies are used to keep contaminants from entering the distribution system from a water consumer's facility. The latter approach would not prevent contamination of the potable water system within the facility where the backflow occurred, but it would prevent the contaminant from leaving the facility and entering the distribution system. Therefore, either type of backflow prevention, isolation or containment, will reduce the number of individuals who are exposed to the contaminant. Backflow preventers must be installed properly and maintained through regular testing and maintenance to ensure that they will continue to limit or prevent the spread of contaminants.

Chemical Pollution/Contamination

Chemicals, like biological agents, can be classified as either contaminants or pollutants. **Chemical pollutants** are substances that do not pose a public health threat, but reduce the aesthetic quality of the water. Therefore, they are low hazards (or non-health hazards). Like biological agents, any chemical that affects the color, odor, or taste of water is considered a pollutant. While chemical pollutants do not themselves cause health problems, they can react with and deplete the chlorine residual in the potable water. Also, as mentioned previously, the aesthetic quality of the water must be maintained to prevent water users from switching to

Table 2-3   Secondary Drinking Water Standards
(Chapter 62-550.320, F.A.C., Table 6, 2/16/2012)

FEDERAL CONTAMINANT FEDERAL ID NUMBER	CONTAMINANT	MCL (mg/L)*
1002	Aluminum	0.2
1017	Chloride	250
1022	Copper	1.0
1025	Fluoride	2.0
1028	Iron	0.3
1032	Manganese	0.05
1050	Silver	0.1
1055	Sulfate	250
1095	Zinc	5
1905	Color	15 color units
1920	Odor**	3 (threshold odor number)
1925	pH	6.5 - 8.5
1930	Total Dissolved Solids	500 (may be greater if no other maximum contaminant level is exceeded)
2905	Foaming Agents	0.5

Abbreviations Used: MCL = maximum contaminant level
mg/L = milligrams per liter

* Except color, odor, corrosivity, and pH.
** For purpose of compliance with ground water quality secondary standards, as referenced in Chapter 62-520, F.A.C., levels of ethylbenzene exceeding 30 micrograms per liter, toluene exceeding 40 micrograms per liter, or xylenes exceeding 20 micrograms per liter shall be considered equivalent to exceeding the drinking water secondary standard for odor.

unsafe water sources. Table 2-3 lists "secondary standards" contained in Florida Drinking Water Regulations, Section 62-550.320, F.A.C. These secondary standards set maximum allowable levels for chemical pollutants commonly found in community water systems, as well as standards for color and odor.

THE HAZARDS OF BACKFLOW

Chemical contaminants are substances that pose a threat to public health; therefore, chemical contaminants are high hazards (or health hazards). Chemical contaminants include both inorganic and organic compounds. Inorganic chemicals are metals or other compounds that do not contain carbon. Organic chemicals are substances that contain carbon. They are often derived from vegetable or animal matter. Maximum contaminant levels (MCLs) for chemical contaminants commonly found in potable water supplies are listed in Section 62-55.310, F.A.C. as "primary standards." Table 2-4 contains a list of primary standards for some organic chemicals.

> Chemical contaminants are substances that pose a threat to public health.

Table 2-4 Maximum Contaminant Levels ("Primary Standards") for Inorganic Compounds
(Chapter 62-550.310, F.A.C., Table 1, 2/16/2012)

FEDERAL CONTAMINANT ID NUMBER	CONTAMINANT	MCL (MG/L)
1074	Antimony	0.006
1005	Arsenic	0.01
1094	Asbestos	7 MFL
1010	Barium	2
1075	Beryllium	0.004
1015	Cadmium	0.005
1020	Chromium	0.1
1024	Cyanide (as free Cyanide)	0.2
1025	Fluoride	4.0
1030	Lead	0.015
1035	Mercury	0.002
1036	Nickel	0.1
1040	Nitrate	10 (as N)
1041	Nitrite	1 (as N)
	Total Nitrate and Nitrite	10 (as N)
1045	Selenium	0.05
1052	Sodium	160
1085	Thallium	0.002

Abbreviations Used: MCL = maximum contaminant level
MFL = million fibers per liter (longer than 10 micrometers)
mg/L = milligrams per liter

Public Health Significance

Just as with biological contaminants, the backflow of chemical contaminants into the potable water system has the potential to detrimentally affect the health of every individual receiving service from that system. The type of chemical contaminant and its concentration will determine just what these detrimental effects will be. For instance, some chemicals such as arsenic are deadly at low levels, with as little as 100 milligrams (mg) of arsenic resulting in severe poisoning. Other chemicals must be present at high levels to have any effect on public health, and even then the effects are much less severe. For instance, while the maximum contaminant level for arsenic is 0.05 pounds per minute (ppm), the maximum contaminant level for fluoride is 80 times greater at 4.0 ppm. The level for fluoride is much higher because fluoride is less dangerous than arsenic. Prolonged exposure to high levels of fluoride does cause mottling of the teeth, and some studies indicate that prolonged exposure to higher levels causes skeletal fluorosis.[10]

Recall that for biological contaminants to have an effect on public health, a certain number of organisms must be present. If this level is reached, the biological agent will have an effect on the individual fairly quickly (within a few hours or days). Chemical contaminants, however, can cause either acute or chronic health problems. An acute effect occurs fairly quickly. High levels of chemical contamination cause acute health effects. For example, as illustrated in Case History 8, if sufficient levels of arsenic are ingested, the individual will die shortly after ingesting the chemical. In contrast, chronic health effects normally develop after long exposure times (10 to 20 years). Long-term exposure to a chemical contaminant at low levels causes chronic effects, such as cancer.

Monetary Cost of Chemical Contaminants

Just as with biological backflows, backflows of chemical contaminants that cause death, disease, or other health problems will probably result in lawsuits. As a result, the same type of costs (cleanup costs, punitive damages, etc.) will be associated with chemical contaminant backflows as with biological contaminant backflows. Chemical contaminants can also cause property damage, as illustrated in Case History 9. Property losses can result in employee layoffs, which will mean lost wages for employees and lost production time for the company. Also, the reputation of a business responsible for a backflow incident will undoubtedly be tarnished by the incident, possibly resulting in lost business.

Another problem is removing the contaminant from the potable water lines. Often, flushing the line will remove most of the contaminant. However, the acceptable levels for some chemicals are so low that flushing the line will not remove enough of the contaminant. For some chemicals the maximum contaminant level is measured in parts per billion (ppb). One ppb of chlorine in water would mean there is one chlorine molecule in one billion

THE HAZARDS OF BACKFLOW

Figure 2-2   Parts per Billion Analogy
Assume that the world population is approximately 6 billion. 1 ppb is equivalent to six people with blue thumbs in a population of 6 billion people with normal thumbs. If these blue-thumbed individuals were then distributed throughout the world, trying to locate them based on that trait alone would be similar to trying to detect 1 ppb of a contaminant in water.

water molecules. Figure 2-2 gives a better indication of just how small a number this is. Assume that the world population is approximately 6 billion; 1 ppb is equivalent to 6 people with blue thumbs in a population of 6 billion people with normal thumbs. If these blue thumbed individuals were then distributed throughout the world, trying to locate them based on that trait alone would be similar to trying to detect 1 ppb of a contaminant in water. For chemicals with very low maximum contaminant levels, the water lines will often need to be removed and replaced, incurring additional expense, as shown in Case History 10.

Chemical Hazards: Case Histories

The following case histories identify a few of the many chemicals that can cause public health problems. Table 2-5 lists organic contaminants, and Table 2-6 lists other chemicals that must be tested in potable water systems. These "other chemicals" do not currently have maximum contaminant levels established because it is unknown whether they present health

Table 2-5   Maximum Contaminant Levels for Volatile Organic Contaminants
(Chapter 62-550, F.A.C., Table 4, 2001) Last modified February 22, 2018

FEDERAL CONTAMINANT ID NUMBER	CONTAMINANT & (CAS NUMBER)	MCL (mg/L)
2977	1,1-Dichloroethylene (75-35-4)	0.007
2981	1,1,1-Trichloroethane (71-55-6)	0.2
2985	1,1,2-Tricholoroethane (79-00-5)	0.005
2980	1,2-Dichloroethane (107-06-2)	0.003
2983	1,2-Dichloropropane (78-87-5)	0.005
2378	1,2,4-Tricholorobenzene (120-82-1)	0.07
2990	Benzene (71-43-2)	0.001
2982	Carbon tetrachloride (56-23-5)	0.003
2380	cis-1,2-Dichloroethylene (156-59-2)	0.07
2964	Dichloromethane (75-09-2)	0.005
2992	Ethylbenzene (100-41-4)	0.7
2989	Monochlorobenzene (108-90-7)	0.1
2968	o-Dichlorobenzene (95-50-1)	0.6
2969	para-Dichlorobenzene (106-46-7)	0.075
2996	Styrene (100-42-5)	0.1
2987	Tetrachloroethylene (127-18-4)	0.003
2991	Toluene (108-88-3)	1
2979	trans-1,2-Dichloroethylene (156-60-5)	0.1
2984	Trichloroethylene (79-01-6)	0.003
2976	Vinyl chloride (75-01-4)	0.001
2955	Xylenes (total) (1330-20-7)	10

Abbreviations used: CAS Number = Chemical Abstract System Number;
MCL = maximum contaminant level;
mg/L = milligrams per liter.

problems or it is unclear what the maximum contaminant level should be to protect public health. Though these lists are fairly lengthy and only include those chemicals commonly found in water (either they naturally occur in water, are added as part of the treatment process, or commonly contaminate the potable water supply). Many other chemicals are not in these lists, and new chemicals are constantly being created. Every one of these chemicals

THE HAZARDS OF BACKFLOW

Table 2-6 Maximum Contaminant Levels for Synthetic Organic Contaminants (Chapter 62-550, F.A.C., Table 5, 2001)

FEDERAL CONTAMINANT ID NUMBER	CONTAMINANT & (CAS NUMBER)	MCL (mg/L)
2063	2,3,7,8-TCDD (Dioxin) (1746-01-6)	3×10^{-8}
2105	2,4-D (94-75-7)	0.07
2110	2,4,5-TP (Silvex) (93-72-1)	0.05
2051	Alachlor (15972-60-8)	0.002
2050	Atrazine (1912-24-9)	0.003
2306	Benzo(a)pyrene (50-32-8)	0.0002
2046	Carbofuran (1563-66-2)	0.04
2959	Chlordane (57-74-9)	0.002
2031	Dalapon (75-99-0)	0.2
2035	Di(2-ethylhexyl)adipate (103-23-1)	0.4
2039	Di(2-ethylhexyl)phthalate (117-81-7)	0.006
2931	Dibromochloropropane (DBCP) (96-12-8)	0.0002
2041	Dinoseb (88-85-7)	0.007
2032	Diquat (85-00-7)	0.02
2033	Endothall (145-73-3)	0.1
2005	Endrin (72-20-8)	0.002
2946	Ethylene dibromide (EDB) (106-93-4)	0.00002
2034	Glyphosate (1071-83-6)	0.7
2065	Heptachlor (76-44-8)	0.0004
2067	Heptachlor epoxide (1024-57-3)	0.0002
2274	Hexachlorobenzene (118-74-1)	0.001
2042	Hexachlorocyclopentadiene (77-47-4)	0.05
2010	Lindane (58-89-9)	0.0002
2015	Methoxychlor (72-43-5)	0.04
2036	Oxamyl (vydate) (23135-22-0) 0.2	0.2
2326	Pentachlorophenol (87-86-5)	0.001
2040	Picloram (1918-02-1)	0.5
2383	Polychlorinated biphenyls (PCBs)	0.0005
2037	Simazine (122-34-9)	0.004
2020	Toxaphene (8001-35-2)	0.003

could potentially contaminate the potable water system through backflow. However, it is simply considered too costly to monitor for every chemical. Therefore, the potential exists for chronic exposure to chemicals that are not monitored. Backflow prevention is necessary to eliminate or reduce this threat.[11,12]

CASE HISTORY 8: A Drink of Arsenic[9]

In California, a man died after drinking from a faucet that had previously been connected to a hose with an aspirator. The aspirator was used to spray a pesticide onto his lawn. The pesticide contained an arsenic compound, and by connecting the hose to the aspirator he created a cross-connection that allowed the backflow of a deadly contaminant into the potable water supply. A backpressure-backflow probably occurred because the hose was allowed to lie in the sun. The water in the hose heated and expanded to a greater pressure than the potable-water system. Therefore, the pesticide in the aspirator backflowed into the potable water line, and when the man took a drink from the faucet he also ingested the pesticide. This backflow incident could have affected the whole neighborhood if the potable water system pressure had decreased while this temporary cross-connection existed.

CASE HISTORY 9: Replacement of Water Lines - 1981[8]

A backflow of pesticide containing chlordane and heptachlor contaminated the water lines to approximately 75 apartments serving 300 people. A pesticide contractor created a cross-connection by allowing a hose connected to the potable water supply to become submerged in the chemicals contained in his tank truck. The hydraulic gradient that caused the backflow was created by a plumber when he cut into a 6-inch water main to put in a gate valve. The water in the line drained from the cut in the pipe and the contaminant backsiphoned into the potable water line. Repeated attempts to flush the chemicals from the water line proved unsuccessful, and the decision was made to replace all the water lines and plumbing that had been contaminated. Water service was restored 27 days later at a cost of $300,000.

CASE HISTORY 10: Propane Gas in Water System - 1982[8]

A work crew that was purging propane from a tank created a cross-connection that allowed the backflow of propane into the potable water system. The pressure in the propane tank (85 to 90 psi) was greater than the water system pressure (65 to 70 psi). Therefore, propane gas entered the potable water system through backpressure-backflow. About 200 cu. ft. of gas was estimated to have entered the potable water supply (this would fill approximately one mile of an 8-inch water main). The gas caused fires in two homes and a washing machine explosion that threw a woman against a wall, and it forced the evacuation of over 500 people from their homes.

CASE HISTORY 11: A Chemical Shower - 1986[11]

J.R. Isbell, a resident of Lacey Chapel, AL, unwittingly took a shower in sodium hydroxide, a caustic substance, when the chemical flowed out of the potable water line one morning. The chemical caused blisters all over his body, creating something like an allergic reaction. Another resident who washed her hair also had a similar reaction. And, other residents who drank the water complained of burned throats or mouths. Apparently, the chemical backflowed from a tank truck located at the nearby Thompson-Hayward Chemical Co. when a water main broke. The tank truck driver created a cross-connection by connecting a hose to the bottom of the tank to fill it. The contaminant backflowed into the potable water line because of backpressure-backflow. When the potable line pressure dropped due to the main break, the weight of water caused the contaminant to move into the potable water line.

Factors That Affect the Magnitude of a Chemical Backflow Incident

The factors that determine how many people will become ill because of a single backflow incident are similar whether the contaminant is a chemical or a biological agent. The size of the distribution system itself will play a significant role. The larger the distribution system, the greater the potential for a large number of people to be affected. Generally speaking, the greater the amount of chemical present, the greater its effect on public health. As with biological agents, dilution of the contaminant reduces the seriousness of illness or injury, as well as the probability that the contaminant will make someone ill. However, recall that even very low levels of some chemical contaminants (for instance, arsenic) can cause severe illness. The other factors that will play a role are the detection time, the onset time, the toxicity of the chemical and the susceptibility of exposed individuals.

Detection/Onset Time/Removal: Many chemical contaminants do not add any odor or color to the potable water, or change the appearance of the water in any other way. Therefore, the water is often ingested or used for other purposes before the contaminant is detected. Obviously, when chemicals do make a detectable change in the characteristics of the water, backflow of the contaminant is likely to do less damage, since it is detected sooner.

Generally, backflows involving chemicals are easier to identify than those involving biological contaminants, because chemical contaminations are only spread through the distribution system and (usually) are not passed from person to person. Therefore, the source of the contamination can be traced more easily and the cross-connection can be eliminated. When chemicals backflow from a manufacturing facility or business, the loss of the chemical will often be noticed, especially when it is used as part of a manufacturing process. When the level of contamination is high, the onset time for most chemical diseases is very short (usually 2 hours at most). The almost immediate appearance of symptoms helps to allow quick detection of a backflow. Thus, the consumers can be quickly notified to use another source of water until the contaminant is removed.

However, if the level of contamination is very low, the contaminant may not cause any acute symptoms, although long-term exposure can produce serious chronic ailments. Very low levels of contamination are most likely detected through testing. Testing frequencies vary depending on the water source and the type of contaminant. Surface water sources are tested for chemicals on a yearly basis, while well water sources are tested every three years. Accurate laboratory analysis is essential for detecting and preventing long-term exposure to low levels of chemical contaminants.

Chemical contaminants will become more and more diluted the longer they remain in the potable water system, but low levels can persist in water for a very long time. The contaminant is usually removed by flushing the potable water lines. However, adequately flushing the contaminant from the water lines becomes exceedingly difficult when its maximum contaminant level is very low (e.g., in the parts per billion range).

Toxicity/Susceptibility: Toxicity refers to the poisonous nature of a chemical and varies widely. Arsenic has already been used as an example of an extremely toxic chemical. Sodium is much less toxic. Therefore, its maximum contaminant level (160 ppm) is much higher than arsenic's (0.05 ppm).

Just as with biological contaminants, the susceptibility of individuals to chemicals is based not only on the amount of the chemical present, but also on their own physical health. Individuals who are already ill or weakened are generally more susceptible to illness from chemical contaminants. Also, chemicals affect groups of people differently. For instance, a very low level of nitrates will cause babies to develop a disease called methemoglobinemia. Methemoglobinemia is a disease that causes babies to turn blue from a lack of oxygen. Babies under the age of three months are susceptible, because they do not yet have the intestinal bacteria that are needed to convert nitrate to nitrite. Lead is another toxic chemical that has the greatest effect on young children (up to 3 years of age), affecting their development and sometimes causing mental retardation.

Backflow Prevention: As with backflow of biological contaminants, the spread of chemical contaminants can be reduced or prevented by utilizing and properly maintaining backflow preventers.

Summary

In this chapter the public health hazards and monetary costs associated with backflow have been discussed in order to illustrate the potential consequences of a backflow incident. Often, funding for prevention programs is insufficient because the costs of the program are more apparent

THE HAZARDS OF BACKFLOW

than the benefits. The case histories illustrate some potential costs of inefficient or non-existent backflow prevention programs and, thus, provide support and justification for a comprehensive backflow prevention program.

This chapter also reviewed the different factors that determine how many people will be affected by a backflow incident. This information will be especially valuable to program managers and water plant operators as they attempt to deal with emergencies resulting from backflows.

CHAPTER TWO

Chapter Two Review

Types of Contaminants or Pollutants:

    Biological Low (Non-health) Hazard
    Biological High (Health) Hazard
    Chemical Low (Non-health) Hazard
    Chemical High (Health) Hazard

2-1   Define a toxic substance. *Something that makes you ill.*

(Please use answers below for questions 2-2 through 2-5.)

LACTOBACILLUS	**GIARDIA LAMBLIA**
LEGIONELLA	**IRON**
MERCURY	**ARSENIC**

2-2   Give an example of a biological pollutant (non-health hazard).
Lactobacillus

2-3   Give examples of biological contaminants (high hazard).
Legionella
Giardia Lamblia

2-4   Give an example of a chemical pollutant (low hazard).
Iron

2-5   Give examples of chemical contaminants (high hazard).
Mercury
Arsenic

REFERENCES

1. "Giardiasis," Hach News & Notes, 1984, 8(3):11.

2. Lawrence, C. H., Gurthie, P. J., and Silberg, S.L. 1987. "Identification of *Legionella pneumophila* in Recreational and Water Supply Reservoirs in Central Oklahoma," *Journal of Environmental Health*. 49(5):274.

3. *Manual of Cross-Connection Control (9th Edition)*, 1994. Foundation for Cross-Connection Control and Hydraulic Research, University of Southern California, Univ. Park, CA.

4. "Drinking Water," Harvard Medical School Health Letter, 1988. 14(1):1-2.

5. Craun, C.F. and McCabe, L.M. 1973. "Review of the Causes of Waterborne Disease Outbreaks," Jour. AWWA 65(1):74. In: *Manual of Cross-Connection Control Procedures and Practices*, 1981. State of California Health and Welfare Agency, Department of Health Services, Sanitary Engineering Section, p.i.

6. Craun, C.F. and McCabe, L.M. 1976. "Waterborne Disease Outbreaks in the U.S. 1971-1974," Jour. AWWA 68(8):420. In: *Manual of Cross-Connection Control Procedures and Practices*, 1981. State of California Health and Health and Welfare Agency, Department of Health Services, Sanitary Engineering Section, p.i.

7. Craun, C.F. and Gun, R.A. 1979. "Outbreaks of Waterborne Disease in the United States: 1975-1976," *Jour. AWWA*, 8(8):422. In: *Manual of Cross-Connection Control Procedures and Practices*, 1981. State of California Health and Welfare Agency, Department of Health Services, Sanitary Engineering Section, p. i.

8. *Cross-Connection Control Training Package*, 1985. American Water Works Assn., New York, NY.

9. *Cross-Connection Control Manual*, 1975, revised 1989. (EPA 430-973-002). U.S. Environmental Protection Agency, Water Supply Division.

10. Heleman, B., 1988. "Fluoridation of Water," *Chemical & Engineering News*, 66(31):32.

11. "Lacy's Chapel Residents Burned by Contaminated Water Supply," *Birmingham News*, October 10, 1986. Birmingham, AL.

CHAPTER THREE
LAWS AND RESPONSIBILITY

In Chapter 2, the hazards of backflow were discussed. Backflows were shown to be the cause of monetary losses, worker lay-offs, disease, and death. Because of these serious impacts on public health, laws have been enacted to prohibit backflow and cross-connections. However, these laws have been slow in development. For instance, it was not until the 1970s that the Federal Safe Drinking Water regulations were written. Moreover, even the latest revision (1996) still does not clearly address the topics of backflow prevention and cross-connection control. Most plumbing laws do address cross-connection control and backflow prevention, but are inconsistent in their treatment of the topic.

The intent of this chapter is two-fold. First, the relevant laws will be examined in an effort to provide a basis for achieving compliance with all pertinent regulations. Second, the responsibilities of everyone involved in the prevention of backflow will be more clearly defined. This is important because compliance with minimum requirements is not necessarily sufficient to limit liability.

Laws

Federal Regulations

On December 16, 1974, Public Law 93-523 established the Safe Drinking Water Act (SDWA). From this, the National Interim Drinking Water Regulations (NIDWR) was promulgated on December 24, 1975 and became effective on June 24, 1977. These regulations replaced the Public Health Service Drinking Water Standards of 1962. While the body of the regulations itself does not clearly address the topic of cross-connections and backflow prevention, the "Statement of Basis and Purpose" does express concerns related to these topics. The "Statement of Basis and Purpose" is the preface that explains the reasoning behind how and why the regulations were developed. It states, "Minimum protection (of the drinking water) should include programs that result in . . . prevention of health hazards, such as cross-connections. . . ."[1] Clearly, backflow prevention and cross-connection control are issues that the developers of the SDWA considered important.

A close reading of the regulations reveals other references to cross-connection control. The NIDWR regulations set maximum contaminant levels for a number of chemicals believed to be harmful to public health. Maximum contaminant levels are set for those substances that are traditionally found in water (for example, substances that occur naturally in the water, chemicals added during the treatment process or substances that can commonly pollute water sources). In addition, these contaminants, as well as many that are not listed, can enter the water through backflow into the distribution system. However, a maximum contaminant level was not established for every toxic or undesirable contaminant that might enter the public water supply, because "standards for innumerable substances which are rarely found in water would require an impossible burden of analytical examination." Therefore, "no attempt has been made to prescribe specific limits for every toxic or undesirable contaminant that might enter a public water supply."[1]

The "Statement of Basis and Purposes" does express concerns about "other" contaminants that do not have established maximum contamination levels. These "other" contaminants could easily enter the drinking water supply through unprotected cross-connections. Further, the preface clearly identifies a need for backflow prevention programs when it states: "Knowledge of physical defects or of the existence of other health hazards in the water supply system is evidence of a deficiency in protection of the water supply system."[1]

Moreover, because the original regulations required that maximum contamination levels be measured at the consumer's tap, the regulations themselves provide an incentive for backflow prevention programs. The regulations established that it was not sufficient to just produce quality water. The water purveyor must also ensure that quality water is indeed what is reaching the consumer. This requirement is contained under the original definition for "maximum contaminant," which is "the maximum permissible level of a contaminant in water which is delivered to the free flowing outlet of the ultimate user of a public water system."[1] Therefore, the purveyor is responsible for the integrity of the entire distribution system, including interior plumbing, as well as the quality of the water itself. A cross-connection that exists inside a private building could potentially allow backflow of hazardous materials into the drinking water supply. If a backflow of contaminants at levels above those allowed by regulations entered the drinking water system within the building, the consumer would not receive water that meets drinking water standards, and the purveyor could be held responsible.

Understandably, being held responsible for conditions not under their control raised some concern among water purveyors. Plumbing inside a commercial or residential establishment is regulated by the local plumbing codes adopted by a city, town or county and therefore is under the control of the local plumbing authority and usually not the water purveyor.

The U.S. Supreme Court limited the water purveyor's control even more when it ruled in *Camara v. Municipal Court of the City and County of San Francisco*[2] that municipal inspectors are required to obtain search warrants, except in a genuine emergency, if they are refused entry to residential or commercial premises for inspection purposes, even if a local ordinance allows entry of such inspectors at reasonable times. Thus, water purveyors were apparently responsible for a situation they had no authority to regulate. This problem was alleviated to some extent when the definition of "maximum contaminant level" was expanded on December 24, 1975[3] making clear that ". . . contaminants added to the water by circumstances under the control of the consumer are not the responsibility of the supplier of water, unless the contaminants result from corrosion of piping and plumbing resulting from the quality of the water supplied."[1] This statement suggests that the consumer must bear some responsibility for preventing the backflow of contaminants into the potable water system. Thus, the purveyor and consumer share some of the responsibility for preventing backflows. This delineation of responsibility was clarified further in 1986 when the definition of a "maximum contaminant level" was modified; "The term 'maximum contaminant level' means the maximum permissible level of a contaminant in water which is delivered to any user of a public water system."[4]

State Regulations

Under Section 1413 of the **SDWA**, states can obtain primary enforcement responsibilities (primacy) for the water quality program. In order to do this, **state regulations** must equal or exceed the federal regulations. The administrator of the Environmental Protection Agency retains authority over states that do not obtain primacy.

Florida was granted primacy over its water program under the authority of Florida Safe Drinking Water Act, Chapter 403.850-403.864 Florida Statutes. In January of 1975, the State of Florida adopted Florida Administrative Code (F.A.C.) Chapter 17-22 (Public Drinking Water Systems) and the regulations went into effect in November of 1977. The Florida regulations revised in November 1987, revised and renumbered in 1989, and revised and renumbered in 1994 (Florida Administrative Codes 62-550, 62-555 and 62-560 have replaced 17-22, revised Aug. 28, 2003) address the topics of cross-connection control and backflow prevention more clearly than the federal regulations. In fact, state regulations clearly prohibit cross-connections. In Rule 62-555.360(1) F.A.C., it states: "Cross-connection as defined in Rule 62-550.200 F.A.C. is prohibited."

A cross-connection is well defined under Rule 62-550.200 F.A.C. (22)

> "'Cross-connection' means any physical arrangement whereby a public water supply is connected, directly or indirectly, with any other water supply system, sewer, drain, conduit, pool, storage reservoir, plumbing fixture, or other device which contains or may contain contaminated water, sewage or other waste or liquid of unknown or unsafe quality which may be capable of imparting contamination to the public water supply as the result of backflow. Bypass arrangements, jumper

LAWS AND RESPONSIBILITY

connections, removable sections, swivel or changeable devices and other temporary or permanent devices through which or because of which backflow could occur are considered to be cross-connections."

In addition, the Florida regulations specifically address "other contaminants" without established maximum contamination standards. Rule 62-550.330 F.A.C. states, "No **contaminant** which creates or has the potential to create an imminent and substantial danger to the public shall be introduced into a public water system." This statement clearly prohibits backflow of any contaminant that is even potentially hazardous.

The Florida Department of Environmental Protection (FDEP) has taken its responsibility for ensuring the quality of the public drinking water one step further by requiring that community water systems establish a routine cross-connection control program. This regulation references the guidelines of the American Water Works Association (AWWA) for the development of the program. Specifically, the 2004 edition of the AWWA manual *Recommended Practice for Backflow Prevention and Cross Connection Control* (Manual M14) is incorporated into the rules of the FDEP by reference. This guideline helps establish the necessary components of a cross-connection/backflow prevention program and is discussed in Chapter 8.

FDEP, through Rule 62-555.360(3) F.A.C., has also given the water purveyor authority and responsibility to discontinue service to any customer who refuses installation of a backflow preventer where a cross-connection exists. The regulation reads: "Upon discovery of a **prohibited cross-connection**, public water systems shall either eliminate the cross-connection by installation of an appropriate backflow prevention device acceptable to the Department or shall discontinue service until the contaminant source is eliminated." This statement is erroneous, however, when it states that the cross-connection will be "eliminated" by the installation of the appropriate backflow preventer. Installation of the appropriate backflow preventer only provides a protected cross-connection, instead of an unprotected cross-connection.

A brief definition of those backflow preventers acceptable to the FDEP is included under Rule 62-555.360(4) F.A.C. (Appendix B). They include: an air gap separation, reduced pressure backflow preventer, atmospheric vacuum breaker, pressure vacuum breaker, dual check valve, and double check valve assembly. As a note, an air gap separation does not utilize any mechanical mechanism to function and is more accurately referred to as a "method" of backflow prevention. This rule also gives the water purveyor the right to control and supervise the installation of an approved device. It states, "They shall be installed in agreement with and under the supervision of the supplier of water or his designated representative (plumbing inspector, etc.) at the consumer's meter, at the property line of the consumer when a meter is not used, or at a location designated by the supplier of water or the Department." This not only gives the water purveyor control

> "No contaminant which creates or has the potential to create an imminent and substantial danger to the public shall be introduced into a public water system."

> "Upon discovery of a prohibited cross-connection, public water systems shall either eliminate the cross-connection by installation of an appropriate backflow prevention device acceptable to the Department or shall discontinue service until the contaminant source is eliminated."

over the installation of the backflow preventers, but also potentially creates some liability for the water purveyor if a backflow preventer is not installed correctly or is not in working order when initially installed.

Regulations in Other States

Each state may or may not have its own regulations related to backflow prevention and cross-connection control, and the department that enforces these regulations also varies from state to state. Enforcement of cross-connection regulations may be the sole responsibility of health authorities (local or state health departments), of environmental regulatory agencies, or of plumbing officials (plumbing inspectors, building inspectors or code enforcement personnel). Enforcement may be shared among the various agencies. Usually, the water purveyor will be involved in the implementation of these programs to some extent.

The lack of specific language in the Federal regulations has resulted in significant differences among various states in the enforcement of their backflow prevention programs. This inconsistency among states creates confusion about the importance of backflow prevention, thereby potentially slowing the implementation or enforcement of backflow programs. Furthermore, individuals involved in backflow prevention who work in several states have to deal with different laws and regulations, thus complicating the implementation of cross-connection control and backflow prevention programs. Training related to a particular state's regulations is necessary to avoid violation of backflow prevention laws (and backflows themselves).

Plumbing Codes

A number of different plumbing codes exist: the International Plumbing Code (IPC), the Uniform Plumbing Code (UPC), the Standard Plumbing Code, the BOCA National Plumbing Code, the National Plumbing Code, the National Standard Plumbing Code, the Florida Building Code, and Florida Administrative Code 10D-9. In addition, many other local or regional codes have been developed. While most of the plumbing codes address the problems of cross-connection and backflow prevention, the extent and depth of coverage varies.

The adoption of many different codes throughout the United States creates a hardship for individuals who install plumbing, as they must be familiar with a number of different regulations in order to comply with the particular codes in effect. This not only makes their job more difficult, but also increases the potential for error. Further, the variation among the codes could create some confusion as to the significance of backflow prevention and cross-connection control.

The State of Florida has implemented the Florida Building Code (March 2002) and has adopted the International Plumbing Code (IPC) as its plumbing code.

LAWS AND RESPONSIBILITY

Responsibilities and Liabilities

So far, the discussion has focused on the various laws, regulations, and codes that are aimed at the prevention of backflow and cross-connections. However, enactment of laws alone will not totally prevent backflows. Fines and penalties are used to induce people to obey the laws; but, when laws are broken or people otherwise act negligently with regard to backflow prevention and people are injured or property is damaged, the court system is called upon to assign civil liability and recover civil damages. The discussion that follows should help to determine some of the responsibilities and liabilities related to backflow prevention and cross-connection control. However, no attempt is being made to give legal advice or delineate definite responsibilities. The former is for lawyers to provide and the latter is for the courts to decide.

The Safe Drinking Water Act establishes some responsibilities for both the water purveyor and the consumer, as well as for those enforcing the regulations. The plumbing codes establish responsibilities for plumbers, contractors, and others who install plumbing, for engineers who design plumbing plans, and again, for those who enforce the plumbing codes. Everyone involved in cross-connection control and backflow prevention has some level of responsibility. It is essential that each individual and agency involved in backflow prevention be aware of the laws, as well as the responsibilities of other individuals and agencies involved in backflow prevention. Not only can this knowledge help reduce the number of backflow incidents, but it can also help to limit individual liability in the event of a backflow incident.

Liabilities and Damages

Water purveyors are legally obligated to provide and distribute safe drinking water. If contaminated water reaches a customer and causes personal or property damage, the purveyor and his/her agents (plumbing contractors, engineers, etc.) may be held liable for those damages. In a potential claim to recover compensation, the injured customer may attempt to establish: that a warranty ensuring safe water existed between the purveyor and the customer, that the purveyor breached this warranty, and that the breach of warranty actually caused the damage.

Existence of a warranty: A warranty is basically a contractual promise that a certain fact is true. In the case of potable water, the promised fact is that the water distributed throughout the water purveyor's system is pure and wholesome. The warranty may be clearly established in the form of a written contract or may be simply implied in the relations between the seller and the buyer. The most common of these implied warranties is the warranty of merchantability, which simply means that the purveyor impliedly promises the purchaser of the water that the water is fit for ordinary purposes such as drinking, cooking, bathing, etc. It may or may not be fit for other purposes such as use as cooling water, making beer, etc. because those are not generally considered ordinary purposes and may require special pretreatment. Contracts can establish conditions of service and may limit liabilities by delineating what the purveyor does not provide or promise.

An **implied warranty** is an assumption made by the customer that water coming out of the faucet is fit for ordinary purposes. The customer naturally assumes this, because laws require that community drinking water meets standards that have been established to protect public health. Court precedent has established that water does carry the implied warranty that it is fit for human consumption: *Hayes v. Torrington Water Co., Hamilton v. Madison Water Co.*[5] Another example of an implied warranty is the assumption that a licensed plumber will install plumbing according to applicable plumbing codes.

An **expressed warranty** is an explicit statement declaring something.[6] For example, an advertisement declaring that water is "safe and pure" might be construed as an expressed warranty.[7]

To recover damages from the water purveyor for breach of warranties, an injured customer must prove in court that an implied, expressed, or written warranty did exist. Determining that an implied or expressed warranty existed is often a point of contention. For instance, the customer's lawyer may argue that an explicit advertisement constitutes an expressed warranty, while the purveyor's lawyer could contend that the advertisement was simply an opinion or sales "puffing."[5] Ultimately, a jury must decide whether a warranty did in fact exist.

Breach of warranty: Once a warranty's existence is established, the injured customer would need to show that the purveyor breached that warranty. A breach of warranty is basically a breaking of the promise that a certain fact (such as fitness for ordinary purposes) is true. Violation of a law or regulation in itself does not establish a breach of warranty, but will likely be evidence of the breach of warranty. If the water purveyor fails to provide water meeting maximum contaminant standards, he/she is in violation of law and would be open to fines (in Florida up to $5,000 per day per offense) in actions brought against her/him by the government even if nobody becomes ill and no property is damaged. Furthermore, even if the purveyor does comply with all regulations, he/she may still be held liable for breach of warranty.[6] However, even if damage does result, the water purveyor cannot be held liable in civil courts for personal injuries or damages just because the law has been violated without additional showing that those violations or the breach of the water purveyor's duty to provide pure and wholesome water actually *caused* the injuries or damages as further discussed below.

Negligence: On the other hand, the customer's lawyer could argue that a "reasonable and prudent" water purveyor would comply with the law because laws help determine the **standard of care** (i.e., what would normally be expected).[3] The duty to adhere to a certain standard of care is the basis for a claim of negligence when the defendant fails to adhere to that standard and that failure causes injury or damage to another. Therefore, even if the purveyor does comply with all regulations, he/she may be held liable for negligence if the accepted standard of care exceeds the government's regulations, and some foreseeable condition causes the purveyor to breach the duty to meet that standard of care. Compliance with the laws alone may not be considered adequate, because laws establish only minimum standards and professionals or experts can "reasonably" be expected to keep up with advances in the industry. For instance, negligence might be found to have occurred if it is shown that the water purveyor should have been aware of the dangers of cross-connections and taken industry-accepted steps to prevent or eliminate cross-connections. This will be of special concern in those states that have relatively weak regulations related to backflow and cross-connection control.

Because laws and regulations are often slow in developing, "standard practices" and "recommended guidelines" should be used when laws are inadequate. For instance, the customer's lawyer might use the AWWA policy statement (Appendix A) on cross-connection control as a "reasonable" expectation of water purveyor responsibilities. Because AWWA is a highly respected organization composed largely of water purveyors, this policy statement could be held as the general consensus of the industry regarding the appropriate approach to backflow prevention.

The last step in a damage recovery suit, and usually the most difficult part for the customer's lawyer, is to prove that the breach of warranty or negligent act or omission was the actual cause of the damage.[5] Many cases have been lost because evidence (e.g., contaminated water) was not handled and tested properly or because the disease creating the injuries could have been caused by other sources such as pools, packaged foods, etc. Sometimes, nobody thinks to save any of the contaminated water at the time of the backflow event. The evidence is usually circumstantial at best. If the evidence is circumstantial, the plaintiff's lawyer will try to show that the backflow event most probably was the cause of the damage.[4] Finally, the lawyer must establish the extent of the damage.

A **personal injury** or "tort" claim can be made on the grounds of negligence or nuisance.[6] Perhaps the water purveyor failed to require the appropriate backflow preventer, or perhaps the correct backflow preventer was installed but failed to work properly because it was not maintained. The courts have ruled that failure to properly inspect the water distribution system to prevent it from becoming contaminated could be negligence.[2] The courts in California have made it clear that the only acceptable defense is evidence that the water purveyor acted in good faith and adopted an acceptable cross-connection control program that was carried out diligently and effectively.[3]

When ascertaining negligence, the court will try to determine whether the defendant took all reasonable steps to avoid the problem or damage. Did the water purveyor properly install, inspect and maintain the backflow preventers? If, by ordinance, the duty of maintaining backflow preventers is delegated to the water customer, did the water purveyor keep records on this maintenance to ensure that it was being completed? Did the plumber install an appropriate backflow preventer and test it to ensure that it was working properly? If the court finds that the defendant (whether plumber or water purveyor) did take all reasonable steps to prevent the problem, or finds that damages resulted from an honest error in judgment, negligence might not be charged. In Florida and most states, if both the water purveyor and consumer were found to be negligent to some degree, the court can assign liability by percentages of negligence, and the damages will be borne by both the plaintiff and the defendant according to those percentages.

In some instances, the water purveyor may have required a backflow preventer (e.g., a vacuum breaker), but a backflow of toxic materials could occur if the device failed. The purveyor's lawyer will try to show that the water purveyor took a "reasonable precaution" by calling expert witnesses to state that the vacuum breaker was adequate protection for the situation. Normally, the defense will also try to show that others were the ultimate, or at least partial, cause of the condition causing the injury or damages. To counter, the plaintiff will try to find expert witnesses to declare that the vacuum breaker was not adequate protection because this type of device cannot be tested, or because this was not the correct device for the particular situation. Further, the plaintiff may try to show that the device was not inspected frequently enough or that the water purveyor failed to adequately inform the consumer of his/her responsibilities.

In addition to negligence, nuisance can be charged in some situations when the act of omission of any person creates an ongoing condition endangering the health and property of water users. Nuisance is defined as an unlawful or unreasonable use of one's own property or unlawful or unreasonable conduct that causes damage.[6] For instance, if a chemical plant allows or creates a condition that threatens to allow a dangerous material to escape and enter the public water distribution system through a backflow, nuisance might be charged.

Clearly, responsibilities extend far beyond the letter of the law. The exact delineation of responsibilities among those who are involved in backflow prevention is not clear. While the following chapters will try to clarify these delineations, those involved in backflow prevention should endeavor to achieve the highest level of protection possible, using sound judgment, in order to avoid the hazards associated with a backflow (i.e., death, disease, and law suits).

Individual Responsibilities

Water Purveyors

In general terms, the water purveyor is responsible for supplying water that is safe and wholesome to the public. First, the water purveyor must produce the water, then monitor the quality of the water and take precautions to protect it from potential sources of contamination through the

Nuisance is defined as an unlawful or unreasonable use of one's own property or unlawful or unreasonable conduct that causes damage.

Those involved in backflow prevention should endeavor to achieve the highest level of protection possible, using sound judgment in order to avoid the hazards associated with backflow.

The water purveyor is responsible for supplying water that is safe and wholesome to the public.

appropriate use of backflow preventers. If these steps are not taken, the water purveyor could be held liable for supplying faulty goods. Legal precedent has established that both private and public water purveyors are subject to the charge of negligence, i.e., public water purveyors are not protected by sovereign immunity.[3]

The responsibilities of the water purveyor have already been discussed to some extent while reviewing Rules 62-550 and 62-555 F.A.C. However, the case has been made that compliance with the regulations alone is not the end of responsibility. Interpretation of the regulations varies significantly among individuals. Therefore, this chapter will identify the various responsibilities that can be imposed on the water purveyor. Again, this material should not be construed as legal advice, and small changes in facts can have a substantial impact on the assignment of liability.

The following example illustrates that compliance with the law alone is not always sufficient. Florida Administrative Code 62-555.360(1) (Appendix B) does allow cross-connections between community water systems that meet water quality standards. No hazard is perceived in this situation, since all community water systems are required to comply with the same standards. However, the installation of a backflow preventer between community water systems provides a good safeguard. It prevents the contamination of one system from affecting the other system(s). For instance, if one community water system's backflow prevention program is in the beginning stages of development, the potential for a backflow incident in that system is much higher. Thus, a reasonable and prudent decision on the part of the water purveyor might be to install a backflow preventer between the two systems to protect against the backflow of contaminants between the systems, even though it is not required by law.

As already mentioned, the regulations have alleviated the water purveyor's responsibility for contaminants added by circumstances under the consumer's control. This change may have resulted from cases such as the 1967 San Francisco court case, that ruled a water purveyor had no right to enter private property when denied entry unless a warrant was obtained.[2] However, the purveyor is still responsible for contamination resulting from the corrosiveness of the water. For instance, the regulation on lead levels in drinking water requires that test samples be taken inside the home at the tap, because it is believed that the corrosiveness of water is leaching lead from the solder used in plumbing. In this case, the purveyor is held responsible for the lead contamination, even though the lead originated in the consumer's plumbing.

The Florida water purveyor is not relieved of all responsibility for backflows that originate inside private property. Florida regulations give the water purveyor the right and responsibility to discontinue water service to anyone who has a cross-connection until the cross-connection is eliminated or protected from backflow (Rule 62-555.360 F.A.C.) (Appendix B) as soon as the water purveyor knew, or should have known, of the cross-connection.[3] In those states where this right has not been established by law or regulation, it can still be acquired by making the elimination of cross-connections a "condition of service." Because maximum contaminant levels are measured at the consumer's tap, a backflow within a private building could easily create an illegal contamination level. As Rule 62-550.300 F.A.C. states: "The ultimate concern of a public drinking water program is the quality of piped water for human consumption when the water reaches the consumer." Recall that the AWWA policy statement (Appendix A) calls upon the water purveyor to make inspections and reinspections of interior plumbing as necessary in order to detect hazardous conditions resulting from cross-connections. Many water purveyors establish their right to inspect interior plumbing within the local ordinance. However, if this right is assumed, the water purveyor might also be considered to assume the responsibilities and liabilities of any plumbing inspector.

The question still remains: where does the water purveyor's responsibility end—at the water meter, the point of delivery from purveyor to consumer, or at the tap? Obviously, there is no definitive answer. Each water purveyor must examine the problem and determine what is "reasonable and prudent" for a particular situation. Generally, each of the essential components of a good backflow prevention program (as discussed in Chapter 8) signify responsibilities the water purveyor could "reasonably" be expected to meet and fulfill. The decision maker must consider what can be achieved financially, what is cost effective, what are normal and accepted practices in the industry, etc. However, the water purveyor should remember to consider the costs of litigation that could result because of a backflow. For many water purveyors, the development of a program is achieved in a series of steps, implementing what is possible under current financial constraints and expanding the breadth of the program as more funds are made available.

Often, the first step is containment (service protection). Containment backflow preventers are installed at the meter in order to stop the spread of a contaminant beyond the consumer who created the problem. This involves surveying present consumers to determine the degree of hazard and then retrofitting the consumers with the highest hazards first. Water purveyors can either perform the retrofit themselves or require by ordinance that the consumer perform the retrofit. Recall that the water purveyor is given authority and thus responsibility to inspect such installations.

Containment backflow preventers are installed at the meter.

Another approach the water purveyor might take is isolation. The isolation method provides internal protection by having a backflow-preventer at each cross-connection within a consumer's property. The water purveyor would need to have inspection privileges written into the ordinance or service contract. Existing and "grandfathered" plumbing is surveyed for cross-connections, and backflow-preventers are installed at each potential point of contamination. In new construction, plan approval by the utility and plumbing inspections by the plumbing authority and/or the water purveyor should prevent the installation of cross-connections. One problem that arises with this method is that alteration of plumbing can easily create new cross-connections or recreate old cross-connections. Ultimately, this

The isolation method provides internal protection by having a backflow-preventer at each cross-connection within a consumer's property.

must be controlled through a routine inspection and testing program. Court precedent was established in *Aronson v. City of Everett*, that simply accepting the promise of the consumer to rectify a hazardous situation is not sufficient. The water purveyor must inspect the facility to ensure that dangerous cross-connections are eliminated.[2] Another method to limit the creation of cross-connections is to designate a "water supervisor" at the larger facilities. The water purveyor generally would be involved in training this supervisor or requiring training through ordinance. The water supervisor would then assume the responsibility, for the consumer, of ensuring that no new cross-connections are created.

In addition to isolation or containment, a third approach consists of a combination of both. This method may be the best approach, because it provides two levels of protection. One level (isolation) deals with each potential source of contamination within the building, and the second level (containment) prevents the spread of contaminants into the public water system if the first level of protection fails. However, this dual approach would definitely be more costly, and it may not be necessary for every facility. The expense of surveying the water customers could be reduced by eliciting the help of others who should be involved in backflow prevention. For instance, the purveyor could involve Health Department officials in the surveying process. Health Department employees could be called upon to look for cross-connections in those facilities that they routinely inspect, e.g., schools, swimming pools, restaurants, hospitals, and industries.

No matter which approach is chosen, the water purveyor should set standards and specifications for backflow preventers. The water purveyor should provide timely testing, repair, and maintenance of backflow preventers or require through ordinance that the consumer perform these duties. If the duty is delegated to the water consumer, the water purveyor should still determine that these requirements are being met (through spot checks, review of test results, etc.). Further, the water purveyor may want to require some minimum qualifications for those performing these duties, such as certification or training in backflow prevention and cross-connection control.

Another responsibility, according to Rule 62-550 F.A.C., includes monitoring the water quality to determine compliance with the regulations. It is reasonable to assume that a prudent water purveyor would perform more than the minimum amount of testing. For instance, the water purveyor may test water quality any time there is a reduction in pressure to assure that a backflow of contaminants has not occurred.

Because the water purveyor could potentially be considered an "expert" on the subject of backflow and cross-connection control, it would seem "reasonable and prudent" that the water purveyor provide some amount of education on the hazards of backflow to the water consumer. In fact, educating consumers, water utility personnel and others is one of the seven essential elements of an effective backflow prevention program, and will be discussed in more detail in Chapter 8.

> **Simply accepting the promise of the consumer to rectify a hazardous situation is not sufficient.**

Most of the water purveyor's responsibilities deal with preventing potential backflow incidents. However, in the event that a backflow incident does occur, the water purveyor is then responsible for confining the spread of the contaminant and notifying the proper regulatory authority, within 2 hours, and the consumer in order to minimize adverse consequences. Mitigation of backflow can more easily be accomplished if the water purveyor has an emergency plan available to deal with the event. A well thought-out written plan can be a great aid when a backflow occurs. One important element of any emergency plan should be accurate record keeping. Records should be kept of conversations, sample collection data (including chain-of-custody records), measures taken to mitigate the backflow of the contaminant, notification procedures, etc. Accurate records could provide valuable evidence that the water purveyor took the appropriate steps both before and after the backflow occurred. Other important elements of an emergency plan will be discussed in Chapter 8.

Finally, the water purveyor has responsibility for cleaning and disinfecting the water distribution system after a backflow event and restoring the quality of water provided to the consumer. Sometimes this is not an easy task, recall the case history from Chapter 2 in which chlordane contaminated the water-distribution system, which required replacing plumbing lines in 75 apartments.

Consumer

The basic responsibility of the consumer is to prevent contaminants or pollutants from entering the public water system. This responsibility begins at the point of service (meter). Recall that if a backflow occurs, the water consumer can be held liable for creating a nuisance. Nuisance can be charged whenever a person allows the escape of dangerous materials from their property. Also, the consumer could be held liable for damages due to negligence. Negligence might be charged if alterations are made in plumbing that result in the creation of a cross-connection, or if plumbing (including backflow preventers) is not maintained as required by local plumbing codes or ordinance.

Generally, the consumer is also responsible for all the duties assigned by the water purveyor through ordinance. Furthermore, the consumer is responsible for the quality of water that reaches individuals inside the facility/premises.[8] Often, new property owners will make an assumption that the building's plumbing meets current standards. A prudent prospective owner should not make such assumptions, but should instead have the building inspected or require a written certification of compliance from the current owner prior to making the purchase. If the consumer has a designated water supervisor, this person as an "expert," would assume an even higher level of responsibility.

If a backflow does occur, the consumer is responsible for confining the further spread of the flow, as well as notifying the water purveyor and proper regulatory officials. Then, either the consumer or the water purveyor (as specified by ordinance) is responsible for removing the contaminant

The basic responsibility of the consumer is to prevent contaminants or pollutants from entering the public water system.

from the private system and restoring water quality. Following a backflow incident, the consumer would be responsible for improving methods of backflow prevention to avoid future accidents.

Regulatory Officials

In general, the responsibility of any regulatory official is promulgation and enforcement of regulations that mandate reasonable and prudent action to prevent backflows.[9] For instance, in addition to the Drinking Water Regulations contained in Rules 62-550, 62-555 and 62-560 F.A.C., the FDEP has promulgated other regulations, such as Rule 62.610 F.A.C. (Reuse of Reclaimed Waters and Land Application) and Rule 62-620 F.A.C. (Wastewater Facilities), which also address the topic of cross-connection and backflow prevention. These regulations deal with substances that could significantly affect water quality in the event of a backflow. Failure to enforce laws or regulations could make a regulatory official partially liable in the event of a backflow.

To some extent, regulatory officials are limited to enforcement of existing regulations and are not allowed the freedom to enforce new or better methods until they are made law or departmental policy. However, the prudent official can make recommendations and record these recommendations in order to better fulfill the responsibility of the position. When doing so, however, the regulatory official should avoid making recommendations that conflict with current law (unless approved by the administration).

In Florida, FDEP is the agency responsible for developing regulations "to ensure that public water systems supply drinking water which meets minimum requirements."[9] Recall also that FDEP is concerned with the quality of water that reaches the consumer's tap (Rule 62-550.300 F.A.C.). It would be reasonable to assume that FDEP is not only responsible for ensuring that the water purveyor prevents contaminants from entering the public water system, but also responsible for the water system within the consumer's premises. However, FDEP does little to assure that the plumbing system within a consumer's property is protected from backflow. Generally, plumbing codes (which are not under FDEP's control) are used to regulate the plumbing inside the consumer's premises.

Because FDEP is granted primacy over the water quality regulations, local health departments generally play a very minor role in the enforcement of backflow prevention programs, except where county public health units have been contracted by FDEP to perform enforcement services. In other states, local health departments may play a much more active role in backflow prevention programs. It seems reasonable to expect that health departments should have some level of responsibility for ensuring that backflows are prevented, as their ultimate goal is protection of public health. Most health departments routinely inspect such facilities as swimming pools, schools, restaurants, and hospitals, where backflow of hazardous materials into either the private or community water system could create highly hazardous situations and have a great impact on public health. Furthermore, health department officials could regulate plumbing systems of such facilities as hospitals, which are not covered under some plumbing regulations, i.e., the International Plumbing Code.

Additional health department functions might include: education of consumers on the hazards of backflow, notification of affected consumers in the event of a backflow incident, collection and analysis of water samples, administration of epidemiological studies to determine the cause of disease and source of contamination, and delivery of safe water to those areas affected by backflow.

Plumbing Inspector

The plumbing inspector is responsible for enforcement of local plumbing codes on the consumer's premises. Generally, the plumbing official's responsibility begins at the meter, the point of delivery. The plumbing official's title may be "plumbing inspector," "building inspector," "code enforcement officer," etc. When reviewing plans for new construction, the plumbing official should be able to prevent the installation of cross-connections by predicting where they may be inadvertently designed or built into the structure. During inspections at the construction or remodeling site, surveys should be conducted to ensure that cross-connections are not installed or that the proper backflow prevention is provided at cross-connections in order to prevent backflows.

Consulting Engineers

Consulting engineers have the responsibility to design plumbing according to all applicable laws, regulations, and codes. However, as experts, these individuals could be charged with negligence for failure to comply not only with current regulations, but also with "standard practices" or updated "best methods." Additionally, consulting engineers would hold some liability for any breach of contractual duties. Therefore, the engineer could be sued not only by the consumer (for damages sustained as the result of a backflow), but also by the plumber with whom the engineer had a contractual agreement. This shifting of responsibility on the part of the plumber will probably be part of the plumber's defense against a charge of negligence.

Contractors, Plumbers, and Testers

Contractors, plumbers, and testers are also responsible for complying with all plumbing codes and local ordinances. The duties of these individuals could involve installation, maintenance, and repair of plumbing such as irrigation systems, fire sprinkler systems, plumbing lines, plumbing fixtures, and backflow preventers, as well as testing of backflow preventers. In some locations, a plumbing license may be required by code or ordinance to

LAWS AND RESPONSIBILITY

do any work on plumbing. Water utility personnel are usually exempt from this requirement when constructing their own facilities. Ordinances may require that only certified backflow prevention technicians test and/or repair backflow preventers. No matter who does the work, they are responsible for maintaining adequate records. For instance, maintenance/repair reports should include a list of all replacement parts or materials used in making the repair. Repairs should be made according to manufacturer's instructions, and any replacement parts must be factory authorized replacement parts.

A case could be made that contractors, plumbers, and testers, like consulting engineers, are experts, since they have received considerable training to maintain their license or certification. They could, therefore, be charged with negligence for failure to comply with updated "standard practices." Negligence could be charged either through general law or breach of contractual agreement.

Approved testers, working in the community, should be required to have a business license and insurance coverage. The local authority should have a "Code of Conduct" document prepared and testers must sign this document before they can be placed on the approved tester list. The local authority should understand that by adding a tester to the approved tester list places some responsibility for the conduct of the tester on the local authority.

Code of Conduct

Certified testers placed on an approved testers list by the local authority should have their credentials checked, obtain a business license, have insurance coverage, and sign a Code of Conduct document.

The local authority should be comfortable that the tester requesting placement on the approved list is a competent tester. Some authorities will require the applicant take a local examination. Not all training facilities teach the same information or require that the tester actually learn the field test procedures to be certified.

It is highly recommended that the local authority create a Code of Conduct for the certified tester to sign. A certified tester who is placed on a list of approved testers that is sent to the customer is a contractor representing the local authority. Testers who do not provide good service to the customers should be removed from the approval list.

Others with Some Level of Responsibility

Wastewater personnel and water reuse personnel also have some responsibility for ensuring that their distribution or collection systems are properly labeled and color-coded. Labeling and color-coding help prevent the accidental creation of cross-connections. In addition, these individuals should be knowledgeable about the hazards of cross-connection and the hydraulic conditions that create backflow, so that cross-connections are not inadvertently created by personnel when performing routine functions.

Summary

This chapter has demonstrated that everyone involved in the provision, distribution, and use of water, as well as those who design, install, modify, and inspect plumbing or distribution systems, share some degree of responsibility for preventing backflows and cross-connections. Because of overlapping responsibilities, the limits of one's liability are often difficult to identify. Therefore, an effort must be made to provide the highest level of protection that is economically feasible and to take all "reasonable" precautions to avoid the hazards of backflow.

Example of a Code of Conduct for Backflow Prevention Assembly Testers

1. A tester must have a current tester certification from an approved training facility or approved certifying agency to test backflow prevention assemblies under the jurisdiction of this authority.

2. A tester must not knowingly falsify the results of backflow assembly field tests performed by him/her.

Examples of Falsification of Field Test Reports

   a. Signing backflow field test reports for tests he/she did not perform.
   b. Making unneeded repairs.
   c. Not having proper backflow certification to perform tests.
   d. Not using proper field test procedures as established by the local authority.
   e. Using unauthorized backflow test equipment.
   f. Using backflow test equipment that is not in calibration or has not been re-calibrated within the last twelve (12) months.

3. A tester must not remove, replace, or relocate a backflow assembly without the approval of the water purveyor or the local authority.

4. All backflow assembly field test reports must be submitted to the water purveyor and/or the local authority within 10 days of the initial test, no matter what the result. If there is a specific problem relating to the test or the field test report form, the tester must call the water agency or the local authority.

5. All backflow reports must be submitted on proper forms. They must be legible and contain all appropriate information pertaining to the field test.

6. A tester must attend a backflow prevention assembly tester update seminar at least once every two years. The seminar must review current field test procedures and be approved by the local authority.

7. It is the tester's responsibility to inform the local authority of any changes in their contact information.

Any tester failing to comply with the provisions of this Code of Conduct is subject to disciplinary action. The results of the action can be the loss of testing privileges in the county or in a water purveyor's jurisdiction. It is a misdemeanor violation to knowingly file a false field test report.

LAWS AND RESPONSIBILITY

Chapter Three Review

3-1 Who has the PRIMARY RESPONSIBILITY if there is a backflow incident and contaminated or polluted water ends up in the distribution system?
Water Purveyor

3-2 What should a utility company be able to do if a customer, with a cross-connection, refuses to install a backflow preventer? *Discontinue Service*

(Please use answers below for questions 3-3 through 3-6.)

WATER PURVEYOR (SUPPLIER)	**TESTER**
PLUMBING INSPECTOR	**CUSTOMER**
HEALTH INSPECTOR	**CONTRACTOR**

3-3 Who might be held responsible for selecting the wrong type of backflow preventer?
Water Purveyor
Customer   Contractor

3-4 Who might be responsible if the backflow preventer was installed incorrectly?
Water Purveyor   Customer
Plumbing inspector   Contractor

3-5 Who might be held responsible if the backflow preventer is not tested at least annually?
Customer
Water Purveyor

3-6 Who might be held negligent if the backflow preventer was not tested properly? (hint: there are multiple answers.)
Tester
Customer

CHAPTER THREE

REFERENCES

1. *National Interim Primary Drinking Water Regulations*. 1976. (EPA 570/9-76-003). U.S. Environmental Protection Agency, Office of Water Supply.

2. Campbell, E.H., "Legal Responsibility," 1969. Presented at the seminar, Cross-Connection Control in Water Supplies, Seattle, WA, April 17, 1969.

3. Federal Register, *Point Measurement*, 1975. Vol. 40, No. 248, pp 59567-59568, December 24, 1975.

4. *Manual of Cross-Connection Control (9th Edition)*, 1994. Foundation for Cross-Connection Control and Hydraulic Research, University of Southern California, University Park, CA.

5. *Cross-Connection and Backflow Prevention (2nd Edition)*, 1974. American Water Works Assn., New York, NY.

6. Davis, J.H., and Murrell, L.R., 1976. "Legal Aspects of Backflow and Cross-Connection Control." *Jour. AWWA*. 68(8):397.

7. Smith, L.Y., and Roberson, G.G., 1977. *Business Law*. West Pub. Co., St. Paul, MN.

8. *Understanding Backflow Prevention Program Development*, (F-BP-MO), Watts Regulator Company.

9. "Drinking Water Standards, Monitoring, and Reporting," *Florida Administrative Code*. 1997, amended April 3, 2003. (62-550, F.A.C.). Florida Department of Environmental Protection, Tallahassee, FL.

CHAPTER FOUR

FUNDAMENTALS OF BACKFLOW

Chapter 2 covered the potential hazards of backflow incidents. The purpose was to identify and describe the hazards, and more importantly, to show the public health consequences and monetary costs associated with these hazards and demonstrate why backflows must be prevented. Recall that there are two basic approaches for preventing backflows that must be implemented. They are cross-connection control and backflow prevention. Cross-connection control involves identification of cross-connections through plan review, inspection of new construction, and surveys of existing facilities. These cross-connection control methods will be more thoroughly discussed in Chapter 8. To conduct these cross-connection control programs, it is necessary to have a thorough understanding of the different types of cross-connections so they can be identified, eliminated, and prevented.

This chapter identifies different types of cross-connections. Examples and illustrations help to demonstrate some common ways that cross-connections are created. Where cross-connections are necessary, installing the appropriate mechanical backflow preventer, thus creating a "protected cross-connection," can still prevent backflows. To understand how backflow preventers work, it is first necessary to understand how and why backflow occurs. Therefore, this chapter also covers the fundamentals of hydraulics and backflow, providing the basic information needed to identify and avoid potential backflow incidents.

Cross-Connections

Cross-connections are the physical links that allow backflows. If cross-connections do not exist, then the backflow of contaminants into the potable water supply cannot occur. Therefore, it is important to identify how and why cross-connections are created.

Types of Cross-Connections

Cross-connections can be classified as temporary or permanent, and direct or indirect. **Temporary cross-connections** are short-term links between potable and non-potable water sources. They are frequently created by a hose or some other type of extension to the permanent water supply line. The cross-connection is created when the end of the hose, or other extension, is submerged in a contaminant or pollutant. Hoses are believed to be the

Cross-connections are the physical links that allow backflow to occur.

Temporary cross-connections are short-term links between potable and non-potable water sources.

Hoses are believed to be the most common temporary cross-connection.

most common temporary cross-connection.[1] Temporary cross-connections can also be created by swing connections, spools, and change-over devices. These types of connections are common where auxiliary water supplies are used for firefighting or irrigation systems.

Temporary cross-connections are commonly created when water is drawn from the potable water system to dilute chemicals in tank trucks or to rinse the inside of the tank. Pesticide companies, lawn maintenance companies, sludge haulers, and milk haulers all commonly create temporary cross-connections.

Figure 4-1　Tank Truck Creates Cross-Connection
Here a hose creates a temporary cross-connection when it is used to dilute toxic pesticides.

Temporary cross-connections are frequently created at piers where large ships dock. When at dock, these ships refill their drinking water tanks from the community water system. If the potable drinking water system is not adequately protected, a hazardous backflow could result, since the potable water lines on board ship are frequently cross-connected to seawater for fire-fighting purposes. Backflows can occur when fire system pumps are exercised while fresh water is added to the potable water tanks. The fire system pumps create a backpressure that exceeds the potable water supply pressure, and a backflow results.

Permanent cross-connections are generally created by solid pipe connections. For instance, a permanent cross-connection is created where water is supplied to a boiler, an air cooling system, or any piece of equipment or holding vessel that could pollute or contaminate the potable water supply. However, a hose can also create a permanent cross-connection if it is used as a fixed extension of a water system.

Cross-connections can also be classified as direct or indirect. **Direct cross-connections** are generally hard pipe connections, thus they always

> Permanent cross-connections are generally created by solid pipe connections.
>
> A direct cross-connection is any arrangement of pipes, fixtures, or devices connecting a potable water supply directly to a non-potable source that is subject to backsiphonage and backpressure.

have the potential for backflow. Often, the terms permanent and direct are used synonymously, but this is not always valid. For instance, an old-fashion bathtub, as illustrated in Figure 4-2, creates an indirect permanent cross-connection. Obviously the plumbing creates a permanent cross-connection, and it is indirect because a cross-connection is only created if the tub is overfilled and the water inlet becomes submerged. **Indirect cross-connections** are "subject to backsiphonage only"[2] and have the potential for backflow only when conditions exist to create a cross-connection, and backsiphonage is possible. If wastewater lines were to leak into a potable water source, a well, or holding tank, this would also create an indirect cross-connection. Recall that this type of cross-connection was responsible for the cholera outbreak that prompted Dr. John Snow to theorize that water spreads disease. Hoses are also often considered indirect cross-connections, therefore, the terms temporary and indirect are often used interchangeably.

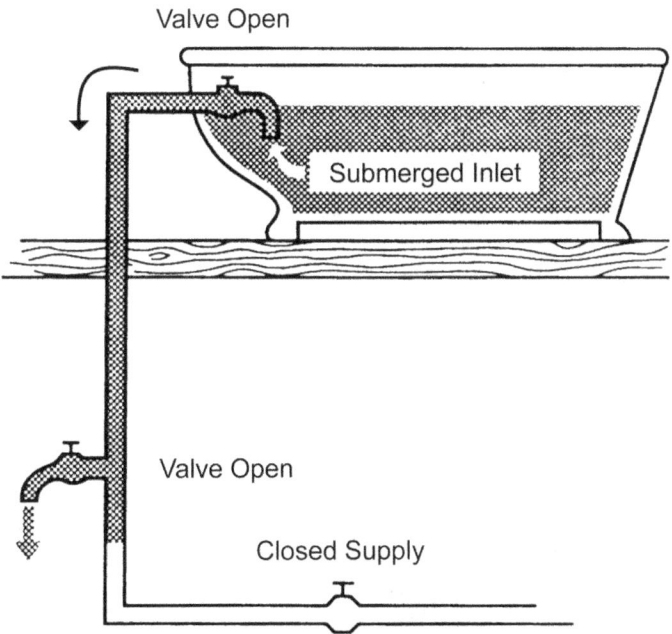

Figure 4-2   **Submerged Bath Tub Inlet Creates a Cross-Connection**
An indirect, permanent cross-connection is created because the bathtub inlet discharges below the flood rim level of the bathtub. The cross-connection is indirect because a physical link exists between potable and not-potable water only under certain conditions, in this case, when water level reaches the inlet.

Why Cross-Connections Are Created

The terminology used to describe cross-connections is meaningful here because it helps to reveal why cross-connections are created. Some cross-connections are created out of necessity. For instance, a direct cross-connection may be created because it is essential that water is directly and

Why Cross-Connections Exist:

1) **People unaware of a problem**
2) **Convenience**
3) **Inadequate devices**

FUNDAMENTALS OF BACKFLOW

continuously supplied to a boiler. Likewise, some laboratory equipment, food processing equipment, and air-to-water heat exchangers also need a direct and continuous supply of water. Although necessary, any of these direct cross-connections can allow backflow unless backflow preventers are in place.

Other cross-connections are created for convenience. The person creating the cross-connection may not be aware of the hazards of cross-connections and frequently does not know about the hydraulic conditions that cause backflow or the methods available for preventing cross-connections.

In Figure 4-3 the utility personnel are using a jet truck to clean out a blockage in the sewer line. If there was a significant reduction in the potable water pressure, the wastewater could backflow into the water supply. Figure 4-4 illustrates a very common cross-connection. In this situation the pool operator has extended the water outlet through the use of a hose down into the filter tank to prevent potable water ("make-up water") from splashing the media off the filters when it is added to the tank. The hose is an extension of the potable water system, and the pool water is now cross-connected to the potable water supply. Any use of a hose must be done with care to ensure that a cross-connection is not created.

Figure 4-3   **Jet Truck Creates a Cross-Connection**
The utility personnel are using the jet truck to clean out a blockage in the sewer line. Because they are not aware that water can move uphill against the force of gravity, they do not realize that they are creating a cross-connection. If there was a significant reduction in the potable water pressure, the wastewater could backflow into the potable water supply.

CHAPTER FOUR

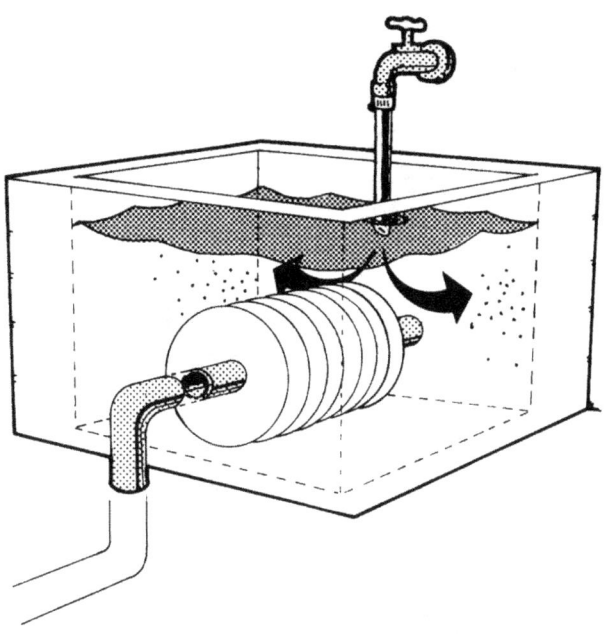

Figure 4-4 **Cross-Connection Created at Swimming Pool**
The pool operator has extended the water outlet through the use of a hose down into the filter tank to prevent potable water ("make-up water") from splashing the media off the filters when it is added to the tank. The hose is an extension of the potable water system, and the pool water is now cross-connected to the potable water supply. Water treatment facilities often have similar cross-connections.

Temporary cross-connections are often created through normal or routine use of products such as "plumber's helpers" (Figure 4-5), siphon mixers (Figure 4-6), toilet tank flush valves, "bath-to-shower adapters" (Figure 4-7), and pool-cleaning devices that operate on water pressure. The problem with these devices is that often a submerged inlet is created through normal use. Thus, an unprotected cross-connection is created. Unfortunately, most manufacturers fail to warn the user that the product will create a cross-connection and that backflow protection should be provided.

One of the most hazardous temporary cross-connections arises through the use of a siphon mixer (Figure 4-6). The siphon itself creates a cross-connection. The movement of water past the opening of the holding vessel creates a vacuum. This vacuum draws the pesticide or chemical fertilizer up the tube and mixes it with water as it is sprayed out onto the lawn. If there is a significant reduction in water-line pressure, a reversal of flow occurs, and the chemical can be drawn up into the potable water system as the water backflows past the opening. Recall, that a California man died after ingesting chemicals containing arsenic that backflowed from a siphon mixer into the potable water line.[3]

FUNDAMENTALS OF BACKFLOW

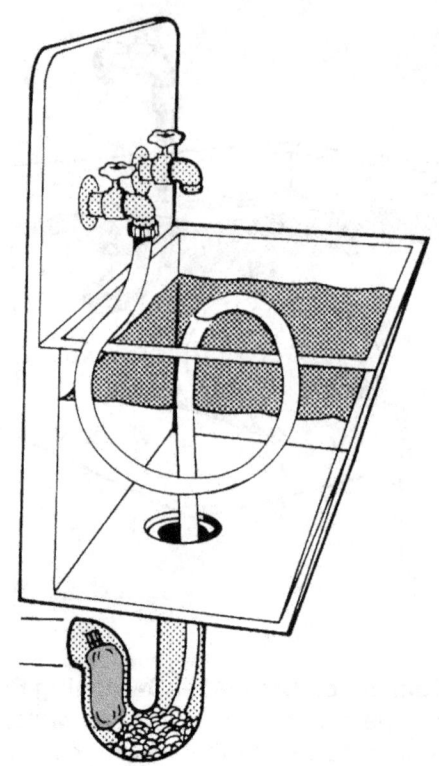

"Plumber's helpers" are used to unclog wastewater lines. The potable water supply is used to inflate the balloon and hold the device in place. Water pressure then removes the blockage. The cross-connection is created because the device is connected to both the potable water supply and to the contaminated water in the waste line.

Figure 4-5   "Plumber's Helper" Creates a Cross-Connection

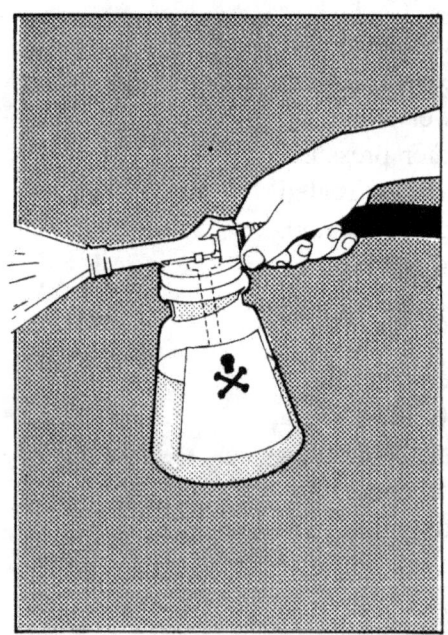

The chemicals in the mixer could be siphoned into the potable water supply.

Figure 4-6   Siphon Chemical Mixer

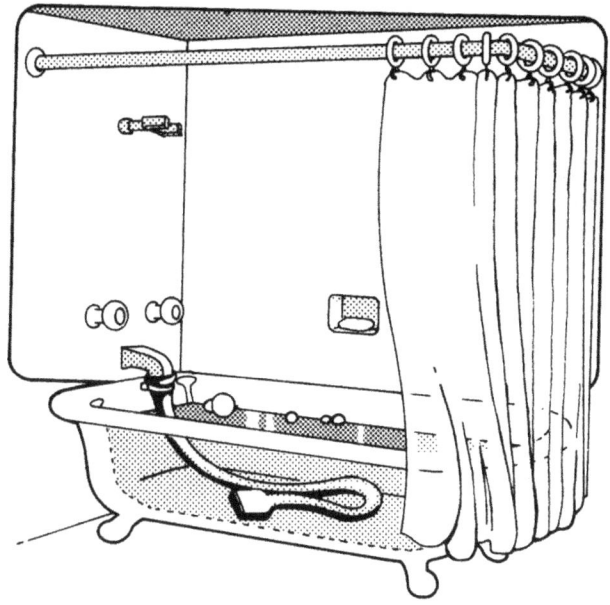

A cross-connection is created when the shower head of a "bath-to-shower adaptor" unit falls down into the bath water. A person taking a shower could easily create this cross-connection as they hastily leave to answer a ringing phone and drop the shower head. If the potable water supply pressure decreases significantly, the bath water could backsiphonage - backflow into the potable water supply.

Figure 4-7   A "Bath-to-Shower" Adaptor Creates a Cross-Connection

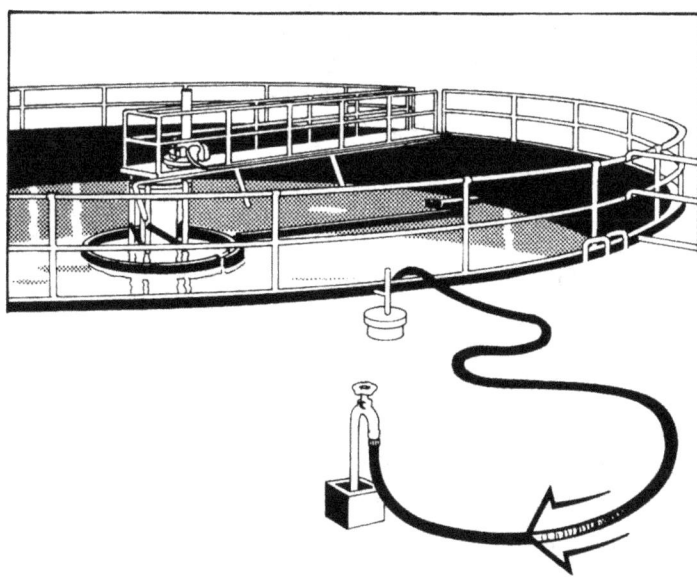

Figure 4-8   **A Hose Creates a Cross-Connection at a Wastewater Treatment Plant**
A hose connected to a potable water-line at a wastewater treatment plant was needed to spray down foam that developed due to a treatment process malfunction. Later that evening the tip of the hose fell down into the treatment tank, creating a cross-connection. If for any reason the potable water pressure dropped below atmospheric pressure, the water in the tank would flow up the hose and into the potable water line.

FUNDAMENTALS OF BACKFLOW

Finally, many cross-connections are created accidentally and unintentionally. Figure 4-8 shows a hose connected to a potable water line at a wastewater treatment plant being used to clean clarifier weirs. If the tip of the hose fell down into the treatment tank it would create a cross-connection. If for any reason the potable water pressure dropped below atmospheric pressure, the water in the tank would flow up the hose and into the potable water line. The kitchen sink spray nozzle in Figure 4-9 can accidentally create a cross-connection if the spray head is left in the sink. This indirect cross-connection could allow any contaminants or pollutants in the sink (e.g., soapy water) to backflow into the potable water system if backsiphonage conditions existed.

The spray nozzle extension commonly provided at kitchen sinks can accidentally create a cross-connection if the spray head is left in the sink. This indirect cross-connection could allow any contaminants or pollutants in the sink (e.g. soapy water) to backflow into the potable water system if backsiphonage conditions existed.

Figure 4-9   Common Kitchen Spray Nozzle Creates a Cross-Connection

There are myriad ways in which cross-connections can be created. As illustrated above, they are often created through the use of commercial products such as the siphon mixer. However, the products themselves are not necessarily the problem. All cross-connections can be protected by installation of the appropriate backflow preventer or sometimes through the proper use of the products themselves. The manufacturers of these products, however, should consider their potential liability if a backflow were to occur due to the use of their product. Failure to provide some warning about the potential for creating cross-connections and the hazards of backflow might be considered negligent.

Cross-connections, whether temporary or permanent, direct or indirect, are generally created out of ignorance, out of convenience or from reliance on inadequate backflow preventers. However, if people are properly educated about the hazards of backflow, most cross-connections can easily be eliminated.

Understanding Backflow

Backflow hazards occur when water moves through a cross-connection in a direction it was not intended to move. The flow reversal is caused by a pressure differential. In order to better understand what causes these pressure differentials, and therefore, what causes backflow, a more thorough understanding of pressure is needed.

Pressure Principles

To explain the causes and consequences of pressure, a review of some basic terminology is needed. Pressure is a measurement of force. Force is any kind of push or pull. Forces may be generated in many ways. A significant type of force for this discussion is weight. Weight is a force generated when the earth's gravity attracts an object that has mass. Mass is the amount of matter making up an object and volume is how much space an object occupies. Density depends on the mass and volume of an object. The greater the mass of an object, the more it weighs. The water in Figure 4-10 weighs more than air. Given the same volume of water and air, water has a greater mass because of its greater density, therefore, it weighs more.

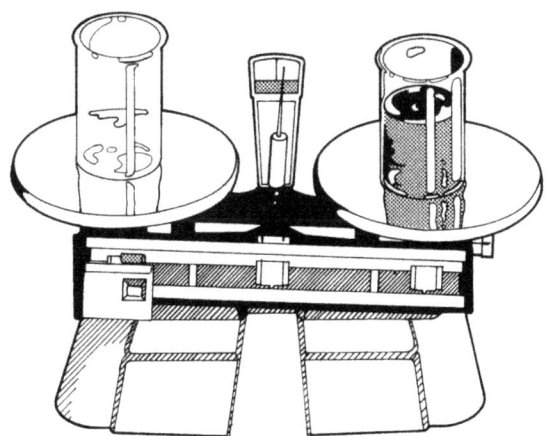

The molecules of water are packed together more tightly than those of air. Given the same volume of water and air, water has a greater mass because of its greater density, therefore, it weighs more.

Figure 4-10   Comparison between the Weight of Water and the Weight of Air

Pressure is a measure of the force acting on a unit of area (e.g., a square inch). Thus, two identical forces acting on different unit areas exert different pressures. Although both blocks in Figure 4-11 have the same mass (same volume and density), they exert different pressures because of the different amount of surface area affected by their weight. Blocks A and B in Figure 4-11 are the same size (4 × 4 × 8-inches) and weight (32 lbs). However, block A exerts more pressure than block B, because block A's weight is exerted over a smaller surface area than B's. In U.S. units the most common measurement for pressure is pounds per square inch (psi).

> Pressure is a measure of the force acting on a unit of area.

FUNDAMENTALS OF BACKFLOW

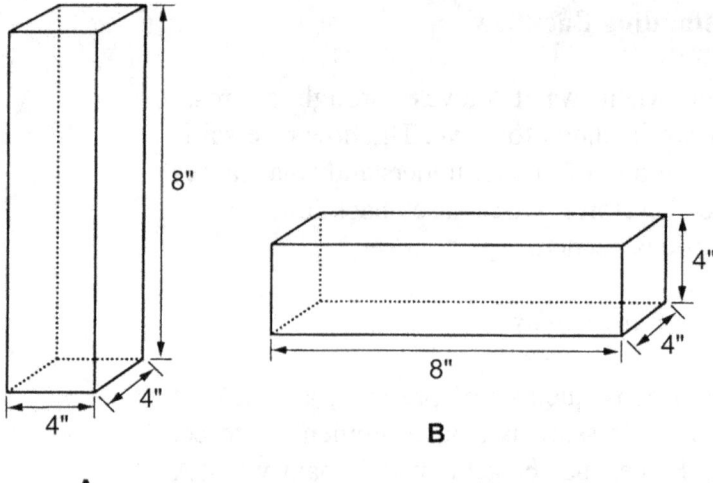

Figure 4-11   **Comparison of Pressures between Two Blocks of the Same Dimensions**
Both blocks are the same size (4 × 4 × 8 inches) and weight (32 lbs.) However, block A exerts more pressure than block B because block A's weight is concentrated on a smaller surface area than B's.

The surface area affected can be found by multiplying the length (L) times the width (W).

$$L \times W = AREA$$
$$4 \text{ inch} \times 4 \text{ inch} = 16 \text{ inch}^2$$

$$PRESSURE = \frac{FORCE}{AREA}$$

$$P = \frac{32 \text{ lb}}{16 \text{ inch}^2} = 2 \text{ psi}$$

$$L \times W = AREA$$
$$4 \text{ inch} \times 8 \text{ inch} = 32 \text{ inch}^2$$

$$PRESSURE = \frac{FORCE}{AREA}$$

$$P = \frac{32 \text{ lb}}{32 \text{ inch}^2} = 1 \text{ psi}$$

Water: Under Static Conditions

Hydraulics is the science that deals with fluids in motion and at rest. Fluids in motion exert pressure. The moving water of a river will propel rocks, tree limbs, and boats. In fact, the pressure that moving water generates is harnessed to create electricity. This is called **kinetic energy**. Fluids at rest (static fluids) also exert pressure. This is called **potential energy**. Water at rest exerts pressure because of the weight of the water and atmospheric pressure.

Water in an open container exerts pressure simply because it has weight. One cubic inch of water weighs 0.036 lbs. In other words, one cubic inch of water exerts a pressure of 0.036 psi. The weight of a column of water, and, therefore, the pressure exerted by the water, is determined by the height of the water column. If the column of water is one foot high (12 inches), it must exert a pressure 12 times as great (Figure 4-12). Therefore, if the height of a water column is known, the pressure it exerts can be calculated. Because the weight of water is constant, a 2-foot column of

The pressure that moving water generates is called kinetic energy.

Water at rest (static) has potential energy.

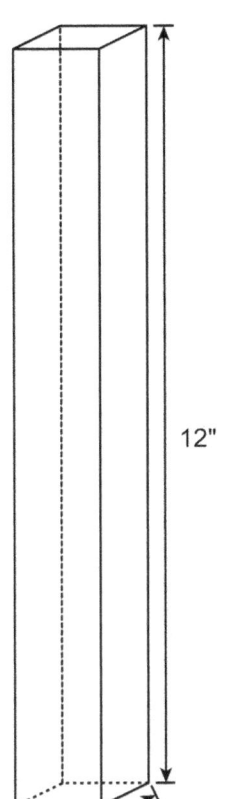

The pressure of a 1-foot column of water acting on a square inch of surface exerts 0.433 psi of pressure.

If: 1 in × 1 in × 1 in = 0.036 psi

Then: 1 in × 1 in × 12 in = 0.433 psi

Since: 0.036 × 12 = 0.433

Thus: A 1-foot column of water acting on a 1 square inch = 0.433 psi

Therefore, if the height of a water column is known, the pressure it exerts can be calculated.

12 inches × 1 inch = 0.433 psi

Figure 4-12   The Weight of a 1-foot Column of Water

water would weigh two times as much as a 1-foot column and thus would exert twice the pressure on a square inch of surface area (Figure 4-13).

Because water pressure is directly related to column height, water pressure can also be measured in terms of "feet of head" or "pressure head." Thus, a 2-foot column exerts a pressure of "two feet of head," which is equivalent to 0.866 psi. Since the weight of water is directly related to height, the shape of the container that holds the water has no effect on the pressure. The pressures at points A and B in Figure 4-14 are equal. At two points of equal depth in water, the pressure is the same, because the earth's gravitational pull on the water is acting equally at both points. However, at different depths the pressure is different, because there is a taller or shorter column of water being acted on by the force of gravity. In the example illustrated in Figure 4-15 assume a pressure gauge determined that the pressure at P1 is 1.0 psi, P2 is 1.0 psi, P3 is 2.0 psi, and P4 is 1.5 psi. The pressure at P1 equals the pressure at P2 because they are at the same depth. The pressure at P3 is greater than at P4, because the column of water above P3 is higher than P4.

If the pressure exerted by a column of water is known, the height of the column can be determined, since 1.0 psi of water pressure is equivalent to 2.31 ft (about 28 inches) of water acting on a square inch of surface area (Figure 4-16).

FUNDAMENTALS OF BACKFLOW

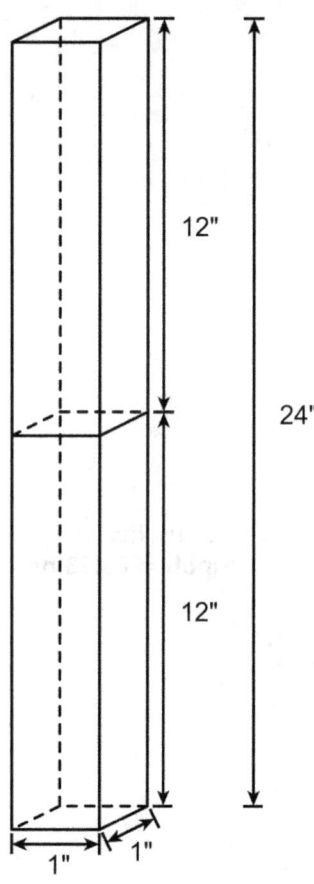

Because the weight of water is constant, a 2-foot column of water would weigh two times as much as a 1-foot column and thus would exert twice the pressure on a square inch of surface area.

12 inches = 1 foot

In a closed system, a change of elevation will change the pressure of the fluid. Each foot of elevation will change the pressure by 0.433 psi.

Figure 4-13   The Weight of a 2-foot Column of Water = 0.866 psi

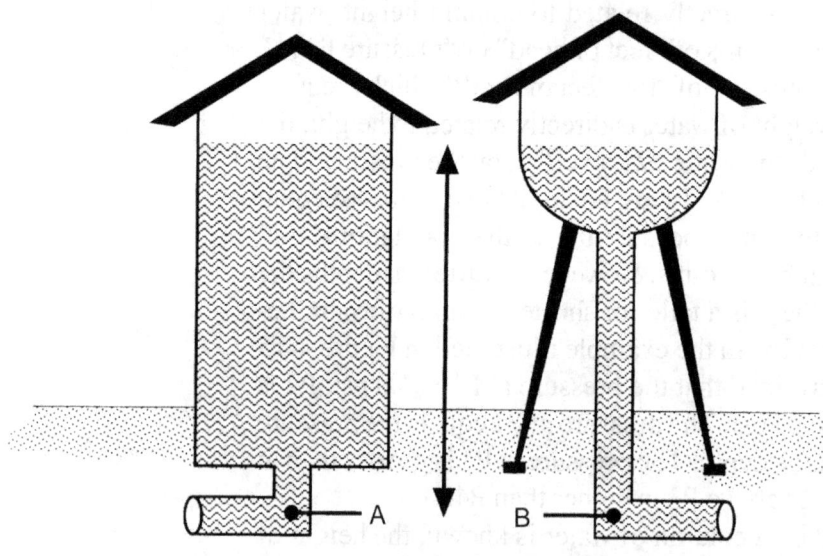

The shape of the container that holds the water has no effect on the pressure.

Figure 4-14   **A Comparison of Water Pressures in Two Different Styles of Water Towers**
The pressure at point A is equal to the pressure at point B.

CHAPTER FOUR

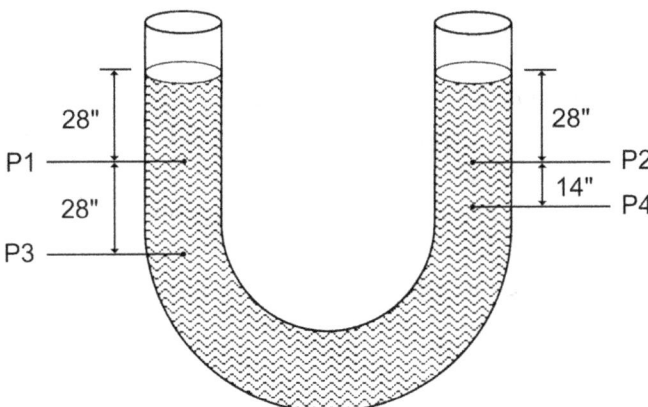

Figure 4-15   **Water Pressure Is Dependent on Depth**
Assume a pressure gauge determined that the pressure at P1 is 1.0 psi, P2 is 1.0 psi, P3 is 2.0 psi, and P4 is 1.5 psi. The pressure at P1 equals the pressure at P2, because they are at the same depth. The pressure at P3 is greater than at P4 because the column of water above P3 is higher than that above P4.

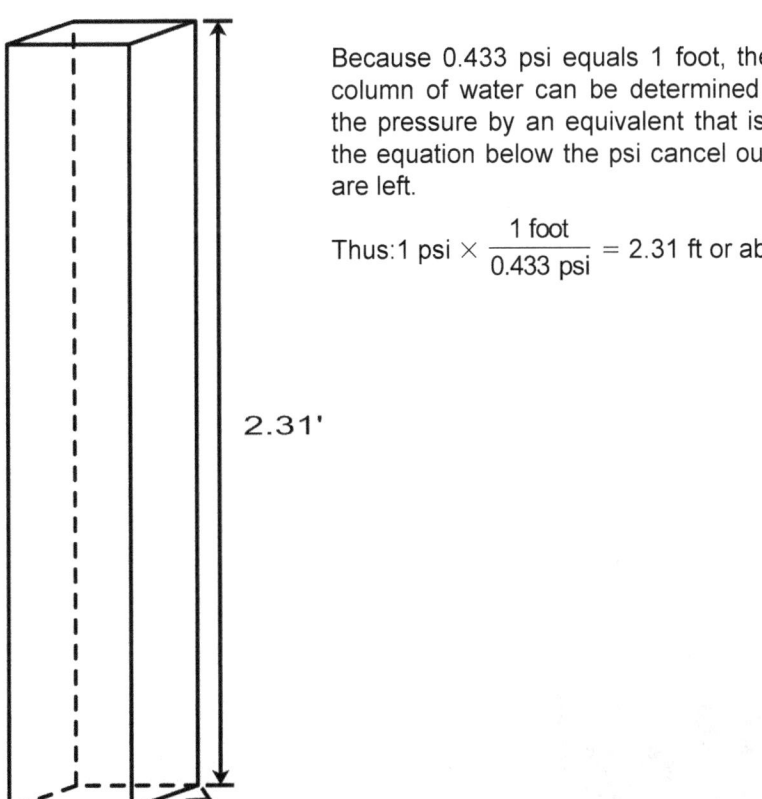

Because 0.433 psi equals 1 foot, the height of the column of water can be determined by multiplying the pressure by an equivalent that is equal to 1, in the equation below the psi cancel out and only feet are left.

Thus: $1 \text{ psi} \times \dfrac{1 \text{ foot}}{0.433 \text{ psi}} = 2.31$ ft or about 28 inches

1.0 psi of water pressure is equivalent to 2.31 feet (28 inches).

Figure 4-16   Determining the Height of a Water Column

FUNDAMENTALS OF BACKFLOW

Atmospheric Pressure

Atmospheric pressure is caused by the weight of air (Figure 4-17). Air weighs very little, however, because the atmosphere is 25 miles thick, the weight of air on the earth's surface is significant. At sea level, the atmospheric pressure is 14.7 psi, meaning the weight of earth's atmosphere on one square inch of surface is 14.7 lbs.

At higher elevations, for instance in the mountains, the atmospheric pressure is less than 14.7 psi, because the column of air is shorter and the air is less dense (Figure 4-18).

At sea level, the atmospheric pressure is 14.7 psi.

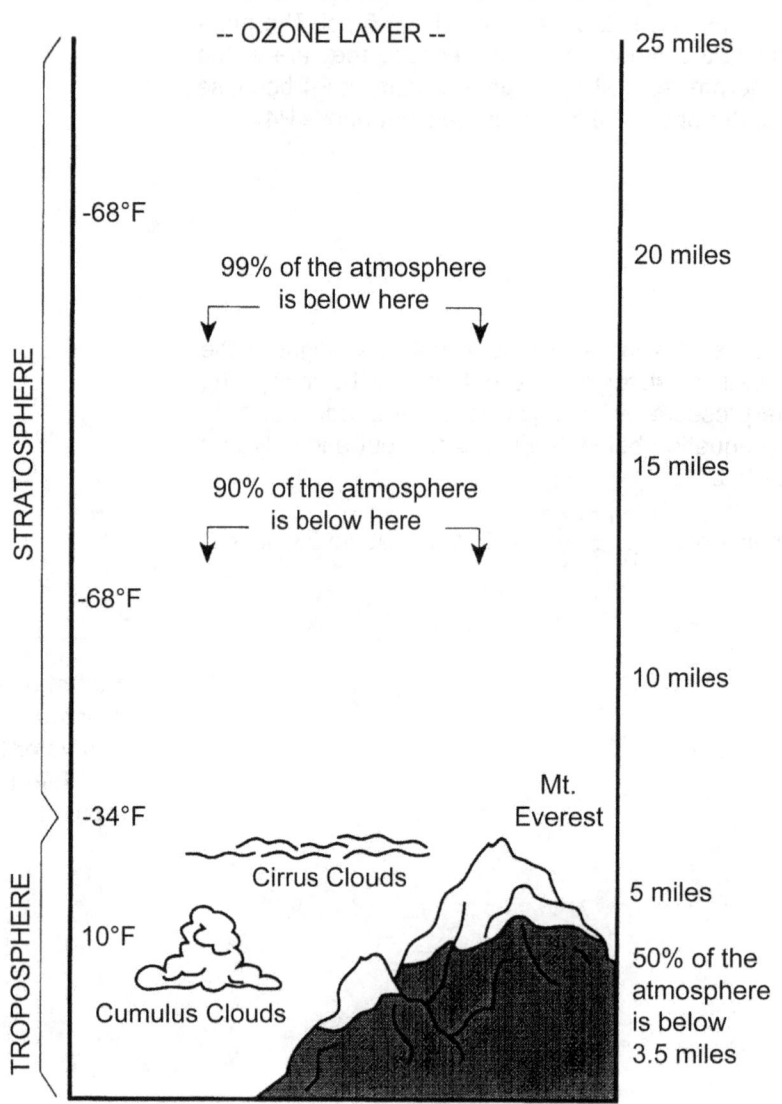

Figure 4-17 **The Weight of Atmospheric Pressure on the Earth's Surface**
Each square inch of the earth's surface has atmospheric pressure acting on it.

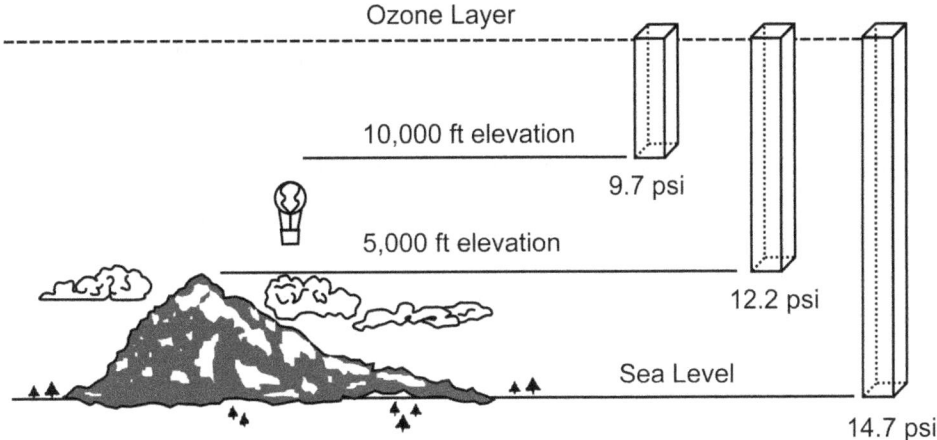

Figure 4-18 Atmospheric Pressure: Sea Level versus the Mountains
The atmospheric pressure at sea level (14.7 psi) is greater than on the mountaintop, because the atmospheric column is higher at sea level than on the mountaintop.

The difference in pressure at different altitudes must be taken into account when calibrating a pressure gauge. Because atmospheric pressure is relatively constant at a given elevation, pressure gauges do not include the atmospheric pressure. Atmospheric pressure is factored out (Figure 4-19).

Gauge pressure readings do not include atmospheric pressure, and are measured in units of pounds per square inch gauge (psig). Gauge pressure readings can be compared with other gauge readings taken at the same altitude. However, since the atmospheric pressure at higher elevations is less than 14.7 psi, the variation in atmospheric pressure must be taken into account when comparing gauge readings taken at different altitudes. If a pressure gauge reads exactly zero, this indicates that only atmospheric pressure is acting on the system. The absolute pressure, however, would be 14.7 psi. This is calculated by adding gauge pressure and atmospheric

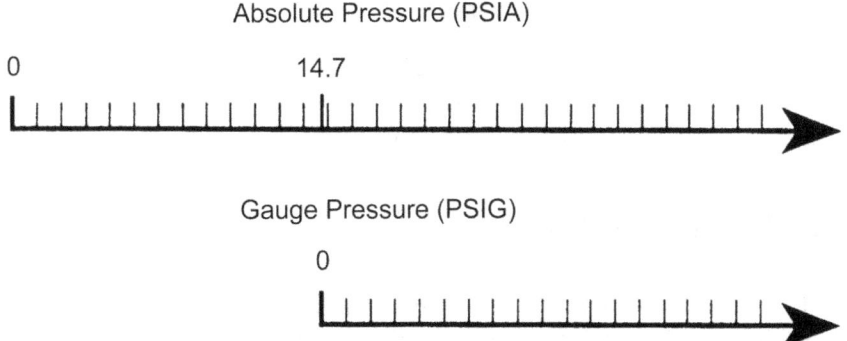

Figure 4-19 Absolute versus Gauge Pressure
If a pressure gauge reads exactly zero, this indicates that only atmospheric pressure is acting on the system. The absolute pressure, then, would be 14.7 psi.

FUNDAMENTALS OF BACKFLOW

pressure at a given altitude, which is absolute pressure. **Absolute pressure** is the gauge pressure plus the atmospheric pressure at a given altitude and it is measured in units of pounds per square inch absolute (psia).

ABSOLUTE PRESSURE = GAUGE PRESSURE + ATMOSPHERIC PRESSURE

Atmospheric pressure is important in hydraulics because the atmosphere exerts pressure on the surface of water. Water at rest redistributes this pressure equally in all directions. Therefore, the total pressure exerted by water at rest depends on both the weight of the water column and the force of atmospheric pressure.

> Absolute pressure is the gauge pressure plus the atmospheric pressure at a given altitude.

Water Movement

Water movement is caused by pressure differentials. Pressure differentials can be created in a number of ways: differences in water weights, differences in atmospheric pressure, differences in water temperature, differences in the velocity of water movement, and differences in mechanical forces. Water will stand still if the pressures in two pipes, tanks, or vats are equal. However, water will move when the pressures are unequal. Water will flow to the point of lowest pressure (Figure 4-20). Recall that pressure is just a measurement of force. Movement will be controlled by the greater force, whether it is a push or pull.

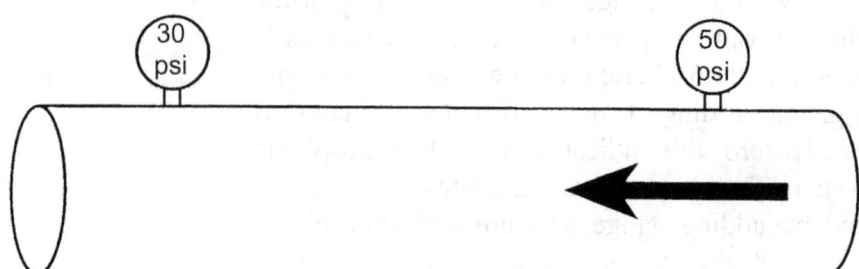

Figure 4-20   **Determining the Direction of Water Movement in a Pipe by Comparing Pressure Gauge Readings**
Because water flows in the direction of lowest pressure, the direction of flow can be determined by comparing the two pressure gauge readings. Here, water will flow from right to left.

Effects of the Weight of Water

Water movement caused by differences in weight can best be demonstrated with a U-tube. The greater pressure on the right side of the valve in Figure 4-21, caused by the higher column of water on the right side, will cause the water to move up the left side of the U-tube as soon as the valve is opened. Here atmospheric pressure is irrelevant, because it is acting on both sides equally.

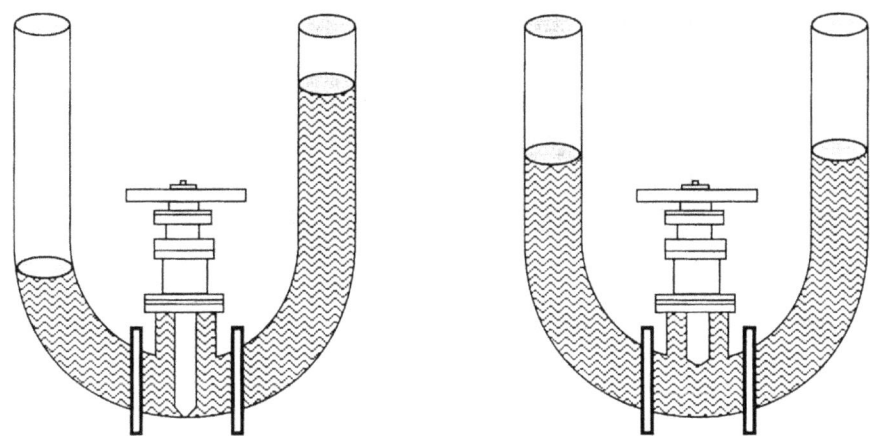

Figure 4-21 Water Movement in a U-tube
The greater pressure on the right side of the valve, caused by the higher column of water on the right side, will cause the water to move up the left side of the U-tube as soon as the valve is opened. (Here atmospheric pressure is irrelevant, because it is acting on both sides equally.)

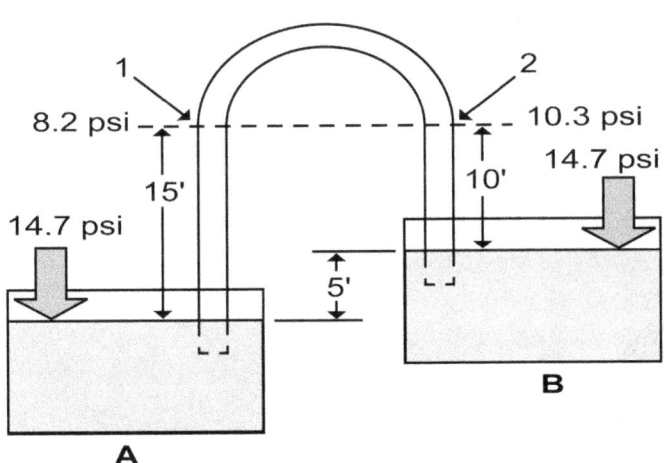

The Pressure at Point 1

$15 \text{ ft} \times \dfrac{0.433 \text{ psi}}{1 \text{ ft}} = 6.5 \text{ psi}$

14.7 psi − 6.5 psi = 8.2 psi

The Pressure at Point 2

$10 \text{ ft} \times \dfrac{0.433 \text{ psi}}{1 \text{ ft}} = 4.3 \text{ psi}$

14.7 psi − 4.3 psi = 10.4 psi

Figure 4-22 Water Movement through a Siphon
Assume that the water level in tank B is 5 feet higher than the water level in tank A, that point 1 is 15 feet above the water level in tank A, and that point 2 is 10 feet above the water level in tank B. And, atmospheric pressure is acting equally on both tanks. Recall also that 0.433 psi equals 1 foot.

FUNDAMENTALS OF BACKFLOW

Figure 4-22 is another example that illustrates a siphon. Again, atmospheric pressure is acting on both tanks equally. It is the unequal weight of water that causes the movement. Assume that the water level in tank B is 5 feet higher than the water level in tank A, that point 1 is 15 feet above the water level in tank A, and that point 2 is 10 feet above the water level in tank B. Remember for this to work atmospheric pressure must be acting equally on both tanks.

Effects of Atmospheric Pressure

Movement of water resulting from atmospheric pressure differentials can be seen when someone sips from a straw (Figure 4-23). As a person starts to sip on the straw, the fluid level rises, because the person is removing some of the atmospheric pressure inside the straw. In effect, the person is creating a low-pressure area. Atmospheric pressure acting on water outside the straw essentially pushes water into the straw.

A partial vacuum or negative pressure is created when atmospheric pressure is reduced. Negative pressure, then, is any pressure less than atmospheric pressure (14.7 psia or 0 psig). This is also called subatmospheric. If water is exposed to full atmospheric pressure at one point and negative pressure at another, water will move in the direction of negative pressure, even if that means moving uphill. If air was induced or allowed into the system, the vacuum would be broken. Likewise, if the straw

Figure 4-23 Drinking from a Straw Illustrates the Effects of Negative Pressure
As a person starts to sip on the straw, the fluid level rises, because the person is removing some of the atmospheric pressure inside the straw. In effect, the person is creating a partial vacuum. Atmospheric pressure acting on water outside the straw essentially pushes water into the straw.

in the above example had a hole in it, the person would not be able to get a drink because atmospheric pressure would enter through the hole as quickly as it is removed.

Theoretically, if atmospheric pressure is totally removed from inside a straw, pipe, tank, etc., water can rise 33.9 ft. Friction opposes this upward movement, however, so water will not actually rise the total 33.9 ft. The vacuum pump in Figure 4-24 is used to remove all of the air inside a pipe. Atmospheric pressure acting on the water outside the pipe forces the water into the pipe. This pressure causes water to move up the pipe 33.9 ft, until the weight of the water column balances the atmospheric pressure.

A **barometric loop** is an old method of backflow prevention that relies on pressure principles to prevent backsiphonage-backflow. A barometric loop is simply a 34-foot vertical loop of piping.

> A barometric loop is simply a 34 or 35 foot vertical loop of piping.

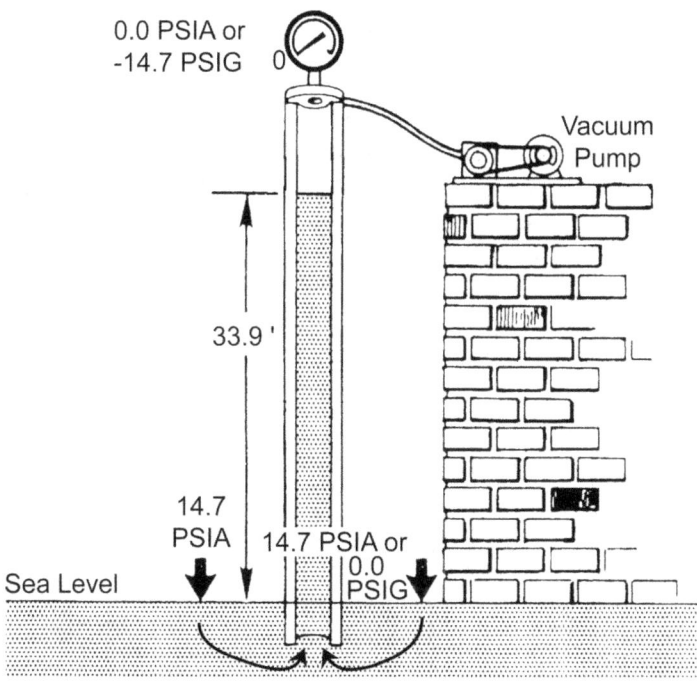

Figure 4-24 **The Creation of a Total Vacuum within a System Theoretically Causes Water to Rise 33.9 Feet**

A vacuum pump is used to remove all of the air inside a pipe. Atmospheric pressure acting on the water outside the pipe forces the water into the pipe. This pressure causes water to move up the pipe 33.9 feet, until the weight of the water column balances the force of atmospheric pressure.

Since: $0.433 \text{ psi} = 1 \text{ foot}$

Then: $14.7 \text{ psi} \times \dfrac{1 \text{ foot}}{0.433 \text{ psi}} = 33.9 \text{ feet}$

Thus: Atmospheric pressure will support a column of water 33.9 feet high.

FUNDAMENTALS OF BACKFLOW

The effect that negative pressure can create is significant to backflow, because negative pressure causes water to move in directions that most people do not expect (i.e., up). Thus, many cross-connections are created unknowingly by persons who install or alter plumbing. Examples of this type of cross-connection will be given later in this chapter when backsiphonage is discussed.

The preceding discussion on the movement of water has shown that water moves in response to differentials in either air or water pressure. Pressure differentials can also be created by changes in the temperature of the water, the velocity of moving water, and mechanical devices such as pumps.

Effects of Water Temperature

As the temperature of water rises, either the volume must increase or the pressure must increase, according to the **Ideal Gas Law**.

This law is depicted by the graph in Figure 4-25. If the volume of water is held constant (as it is in a boiler), and the temperature increases, the pressure inside the boiler must increase (Figure 4-26). As the temperature inside the boiler increases, the increase in pressure will cause water to move out of the boiler and to backflow into a feeder pipe. Just how far the boiler water will backflow depends on the water supply pressure. The potential hazards caused by such a backflow are discussed later under backpressure.

As the temperature of water rises, either the volume must increase or the pressure must increase, according to the Ideal Gas Law.

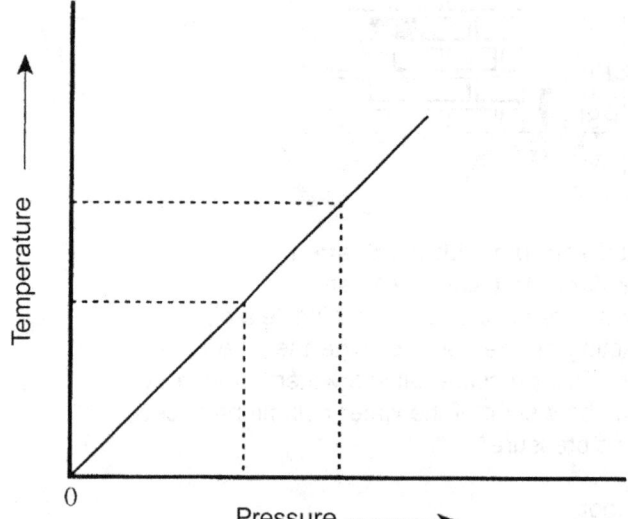

Figure 4-25   A Graph of the Relationship between Temperature and Pressure at a Constant Volume
If the volume is held constant and the temperature increases, the pressure will increase.

CHAPTER FOUR

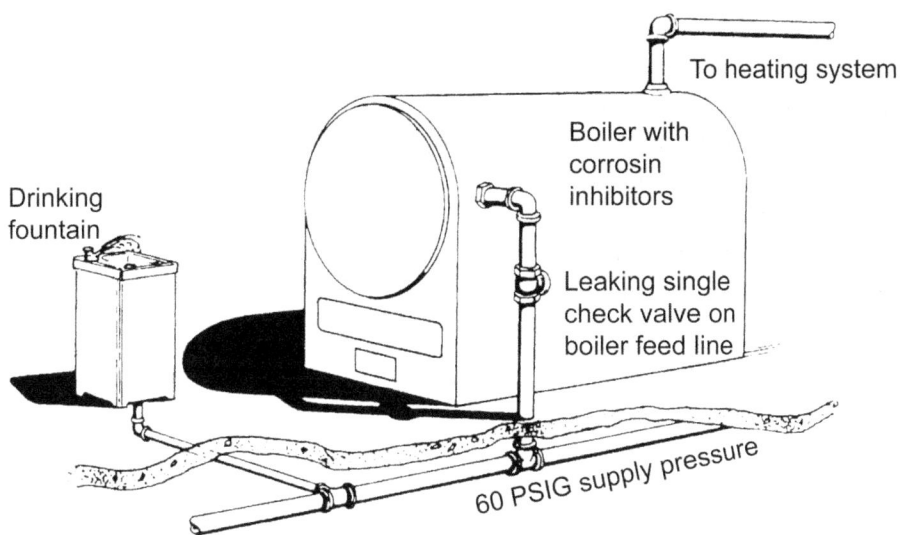

Figure 4-26 **How the Effects of the Ideal Gas Law Could Cause Backflow from a Boiler**
Because the volume that the boiler can hold is constant, as the temperature in the boiler increases, so will the pressure. If this generated pressure is greater than the potable supply pressure, the hot boiler water can backflow into the potable water supply.

Effects of Water Velocity

Where the velocity of a fluid is high, the pressure is low. Figure 4-27 depicts this phenomenon, known as Bernoulli's Principle or **Venturi Effect**.

The Venturi Effect is caused by water traveling at high velocity past or through a small opening.

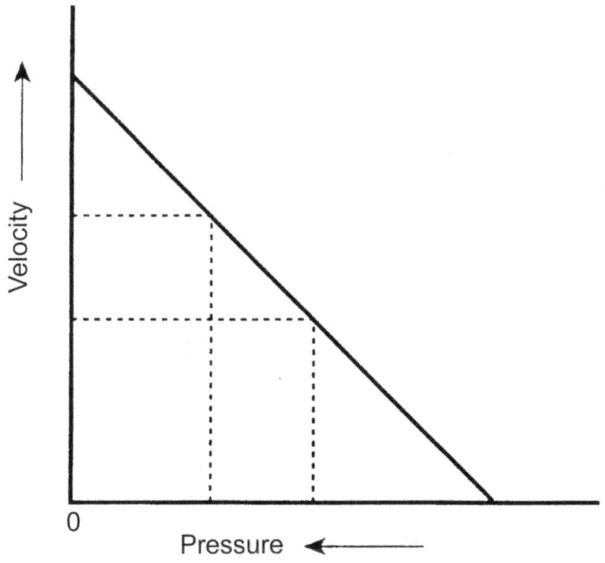

Figure 4-27 **A Graph of the Relationship between Velocity and Pressure**
This graph demonstrates that as the velocity increases, the pressure decreases.

FUNDAMENTALS OF BACKFLOW

When water is forced to flow through a smaller diameter pipe, its velocity must increase, and therefore, the pressure at that point must decrease. Figure 4-28a illustrates the effects of this principle on lateral connections. Because the same amount of water is passing through both points 1 and 2, the water must flow faster through point 2. In order for the water to move faster, it must be under less pressure. The gauges show the pressure at each point. Since more pressure is at point 1, the water pressure will be higher at point 1. As the system pressure drops (Figure 4-28b), the levels in the service connection fall. If the system pressure becomes low enough, a negative pressure is created at point 2.

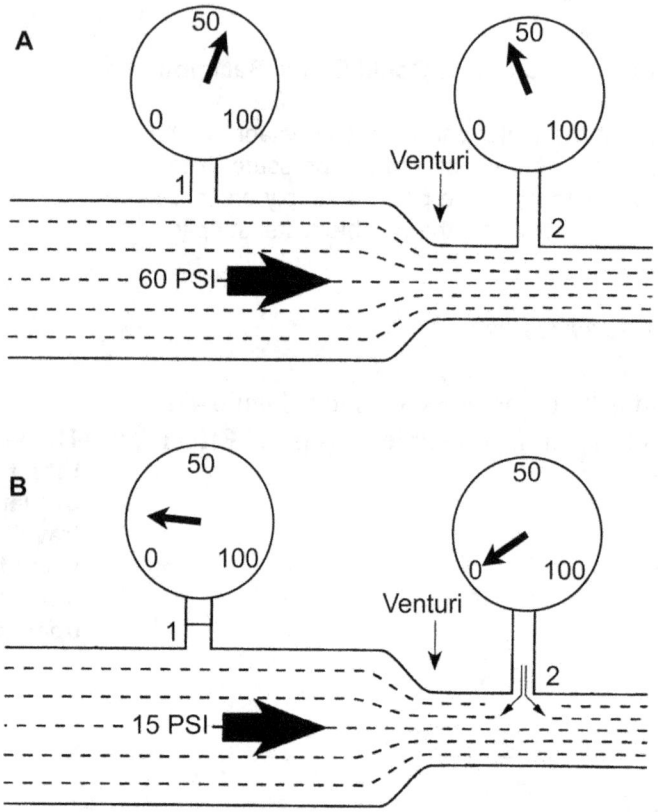

Figure 4-28a   **An Illustration of How a Venturi Can Cause Backflow**
Because the same amount of water is passing past both points 1 and 2, the water must flow faster past point 2. In order for the water to move faster, it must be under less pressure. The gauges show the pressure at each point. Because there is more pressure at point 1, the water pressure will be higher at point 1.

Figure 4-28b   **An Illustration of How the Water Pressure Falls as the System Pressure Drops**
As the system pressure drops, the levels in the service connections fall if the system pressure becomes low enough that a negative pressure is created at point 2.

Because of Bernoulli's Principle, pressure differentials can be created whenever a large pipe feeds a smaller pipe, when there are obstructions in pipes, or when devices or fittings are placed in line to reduce pressure. A negative pressure at point 2 would allow contaminants or pollutants to back-siphonage backflow into the potable water line if a cross-connection existed. Likewise, a venturi meter (which measures flow rate) restricts the flow of water, and therefore, reduces the pressure. A venturi meter can thus create a pressure differential and cause a backflow. The lower pressure at the meter could draw non-potable water backwards, since water always flows in the direction of lowest pressure.

> **Pressure differentials can be created whenever a larger pipe feeds a smaller pipe.**

Mechanical Devices

Pressure differentials can also be created by mechanical devices, such as pumps. Pumps can increase the pressure in a non-potable supply over that of a potable supply, thus creating pressure differentials that could cause a hazardous backflow. The pump in Figure 4-29 creates and maintains a higher pressure than the potable-water line pressure. The higher pressure may be needed to properly dispense the chemicals or to keep them in solution. However, if the shut-off valve is leaking or left open, the higher pressure generated by the pump may force the contaminants into the potable water system.

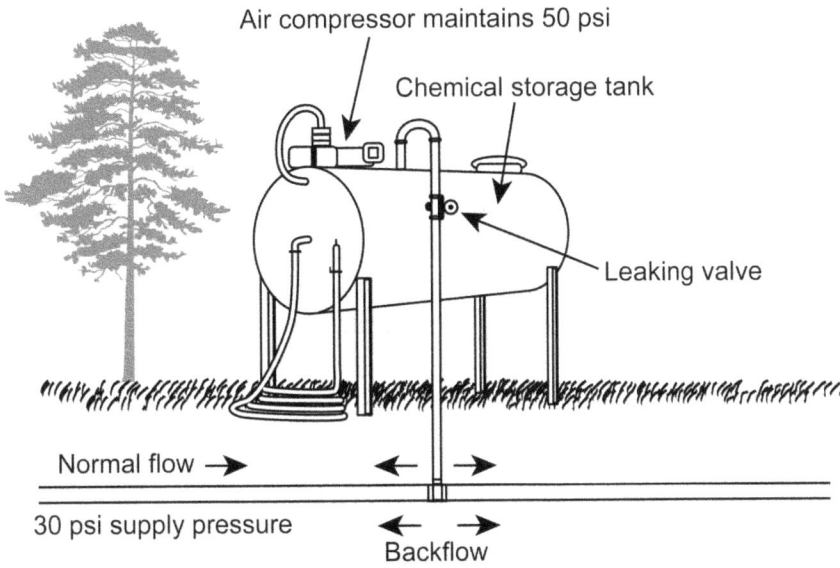

Figure 4-29 **An Illustration of How a Pump Can Cause Backpressure-Backflow**
The pump creates and maintains a higher pressure than the potable-water line pressure. The higher pressure may be needed to properly dispense the chemicals or to keep them in solution. However, if the shut-off valve is leaking or left open, the higher pressure generated by the pump may force the contaminants into the potable water systems.

FUNDAMENTALS OF BACKFLOW

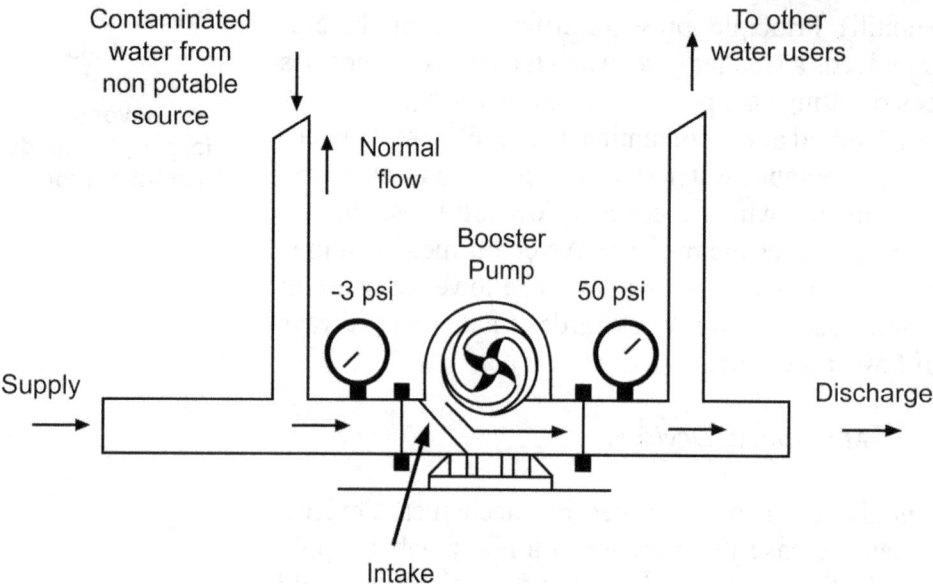

Figure 4-30 **An Illustration of How a Pump Can Cause Backsiphonage-Backflow**
The intake side of the pump is creating a reduction in pressure. If the pressure is reduced sufficiently, water from the non-potable source could backflow into the potable water supply.

Pumps can also create a reduction in pressure. The reduction in pressure occurs on the intake side of the pump. The reduction in pressure occurs because the pump is trying to move the water through itself faster than the water is arriving at the intake (Figure 4-30). If the pressure is reduced sufficiently, water from the non-potable source could backflow into the potable water supply.

Any factor that creates a pressure differential—pumps, water temperature, water velocity, atmospheric pressure or the weight of water itself—can cause a reversal of the normal flow and a backflow of hazardous materials if a cross-connection exists. Two general types of backflow are recognized, based on the nature of the pressure differential that is generated. These are backpressure and backsiphonage.

The two general types of backflow are backpressure and backsiphonage.

Backpressure

Backpressure, one type of backflow, occurs whenever the pressure on the non-potable system is greater than the pressure in the potable water supply system. It occurs due to;

- a decrease in the potable supply pressure (below that of the non-potable source),
- an increase in the non-potable supply pressure above the potable supply pressure; or,
- a combination of both events.

Backpressure is also referred to as superior pressure,[1] since a higher pressure on the non-potable system side causes the backflow.

Any water demand that is larger than what the water purveyor can supply creates a reduction of pressure in the potable supply. Examples of large water demands include fire fighting, water-line flushing, and water-line or fire hydrant leaks.

If the pressure on the potable-water side falls below that of a non-potable source, a backpressure-backflow can occur. An example is irrigation wells that are cross-connected to drinking water supplies. Because irrigation wells are not required to meet potable drinking water standards, the water from the irrigation well is considered non-potable. Irrigation wells have their own pumps; therefore, the non-potable irrigation water can maintain some constant pressure. If a fire occurred, the potable water supply pressure would be reduced. If it falls below that maintained by the pump on the irrigation system, non-potable water will flow into the potable water lines.

It is important to be aware that most air-to-water heat exchangers use private wells as their source of water, and many are also cross-connected to the community water system to ensure that water is supplied in case the private well fails (Figure 4-31).

Backpressure can also be created by an increase in the non-potable system pressure. Recall, for example, how the pump on a chemical storage tank can maintain a higher pressure in the tank than the potable supply pressure (recall Figure 4-29). The higher pressure may be needed to properly dispense the chemicals or to keep them in solution, but because the water supply line to the chemical storage tank creates a cross-connection,

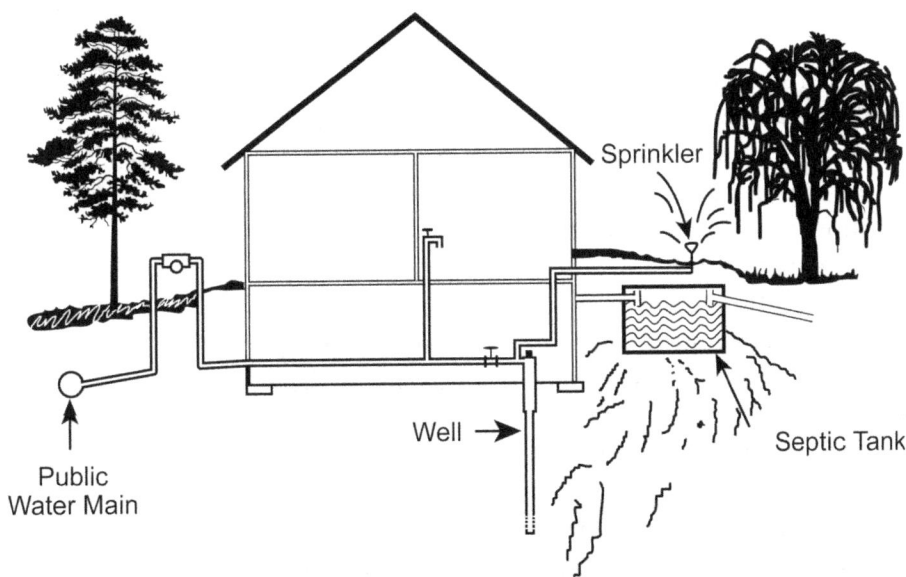

Figure 4-31   **A Cross-Connection to an Irrigation Well**
Irrigation wells are commonly cross-connected to drinking water supplies. The well water in this case has become contaminated with wastewater from the septic system.

backpressure-backflow becomes possible. If the water supply valve to the tank was accidentally left open while the pump was running, the chemicals in the storage tank would backflow into the water supply line.

Boilers are also prime candidates for backpressure-backflow problems (recall Figure 4-26). Increased pressure inside the boiler as a result of higher temperatures can force boiler water to backflow into the potable water supply. This is potentially hazardous, because boric nitrate is often added to boiler water to prevent calcium scale buildup. Boric nitrate is toxic, and can have a significant effect on children and pregnant women. Therefore, a cross-connection between the boiler and the potable-water supply would create a health hazard. As a precaution, coloring agents (e.g., phenols, which act like a laxative if consumed in large dosages) are often added to descalers to prevent people from drinking contaminated water. However, it is possible for people to consume the "colored" water before noticing the color, and children in particular could be attracted to colored water.

These are just a few scenarios of conditions and events that could create backpressure-backflow. Bernuolli's Principle could also contribute to pressure differentials, which would lead to a backpressure-backflow. An example of this type of situation is backsiphonage that follows.

Backsiphonage

Backsiphonage-backflow can occur whenever a cross-connection exists and a negative or sub-atmospheric pressure is generated in the potable water supply. Therefore, any event that reduces the potable supply pressure below atmospheric pressure can create **backsiphonage-backflow.** A number of events that reduce potable supply pressure were listed above under backpressure, e.g., broken water mains, firefighting, water-line flushing, etc. Any of these events could also cause a backsiphonage-backflow, if the potable supply pressure is decreased below atmospheric pressure. If the potable supply pressure is reduced below atmospheric pressure, normal flow would be reversed simply as a result of atmospheric pressure acting on the non-potable source. The following examples illustrate events or situations that could cause backsiphonage-backflow.

A temporary cross-connection is created in a restaurant kitchen (Figure 4-32) when the hose connected to the potable-water system is left lying in the dirty mop water. This faucet just happens to be leaking. The mop water flows into the hose (just as water rises inside a straw that is partially submerged in water). When an automobile runs into a fire hydrant, causing it to burst, it suddenly creates a large loss of water, and therefore, a large decrease in water pressure. In fact, a negative pressure is created in the water-line servicing the restaurant. Because the dirty mop water surrounding the hose is exposed to atmospheric pressure, and the pressure inside the hose is now less than atmospheric, the mop water flows up the hose. The mop water flows through the leaky faucet and into the potable drinking water supply. Ultimately, the restaurant dishwasher is supplied with dirty mop

Any event that reduces the potable supply pressure below atmospheric pressure can create backsiphonage.

CHAPTER FOUR

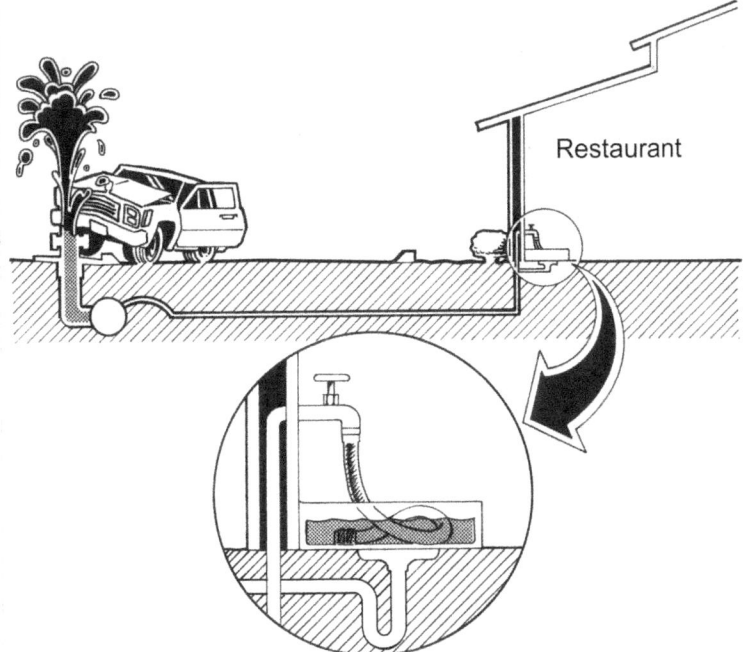

Figure 4-32   Backsiphonage at a Restaurant
The leaking fire hydrant reduces the potable water supply pressure below atmospheric in the water main that supplies the restaurant. Therefore, atmospheric pressure pushes the dirty mop water into the potable water supply.

Broken water mains seem to be the most common cause of backsiphonage-backflow.

water. Any restaurant patron who eats from the dishes could potentially be exposed to a number of pathogenic organisms. The type of temporary cross-connection illustrated in this example is common, because most people mistakenly believe that water cannot flow upward.

Pumps can also create negative pressures. Recall from Figure 4-30 that the intake side of the pump created a negative pressure that drew contaminants from a non-potable source into the pump.

Negative pressures also occur in plumbing systems because of Bernoulli's Principle. Recall that where velocity of a fluid is high, the pressure is low. Therefore, any type of fixture arrangement that produces negative pressure (e.g., venturi meter, large pipe feeding smaller pipe, etc.) can potentially create backsiphonage-backflow if a cross-connection exists. Bernoulli's Principle is illustrated by an aspirator Figure 4-33. As water flows through the aspirator, the restriction in the aspirator causes it to flow very fast. This fast flow of water creates such a reduction in pressure that a vacuum is created. Connecting the aspirator to the flask creates a vacuum in the flask. A bacteria-laden water sample is dumped into the funnel, and the vacuum is used to pull the water sample through the filter and into the flask, trapping bacteria on the surface of the filter.

However, if the filter tears, the bacteria will pass through the filter apparatus into the flask. The aspirator, itself, provides a cross-connection

FUNDAMENTALS OF BACKFLOW

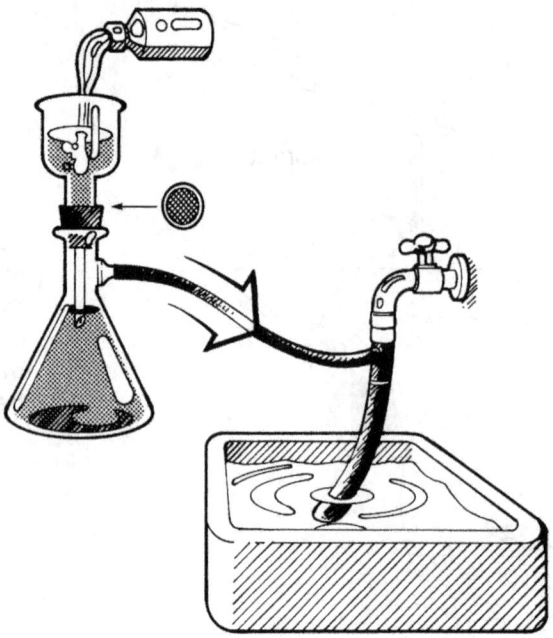

Figure 4-33 **Backflow in a Laboratory Caused by an Aspirator**
The overfilled receiving flask provides a source of contamination; the aspirator provides the cross-connection between the potable water supply and the contaminant; and Bernoulli's Principle reveals why the flask water will backflow into the potable water supply.

between the potable water supply and water in the sink as well as in the flask. An indirect cross-connection between the flask, laden with bacteria, and the potable water supply is created if the water is not dumped out of the receiving vessel and the vessel overfills. If the potable water supply pressure falls below atmospheric pressure for some reason, not only will water from the sink be pushed into the potable water supply by atmospheric pressure, but also the bacteria-laden flask water will be drawn up into the potable water supply. This occurs because the aspirator will create a vacuum no matter which way the water is flowing.

Summary

Understanding how and why backflow occurs is essential if it is to be prevented. To prevent backflows, it is necessary to identify the hydraulic conditions that will create them. Pressure differentials between potable water supplies and non-potable supplies are the cause of backflow events.

Potable water systems must be scrutinized both for cross-connections, which provide the physical links that allow backflows, and also for conditions that could potentially create pressure differentials. The methods used to survey and protect potable water systems are addressed in the following chapters.

Chapter Four Review

4-1 Atmospheric pressure plays a major role in which type of backflow? (Backsiphonage) or Backpressure?

4-2 According to the ideal gas law, if the volume is held constant and the temperature increases, what will happen to pressure? (Increase) or decrease?

4-3 How far can water be backsiphoned up a column if you have a perfect vacuum?
    33.9 ft

4-4 Water traveling at high velocity past or through a small opening can cause backsiphonage by what effect? Venturi

4-5 How can you break or stop backsiphonage?
    add air

4-6 A water column one foot in height equals how many psi?
    .433 psi

4-7 One psi equals how many feet? How many inches?
    2.31 ft   28 in

4-8 Thirty-seven psi equals how many feet or inches?
    37 × 2.31 = 85 ft   1 psi = 2.31 ft.

4-9 A water column 86 feet in height equals how many psi?
    86 × .433 = 37 psi
    1 ft = .433 psi

REFERENCES

1. *Cross-Connection Control Training Package*, 1985. American Water Works Association, New York, NY.

2. *Cross Talk*, Foundation for Cross-Connection Control and Hydraulic Research at the University of Southern California, Autumn 1998, page 6.

3. *Cross-Connection Control Manual*, 1975, revised 1989. (EPA 430-9-73-002). U.S. Environmental Protection Agency, Water Supply Division.

CHAPTER FIVE
METHODS AND MECHANISMS FOR PREVENTING BACKFLOW

Chapter 4 identified the different types of cross-connections that can be created. Learning this information is crucial to preventing the installation of cross-connections and eliminating existing cross-connections. Also, the hydraulic conditions that cause backflow were discussed. Understanding these conditions not only helps to explain why backflow occurs, but also aids in the identification of actual and potential cross-connections that might otherwise be overlooked.

This chapter focuses on methods and mechanisms for preventing backflow. The provision of an air gap between potable and non-potable sources is a **method** for preventing backflow. Air gaps prevent backflow by eliminating the cross-connection. Where cross-connections cannot be eliminated, mechanical backflow preventers (atmospheric vacuum breakers, pressure vacuum breakers, spill-resistant vacuum breakers, double check valve assemblies, and reduced-pressure principle assemblies) provide a mechanism for preventing backflow. Mechanical backflow preventers provide a physical barrier to backflow.

Standards and specifications for the mechanical backflow preventers and air gaps will be reviewed. Adhering to these standards is necessary to ensure that any backflow preventer provides reliable protection. Detailed information will be given on how each method or mechanism works to prevent backflow, when each can be used, how each should be installed, and some of the advantages and disadvantages of each.

In addition, other backflow prevention methods will be discussed, as will the rationale for why these devices should not be used as backflow preventers.

Devices and Approved Assemblies

Before discussing the different mechanical backflow preventers, it is important to clarify the terms device and assembly. The term **device** refers to the backflow prevention mechanism only. It does not include the shut-off valves located on the inlet and outlet of the backflow preventer. An **assembly** is the entire backflow prevention unit, including not only the mechanism that actually prevents backflow, but also the shut-off valves and test cocks.

A method is a non-mechanical way of eliminating a cross-connection, such as an air gap or a barometric loop.

A device is a mechanical way to protect against backflow.

An assembly is the entire backflow prevention unit, including not only the mechanism that actually prevents backflow, but also the shut-off valves and test cocks.

METHODS AND MECHANISMS FOR PREVENTING BACKFLOW

Although several agencies "approve" or set standards and specifications for backflow prevention assemblies, only the Foundation for Cross-Connection Control & Hydraulic Research (FCCC & HR) at the University of Southern California, field tests backflow assemblies. Therefore, its guidelines are being used to define "approved assembly."

To be an **approved assembly**, the backflow preventer must be shipped from the manufacturer as one unit. It should not be assembled in the field. However, this does not mean that the whole assembly must be replaced whenever a part wears out. Worn out parts should be replaced with genuine manufacturer replacement parts.

The backflow preventer must also have shut-off valves that have a **full-flow characteristic** (Figure 5-1). Normally, full-flow valves have smooth interior walls, unlike the older types of gate valve. The terminology has been revised from fully-ported valves to full-flow valves. Fully-ported valves are valves with inside dimensions that match the size of the valve. A valve that is not fully ported can, however, still provide full flow. Full-flow potential is the essential factor. The purpose is to eliminate unnecessary reduction in pressure, since backflow preventers themselves already reduce line pressure.

> **Approved assemblies must be shipped from manufacturers as one complete unit.**
>
> **Shut-off valves *must* have a full-flow characteristic.**

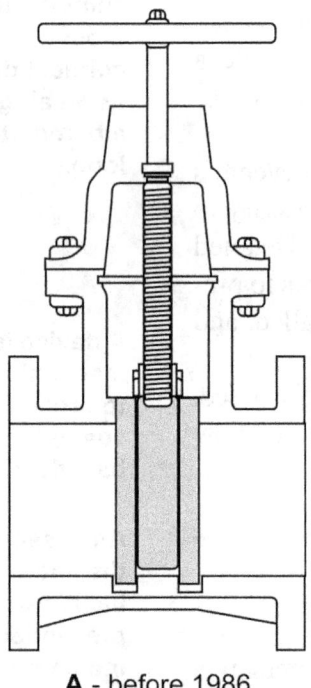
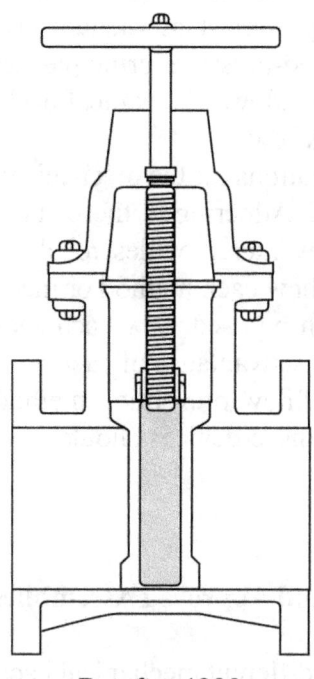

A - before 1986
Gate Valve

B - after 1986
Resilient Wedge Valve

Figure 5-1   Shut-Off Valves
Figure 5-1A   Grooves and notches inside the gate valve used before 1986 tend to collect foreign materials that disrupt the normal functioning of the backflow preventer.
Figure 5-1B   Normally, full-flow valves that were produced after 1986 have smooth interior walls, unlike the older types of gate valves.

CHAPTER FIVE

The valve itself must also be **resilient seated**. Resilient is defined in the *AWWA Water Distribution Operator Training Handbook* as "capable of withstanding shock without permanent deformation or rupture."[1] A resilient seated valve usually contains some soft rubber or other flexible material that can withstand shock. A valve with metal seated against metal is not resilient seated. Some resilient seated valves are ball valves. (One manufacturer makes a resilient seated valve that looks very much like a gate valve.) Valves that are 2½ inches or larger are usually resilient wedge valves. Butterfly valves may also be resilient seated.

Additionally, since July 1988, shut-off valves must provide the following minimum markings: manufacturer's symbol or name, nominal size of valve, model number, and working pressure. The model number should indicate whether the particular model is resilient seated. This is usually designated with an "R" or "RS."

Since June 1988, test cocks used on all approved backflow prevention assemblies must be: resilient seated, fully ported, operating stem must indicate if test cock is open/closed, operating stem on a ball valve type of test cock must be "blow-out proof," and body materials must be specified as bronze, stainless steel, or engineered plastic. A "blow-out proof" stem is required on the test cock to ensure that the stem does not come out of the body (Figure 5-2). An internally loaded stem prevents it from blowing out.

Shut-off valves must be resilient seated and have full-flow characteristics.

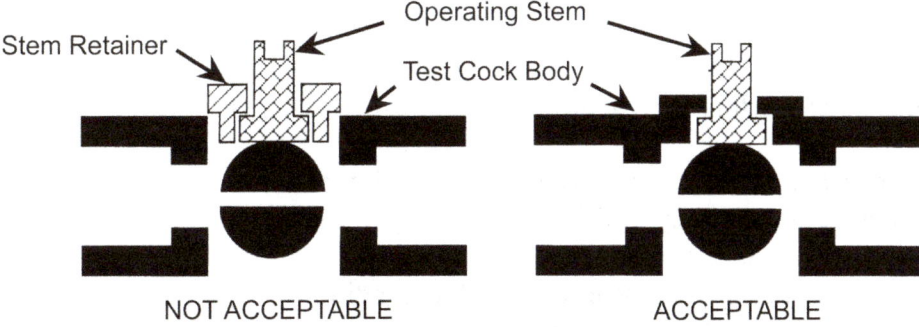

Figure 5-2 Test Cock "Blow-out Proof" Stem
An internally loaded stem prevents it from being blown out.

The assembly itself is also required to be labeled with the following minimum markings: manufacturer's symbol or name, type of assembly, size and model number, serial number, rated working water pressure, rated working water temperature, and direction of flow.

To be "approved," a backflow preventer must be testable and repairable in-line. In other words, the backflow preventer must be able to stay in place on the potable water line when it is tested and repaired (with the exception of the atmospheric vacuum breakers). Table 5-1 provides a summary of the requirements for an approved assembly.

METHODS AND MECHANISMS FOR PREVENTING BACKFLOW

Table 5-1 Summary of Requirements for an Approved Assembly

1. Shut-off valves located before and after the backflow prevention device.
2. Shut off valves must have full flow characteristics.
3. Shut-off valves must be resilient seated.
4. The assembly can be tested and repaired in-line.
5. Assemblies are properly labeled with relevant information.
6. Assembly must ship as a complete assembly.

Since backflow prevention assemblies are mechanical, it is important to have established standards and specifications to ensure that the assembly will function properly. For instance, the specifications for backflow preventers used on hot water lines naturally require parts that can withstand the high temperatures. Table 5-2 lists organizations that set standards and specifications for backflow prevention assemblies.

Table 5-2 Organizations That Establish Standards for Backflow Prevention Assemblies

ANSI (American National Standards Institute)
ASME (American Society of Mechanical Engineers)
ASSE (American Society of Sanitary Engineering)
ASTM (American Society for Testing and Materials)
AWWA (American Water Works Association)
CSA (Canadian Standards Association)
FCCC & HR (Foundation for Cross-Connection Control and Hydraulic Research)
FM (Factory Mutual)
NSF (National Sanitation Foundation)
UL (Underwriters Laboratory)

Mechanical Backflow Preventers

For each type of mechanical backflow preventer discussed here, a basic description of the device or assembly will be given, along with an explanation of how it operates. This discussion will include operation under normal flow conditions as well as during both types of backflow conditions. The proper application for each backflow preventer will be given, i.e., whether it can be used for low or high hazards and whether it can be used in backpressure as well as backsiphonage situations. In addition, special limitations or disadvantages associated with the backflow preventers will be discussed, as well as common applications. Also, an indication of whether the assembly is most frequently used for isolation or containment purposes will be given. Finally, installation recommendations will be given for each backflow preventer.

Atmospheric Vacuum Breaker

Atmospheric vacuum breakers (AVBs) are the simplest type of mechanical backflow preventer that are currently approved by the A.S.S.E. AVBs are not approved by the FCCC & HR since no method exists to fully

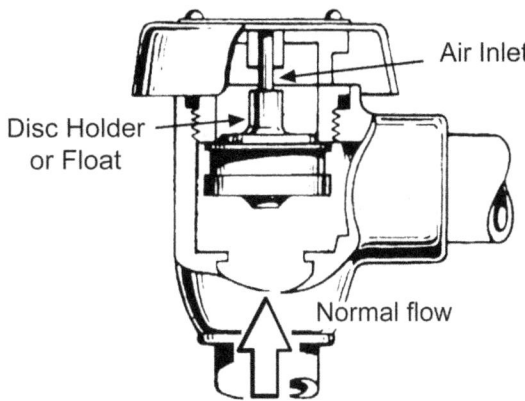

Figure 5-3 **Atmospheric Vacuum Breaker Normal Flow**
Water pressure pushes the disk float up against the air inlet port, no matter which direction the water is flowing.

test these devices in-line. However, the FCCC & HR does include these devices on its approved list, because they have been tested and approved by the Los Angeles City Mechanical Testing Laboratory.

An AVB contains a single disk float (also know as a "poppet" valve) that operates on atmospheric pressure (Figure 5-3) and an air inlet port, which is protected by a canopy. The canopy on the atmospheric vacuum breaker prevents debris from falling into the device. The AVB does not possess any test cocks or shut-off valves; however, a shut-off valve occasionally is located just prior to the device. This is one backflow preventer listed by FCCC & HR that cannot be accurately tested in-line.

When water is flowing normally through the device, the disk float is pushed upward against the air inlet, blocking it, and the water flows past (Figure 5-3). Under backpressure conditions, the disk float will remain pushed up against the air inlet port, allowing contaminants to backflow into the potable water supply. Thus, the AVB can be used for protection against backsiphonage, but NOT against backpressure.

In a backsiphonage-backflow condition, the drop in water pressure creates a partial vacuum that tends to reverse the flow, "pulling" contaminants into the potable water supply. However, when the water pressure falls below atmospheric pressure (14.7 psi), the disk float on the AVB falls. As the disk float falls, it allows air into the system and thus breaks the vacuum (Figure 5-4).

Essentially, when the air inlet port opens, an air gap is created. When air is allowed into the system or pipe, the vacuum that would have moved the contaminant into the potable water supply line is broken. Therefore, the contaminant does not move any further into the potable water supply. Recall, this is similar to trying to drink soda through a straw that has a hole in it. Even though the pressure inside the straw is reduced by sipping, air continues to enter through the hole.

AVBs are NOT approved for use where they will be subjected to continuous pressure in the direction of flow. In other words, shut-off valves

METHODS AND MECHANISMS FOR PREVENTING BACKFLOW

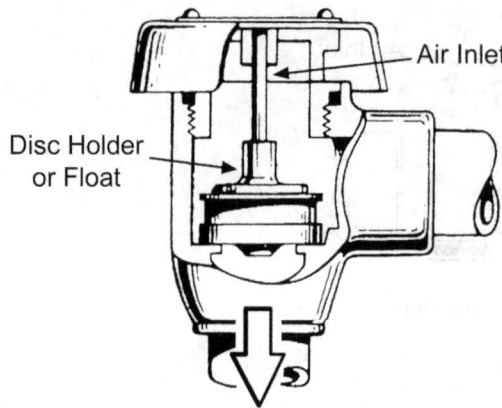

Figure 5-4 Atmospheric Vacuum Breaker During Backsiphonage
When the water pressure falls below atmospheric pressure (14.7 psi), the disk float on the AVB falls and lets air enter the top to form an air gap.

are not allowed downstream from this device. After 12 to 14 or more hours of continuous pressure, the disk float can become permanently stuck to the air inlet port. To prevent this, experts recommend that these devices be exercised at least once every 12 hours. The water should be turned off to allow the valve to drop down from the air inlet port.

Another major drawback of the AVB is that no approved method of testing the device exists. However, simply turning off the water at the shut-off valve just prior to the device can determine whether the device is working. The disk float can usually be heard as it falls. A visual inspection can also be made by removing the canopy. However, in spite of this easy method for monitoring, AVBs are not commonly inspected or replaced. Most water users are unaware of how the device works even if they know its purpose. As a result, the device often becomes nothing more than an expensive elbow in the water line.

According to installation specifications, an AVB must be installed at least 6 inches higher than the highest point in the system. This requirement prevents backpressure from the weight of water in piping downstream. Keep in mind that AVBs do not provide protection against backpressure.

The AVB must be installed in an upright position in order for the disk float to work properly. Moreover, it is essential that all backflow preventers be examined closely to determine the direction of flow as disk floats only function properly in one direction. "Approved" backflow preventers are required to include an arrow that indicates the direction of flow. Shut-off valves should be avoided downstream of the AVB. To be an "approved device" a mark must be provided to designate the critical level on the backflow preventer indicating from where the elevation should be measured and an arrow should indicate the direction of flow. Figure 5-5 illustrates the use of an atmospheric vacuum breaker in an irrigation system.

The AVB must never be subjected to flooding, since it relies on atmospheric pressure to function. For the same reason, the air inlet port must

AVBs do not provide protection against backpressure.

There must not be any shut-off valves downstream of the AVB.

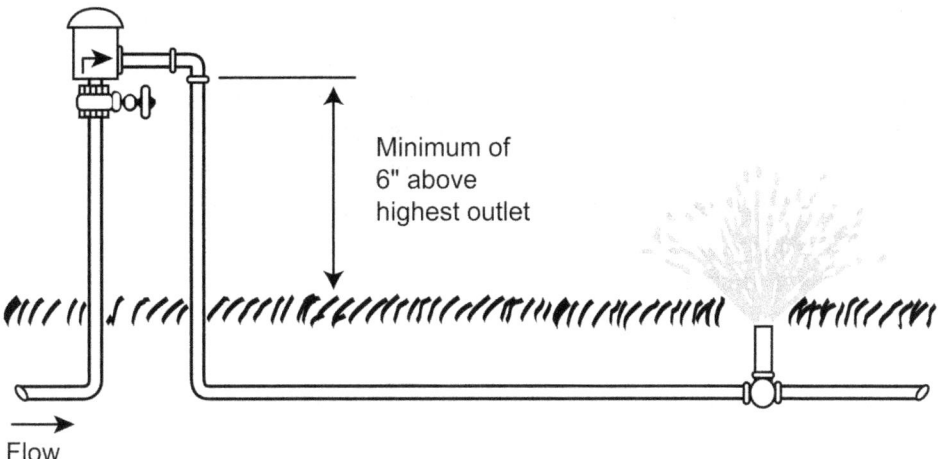

Figure 5-5   Atmospheric Vacuum Breaker In-line on an Irrigation System
An atmospheric vacuum breaker must not have any shut-off valves down stream. To be an "approved device" a mark must be provided to designate the critical level on the backflow preventer indicating from where the 6 inches should be measured and an arrow should indicate the direction of flow.

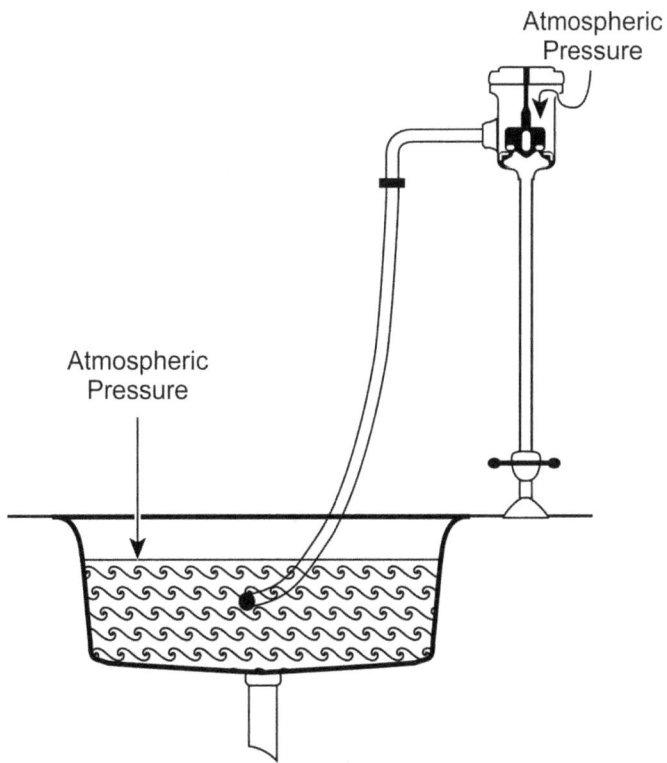

Figure 5-6   Atmospheric Vacuum Breaker Installed on a Laboratory Sink
As the pressure within the potable water system drops below atmospheric, a vacuum is created within the potable water lines. Atmospheric pressure acts on the contaminant and pushes it backward toward the potable water system. However, because atmospheric pressure acts equally and simultaneously on both the contaminant and the disk float, the disk float will fall and allow air into the water line, breaking the vacuum. The air breaks the siphon inside the water line, which prevents the contaminant from backsiphoning into the potable water system.

METHODS AND MECHANISMS FOR PREVENTING BACKFLOW

never be blocked or sealed. Often, individuals who are unaware of how the device functions will wrap the air inlet with cloth or tape to prevent water from spraying out. Water may spray out of the air inlet when the device is subjected to backpressure, e.g., when a hand-operated sprayer is attached to a hose after the device. This occurs because the amount of backpressure in the hose fluctuates as the shut-off valve within the sprayer is opened and closed. When the backpressure drops, the valve opens to the atmosphere. When the backpressure increases, the valve slams closed, forcing out a small amount of water.

Hose Bibb Vacuum Breaker

A **hose bibb vacuum breaker** (HBVB) is an atmospheric vacuum breaker that has been adapted to fit on a hose bibb or sill cock (Figure 5-7). It is commonly installed on hose bibbs where a cross-connection could easily be created by attaching a hose to a faucet. Again, as with the AVB, in order for the hose bibb vacuum breaker to work properly shut-off valves should be avoided downstream from the device. No method to adequately test hose bibb vacuum breakers exists.

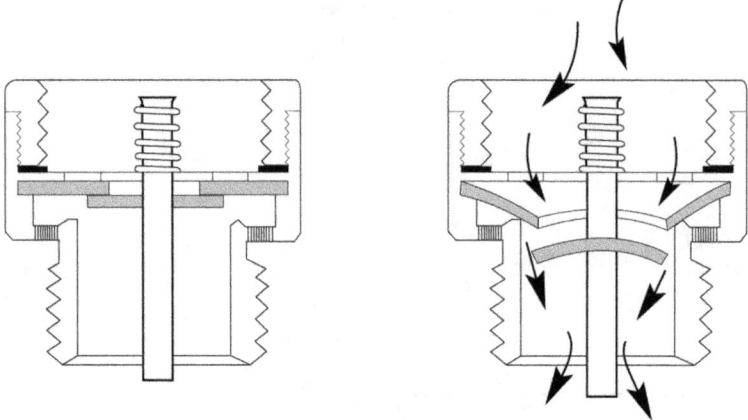

Figure 5-7   **Hose Bibb Vacuum Breaker**
The hose bibb vacuum breaker can be used to prevent backsiphonage but not backpressure, and it can be used as protection against both high and low hazards.

Pressure Vacuum Breaker

Pressure vacuum breakers (PVBs) are very similar to AVBs in design. However, the PVB is an assembly instead of a device; it has two shut-off valves and two test cocks. Also, the air inlet valve and the check valve operate independently of each other (Figure 5-8).

The spring on the check valve keeps the check valve closed, unless water is flowing through the assembly in the proper direction under suffi-

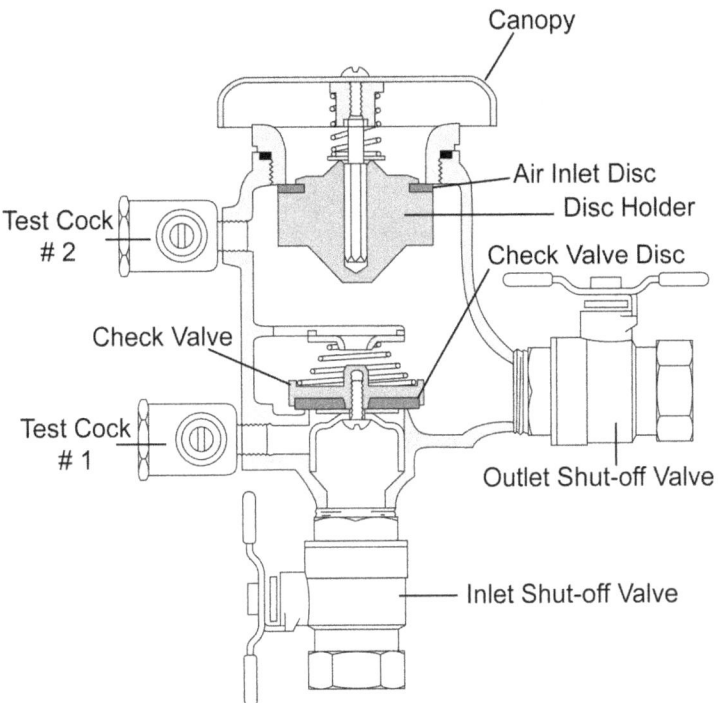

Figure 5-8   Pressure Vacuum Breaker
The check valve must be forced open in the normal direction of flow. During backsiphonage, the air inlet opens and allows atmospheric pressure into the assembly.

cient pressure. Since the check valve must be forced open in the direction of flow, it provides even better protection against backflow than the float-type check. The AWWA specification for a spring-loaded, soft-seated check valve is that it must be drip tight and hold 1.0 psi in the direction of flow. Because the air inlet is also loaded, it will open in response to backsiphonage before the water pressure falls to within 1.0 psi of atmospheric pressure. In other words, it will open before the water line pressure falls below 15.7 psi.

While the springs provide slightly more protection, they can not be relied on to prevent backpressure-backflow. Therefore, like the AVB, the PVB is only approved for backsiphonage conditions. This assembly can be used under either low or high hazard situations. Generally, the PVB is used for isolation purposes. For instance, it is commonly used on lawn irrigation systems (Figure 5-9).

Unlike the AVB, the PVB can be used where it will be subjected to continuous pressure. However, while the air inlet loading helps prevent the air inlet valve from becoming sealed closed, the loading does not totally mitigate the problem, and the air inlet valve will often stick in the closed position if the PVB is subjected to long periods of continuous pressure. Because the PVB can be subjected to continuous pressure, it is acceptable to have a shut-off valve downstream of the assembly.

The PVB cannot be subjected to backpressure.

METHODS AND MECHANISMS FOR PREVENTING BACKFLOW

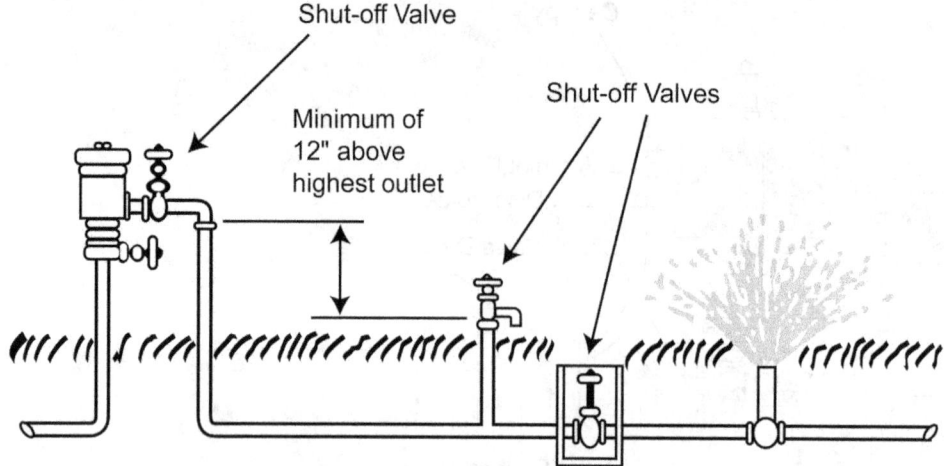

Figure 5-9   Shut-off Valves Are Allowed Downstream on Pressure Vacuum Breakers
Because the pressure vacuum breaker can be subjected to continuous pressure, it is acceptable to have a shut-off valve downstream.

Since the PVB cannot be subjected to backpressure, the installation specifications require that it be installed so that the critical level is 12 inches above the highest point in the system after the vacuum breaker. The **critical level** of a PVB is usually located at the opening of the check valve. As with the AVB, the air inlet port should not be blocked as the assembly depends on the inflow of air to work properly. The assembly must not be subjected to flooding, and it must be installed in the direction of flow. Since this assembly can be tested, it is essential that it be installed in a location where it can be reached.

The PVB must be installed so that the critical level is 12 inches above the highest point in the system.

Backflow Protection for Irrigation Systems

Lawn irrigation systems are classified as a high hazard. Obviously, if chemicals, such as, fertilizers or pesticides are injected into the system, a greater hazard is created. Pop-up sprinkler heads or other ground-level sprinklers also present a high hazard, because irrigation lines can become contaminated with fertilizers, pesticides and other substances (Figure 5-10). Recall the Holy Cross football team case, where the irrigation lines were contaminated with hepatitis after children played near the irrigation heads.

Lawn irrigation systems are classified as a high hazard.

Spill-resistant Vacuum Breaker Assembly

Spill-resistant vacuum breakers (SVBs) are similar to PVBs in design. The PVB has two shut-off valves and two test cocks. The SVB has two shut-off valves, one test cock and one bleed screw (Figures 5-12 to 5-16).

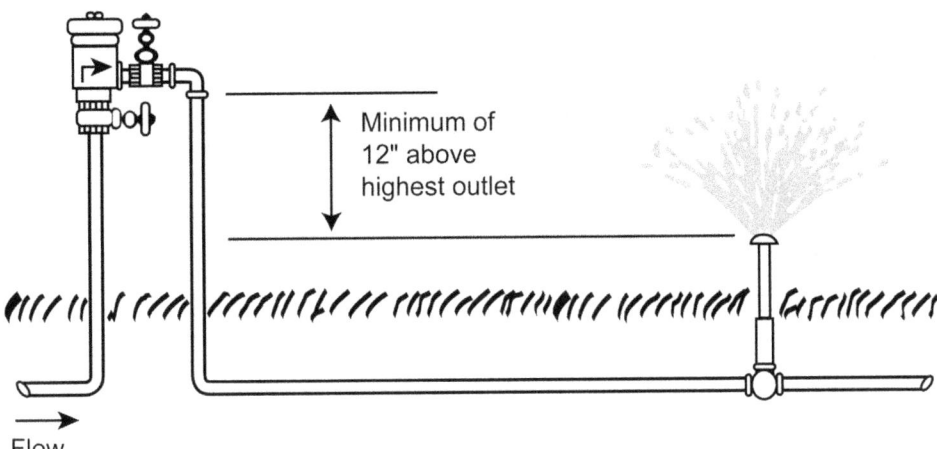

Figure 5-10   Pop-up Irrigation Heads
This irrigation system presents a high degree of hazard. Pop-up irrigation heads or heads installed at ground level allow fertilizers and pesticides to backflow through them into the potable water system if there is a reduction in potable water pressure.

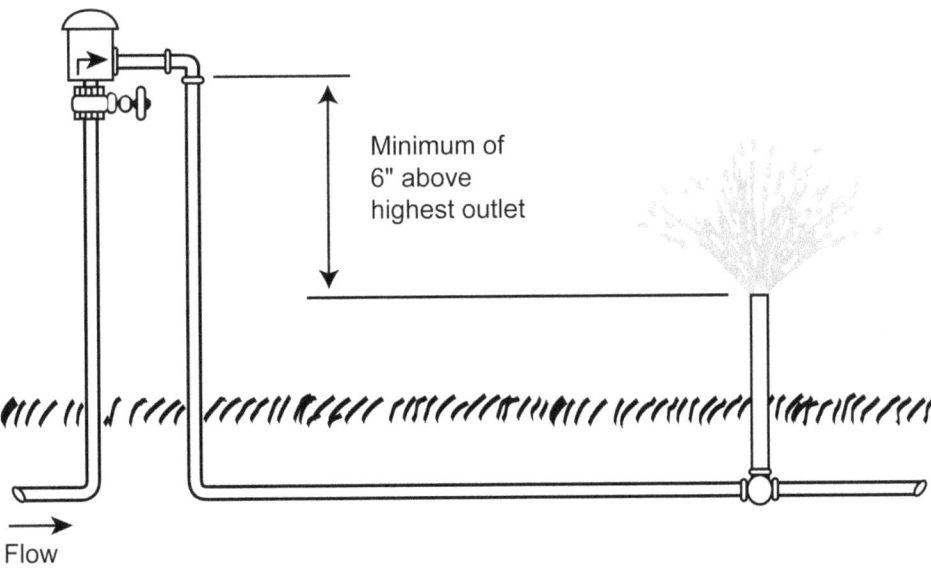

Figure 5-11   Elevated Irrigation Heads
Elevated irrigation heads prevent backflow of fertilizers and pesticides; therefore, they present a low hazard. An AVB can be used as long as there are no shut-off valves downstream from the unit.

METHODS AND MECHANISMS FOR PREVENTING BACKFLOW

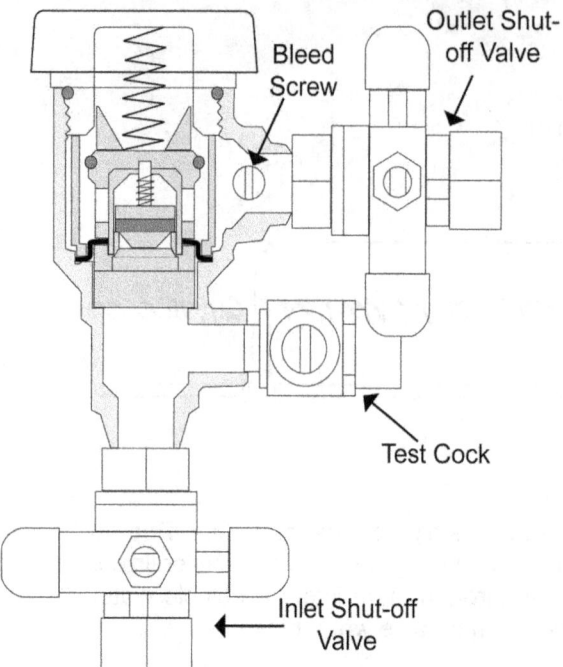

Figure 5-12 **Spill-resistant Vacuum Breaker With No Flow**
When the supply pressure drops below 15.7 psi, the air inlet valve should open to let air inside the assembly.

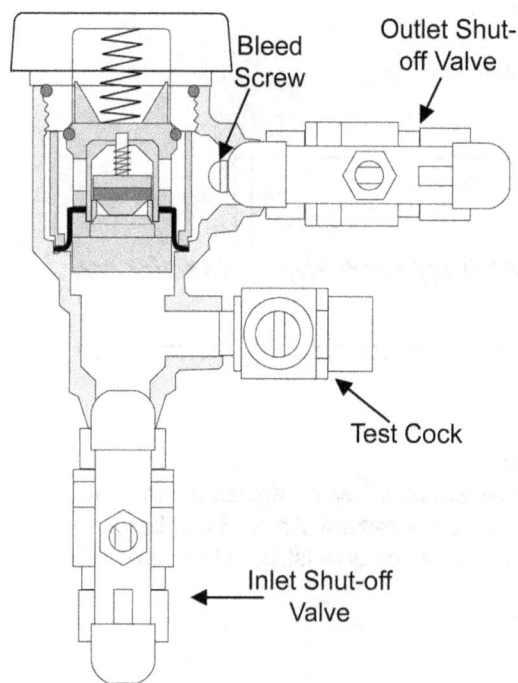

Figure 5-13 **Spill-resistant Vacuum Breaker during Normal Flow Conditions**
The air inlet is sealed shut then the check valve opens to supply water to the customer's equipment.

The spring on the check valve keeps the check valve closed, unless water is flowing through the assembly in the proper direction under sufficient pressure. Since the check valve must be forced open in the direction of flow, it provides even better protection against backflow than the float-type check in the AVB.

The AWWA specification for a soft-seated, spring-loaded check valve is that it must hold drip tight 1.0 psi in the direction of flow. Because the air inlet is also loaded, it will open in response to backsiphonage before the water pressure falls to within 1 psi of atmospheric pressure. In other words, it will open before the water line pressure falls below 15.7 psi.

While the springs provide slightly more protection, they can not be relied upon to prevent backpressure. The SVB is only approved for backsiphonage conditions. This assembly can be used in either low or high hazard situations.

The SVB can be subjected to continuous pressure. Since the spill-resistant vacuum breaker can be subjected to continuous pressure, it is acceptable to have a shut-off valve downstream of the assembly.

Because the SVB cannot be subjected to backpressure, the installation specifications require that it be installed so that the critical level is 12 inches above the highest point in the system after the vacuum breaker. The critical level of a SVB is usually located at the opening of the check valve.

The SVB does not discharge water when the assembly is pressurized.

The air inlet port should not be blocked, since the assembly depends on the inflow of air to work properly. The assembly must not be subjected to flooding, and it must be installed in the direction of flow. Because this assembly can be tested, it is essential that it be installed in a location where it can be reached.

The SVB should be installed in any location that currently utilizes the AVB with a downstream shut-off valve.

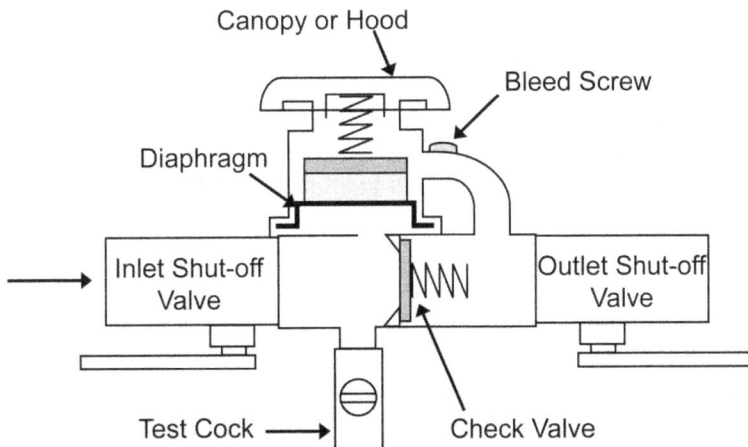

Figure 5-14 Spill-resistant Vacuum Breaker
The Spill-resistant Vacuum Breaker is very similar to the Pressure Vacuum Breaker. The SVB has only one test cock, and the bleed screw replaces test cock #2 on the PVB.

METHODS AND MECHANISMS FOR PREVENTING BACKFLOW

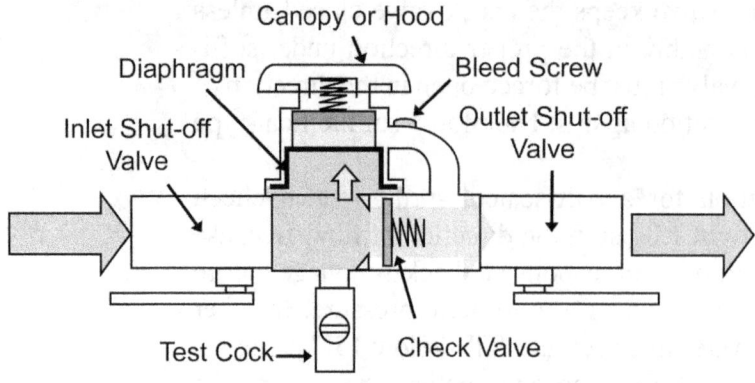

Normal Flow

Figure 5-15   Spill-resistant Vacuum Breaker during Normal Flow
As the assembly is pressurized, the check valve opens after the air inlet closes. The diaphragm pushes the air inlet closed as the assembly is pressurized. This prevents water from discharging out the air inlet valve.

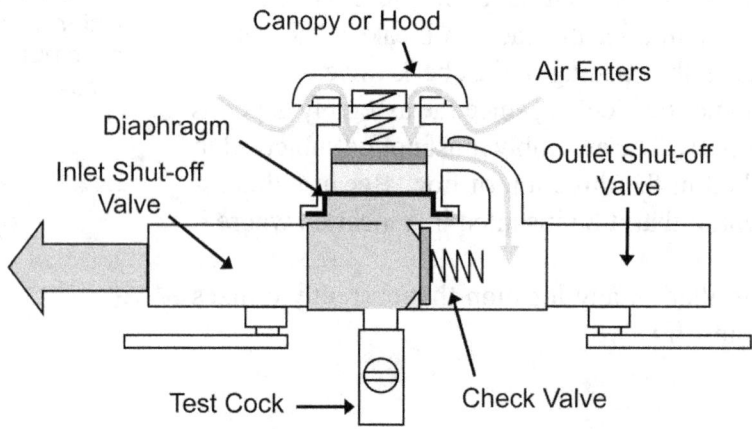

During Backsiphonage

Figure 5-16   Spill-resistant Vacuum Breaker
During backsiphonage or loss of system pressure below 15.7 psi, the check valve closes while the air inlet opens to let in air forming an air gap.

Double Check Valve Assembly

A **double check valve assembly** (DCVA) is basically what its name implies - two independently-operating check valves. The springs in the first and second check valve are identical and, sometimes, interchangeable. The DCVA also has two shut-off valves and four test cocks (Figure 5-17). The check valves will stay closed unless forced open by water in the normal direction of flow. According to AWWA specification, they should hold drip

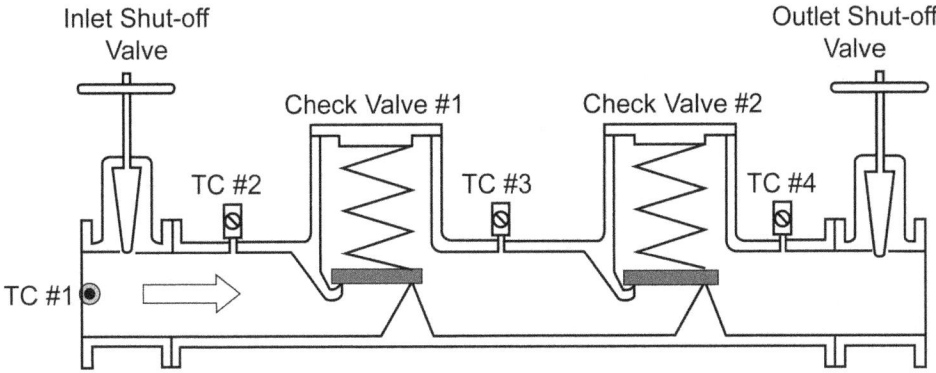

Figure 5-17   Double Check Valve Assembly
The springs in the first and second check valve are identical and sometimes interchangeable.

tight against 1.0 psi in the normal direction of flow. The maximum pressure drop allowed across the assembly is 10 psi. The maximum pressure drop of 10 psi allowed on a DCVA is significant to engineers who design plans for fire sprinkler systems. This maximum pressure drop provides a known quantity to manage when designing fire-sprinkler systems.

Double check valves provide adequate protection against both types of backflow: backsiphonage and backpressure. They can also be subjected to continuous pressure. Under normal flow conditions, the first check valve effectively reduces the supply line pressure by 1.0 to 5.0 psi. This reduction in pressure prevents water from flowing back through the check valve as water always flows in the direction of lowest pressure. The second check valve operates in the same manner.

Both check valves should remain closed during either backsiphonage-backflow or backpressure-backflow. If one check valve fails to close because it is fouled with grit or other debris, the other check valve functions as a backup. Unlike the vacuum breakers, the double check valve assembly does not utilize atmospheric pressure to provide protection. In the rare event that both check valves are fouled at the same time, the assembly no longer provides protection against backflow. For this reason, double check valve assemblies are only approved for low hazard applications.

DCVAs are traditionally used for containment and are installed after the meter or at the point of service. Parallel installations are frequently made where a continuous supply of water is needed (Figure 5-18). Parallel installation allows one backflow preventer to function and supply water to the facility while the other one is being tested or repaired.

DCVAs should be installed at least 12 inches and at most 36 inches above the ground, drainage system, or flood elevation. Installation in a pit or vault is allowed if the pit is not subject to flooding. **Pit installations** are not recommended. Many water purveyors have very stringent specifications for pit or vault installation in order to discourage this type

The maximum pressure loss allowed across the DCVA cannot exceed 10.0 psi during normal flow conditions.

Double check valve assemblies are only approved for low or non-health hazard situations.

Pit installations are not recommended.

Figure 5-18   Double Check Valve Assemblies in Parallel
Parallel installation allows one backflow preventer to function and supply water to the facility while the other one is being tested.

of installation (Figure 5-19). Commonly encountered restrictions include: 1) the top of the vault must extend above grade to prevent flooding or sand intrusion into the pit, 2) the pit must be water-tight (including the bottom) to prevent intrusion of sand or ground water, 3) the backflow preventer must be easily tested from above, 4) the assembly must be within 6 inches of the lid, and 5) brass plugs must be installed in the four test cocks. Many additional requirements can be established. Keep safety in mind when installing DCVAs. **Vault installations** are considered confined spaces. DCVAs that are installed too high off of the ground may require permanent scaffolding.

As with vacuum breakers, it is important that the assembly be examined to determine the direction of flow. This assembly will sometimes be installed vertically because of space limitations. However, manufacturer's recommendations should be examined to determine if vertical installation is approved. Many vertical installations have been approved by FCCC & HR.

Vault installations are considered to be confined spaces.

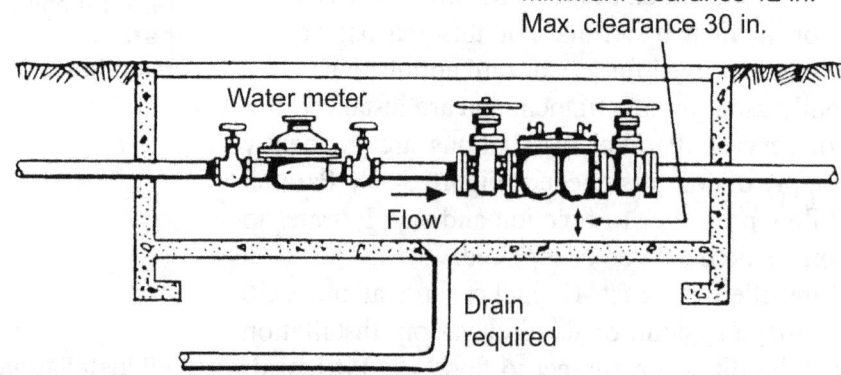

Figure 5-19   Double Check Valve Assembly Installed in a Vault
Vault installations are considered to be confined spaces.

Double Check Valve Assemblies are the only approved backflow preventers that can be installed below ground. This installation is usually discouraged.

CHAPTER FIVE

Reduced Pressure Principle Assembly

The **reduced pressure principle backflow prevention assembly** (RP) consists of two independently operating check valves and a hydraulically operating, mechanically independent pressure differential relief valve, as well as the two shut-off valves and four test cocks (Figure 5-20).

The first check valve is designed to reduce water pressure by a minimum of approximately 5.0 psi. The second check valve has a weaker spring and is only designed to hold back a minimum of 1.0 psi in the direction of flow. As with the DCVA, the two check valves reduce the water pressure so that under normal flow, the pressure on the downstream (customer's) side will always be less than the upstream (supply) side. A hydraulic gradient is thus created that will not allow water to backflow. Moreover, the check valves act to prevent backflow.

The RP has an additional important feature—it opens to the atmosphere through the relief valve. The relief valve is located below or adjacent to the reduced pressure zone. The reduced pressure zone is the area between the first check valve and the second check valve. When the relief valve opens to the atmosphere, it provides an air gap between the two check valves. An air gap is the best method of backflow prevention.

The relief valve opens or closes because of pressures that are applied to both sides of a diaphragm. The water inside the reduced pressure zone applies pressure to one side of the diaphragm and the water on the high pressure side (the supply side) of the first check valve supplies

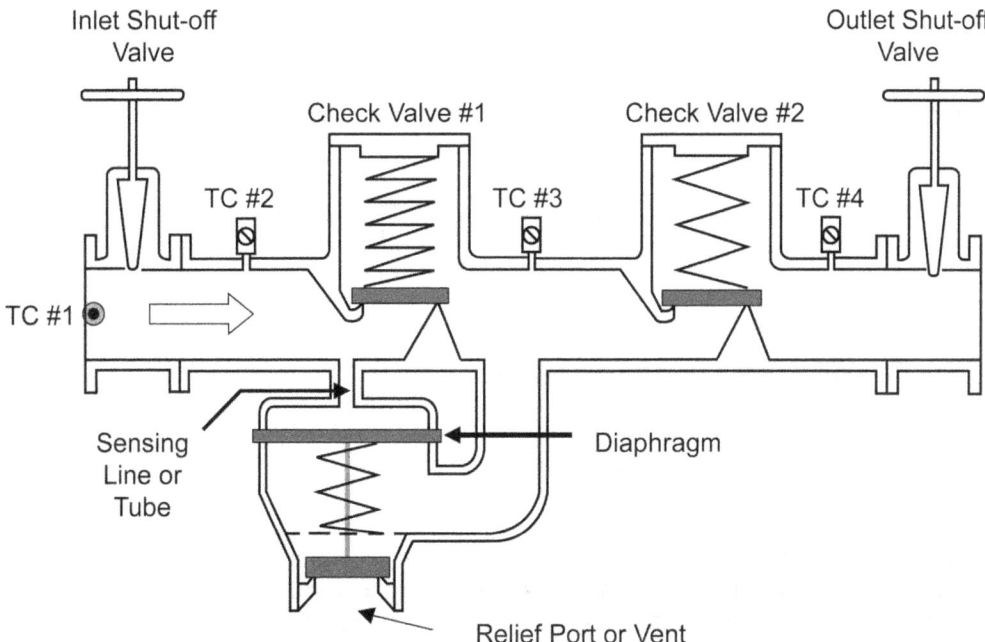

Figure 5-20   Reduced Pressure Principle Assembly
This backflow preventer is also frequently called a Reduced Pressure Zone Assembly (RPZ).

METHODS AND MECHANISMS FOR PREVENTING BACKFLOW

pressure to the diaphragm through the sensing tube. The pressure from the high side acts to keep the relief valve closed. The water pressure inside the reduced pressure zone works to open the relief valve. In addition, a spring attached to the relief valve also works to open the relief valve. The spring is designed to provide a minimum pressure of 2.0 psi.

Figure 5-21 illustrates the normal operation of an RP. Assume a supply pressure of 100 psi and a pressure reduction by the first check valve of 5.0 psi, creating a pressure of 95 psi in the reduced pressure zone. If the relief valve has a 2.0 psi spring, the pressure working to open the port would be 97 psi (95 psi + 2.0 psi) and the pressure working to keep the port closed is 100 psi. Thus under normal flow conditions, the relief valve remains closed.

Under normal operating conditions, the pressure inside the reduced pressure zone is approximately 5.0 to 12.0 psi (at least 5.0 psi) less than the pressure on the supply side of the first check. Even with the additional pressure supplied by the relief valve spring, at least 2.0 psi, the pressure supplied to the diaphragm from the zone is less than that needed to open the relief valve. Figure 5-21 clarifies why the relief valve remains closed under normal conditions.

The 5.0 psi deficiency between the high side pressure and the zone pressure allows for some fluctuation in the line pressure. In other words, if

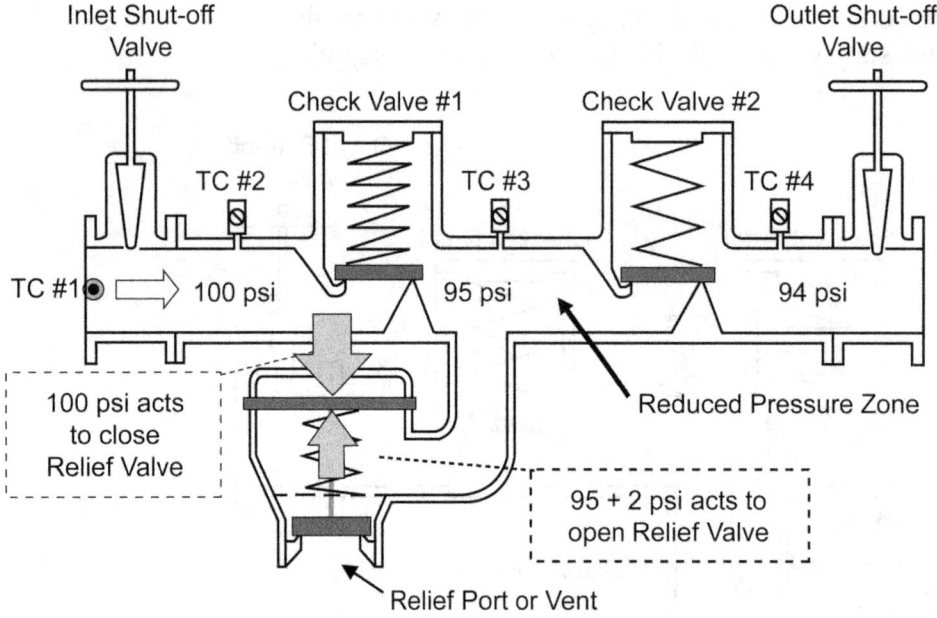

Figure 5-21   **RP: Normal Flow**
Assume a supply pressure of 100 psi and a pressure reduction by the first check valve of 5.0 psi, creating a pressure of 95 psi in the reduced pressure zone. If the relief valve has a 2.0 psi spring, the pressure working to open the port would be 97 psi (95 psi + 2.0 psi) and the pressure working to keep the port closed is 100 psi. Thus under normal flow conditions, the relief valve remains closed.

the supply pressure fluctuates from 100 to 97 psi, the relief valve would still remain closed.

Under a backpressure condition in which there is an increase in pressure on the customer's side, the second check valve should hold tight to prevent backflow. Figure 5-22 illustrates a backpressure from the downstream side of 120 psi. If the second check valve is working properly the pressure in the reduced pressure zone will not increase. However, even if the second check valve is fouled, the assembly will still provide protection against backflow (Figure 5-26).

Under backsiphonage conditions in which the supply line pressure drops toward 0 psi, the pressure exerted on the supply side of the diaphragm falls below that required to keep the relief valve closed. Therefore, the relief valve opens and an air gap is created that prevents backflow. The example depicted in Figure 5-23 assumes that the assembly is under static conditions, i.e., water is not moving through the assembly (perhaps because the customer is not using water or perhaps because the assembly is being tested). If the supply pressure falls to 96 psi because the assembly is under static conditions, the pressure in the zone will remain at 95 psi because the first check prevents backflow of water out of the zone, and water in the zone will not move downstream until there is a demand for water from the customer. The pressure of 95 psi within the zone combined with the pressure exerted by the spring, 2.0 psi, will cause the relief valve to open and **dump** the water in the zone (Figure 5-24).

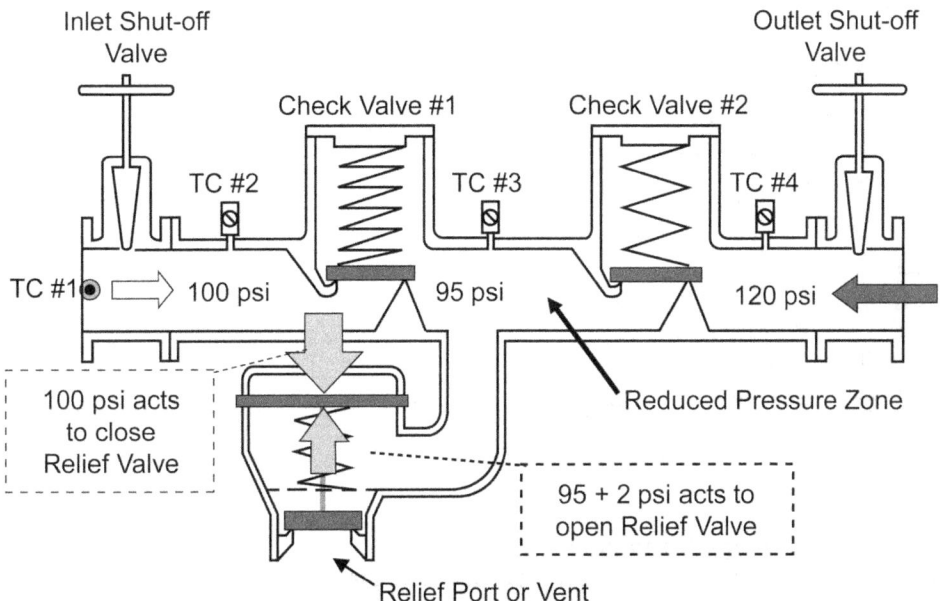

Figure 5-22 RP during Backpressure Conditions
Assume a supply pressure of 100 psi, a relief valve spring pressure of 2.0 psi, a spring pressure of 5.0 psi for the first check valve and backpressure from the downstream side of 120 psi. If the second check valve is working properly the pressure in the reduced pressure zone will not increase.

METHODS AND MECHANISMS FOR PREVENTING BACKFLOW

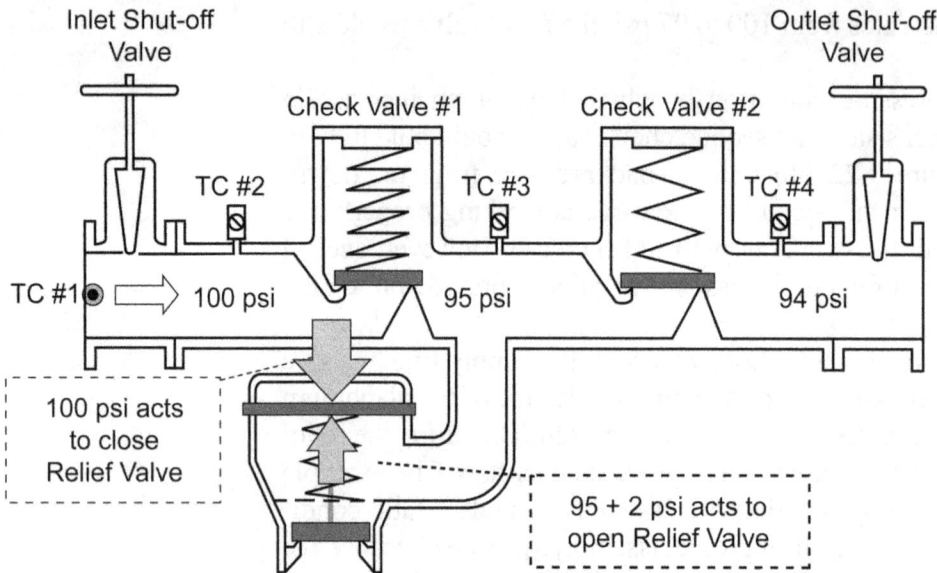

Figure 5-23 **RP Before Backsiphonage Conditions**
Assume a supply pressure of 100 psi, a relief valve spring pressure of 2.0 psi and a spring pressure of 5.0 psi for the first check valve.

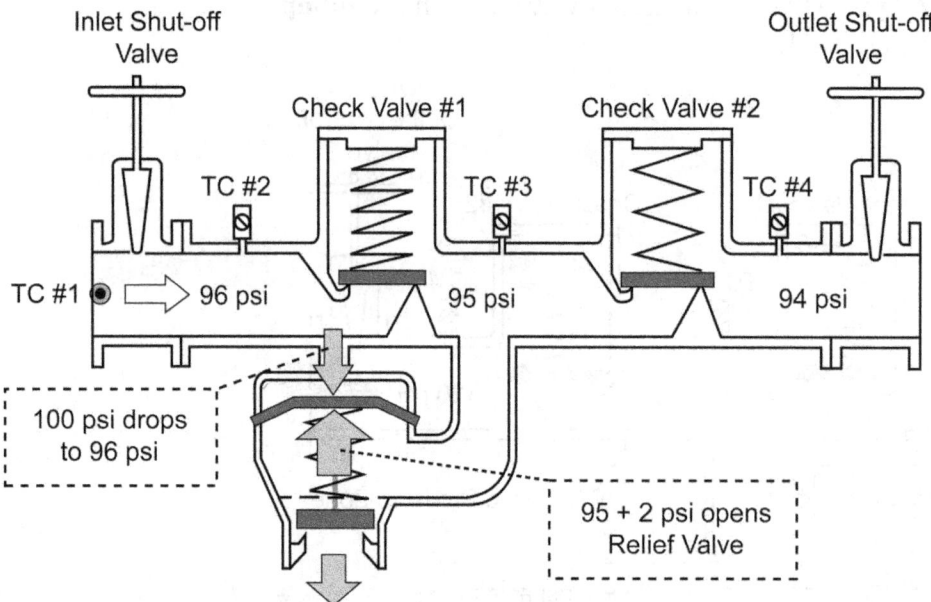

Figure 5-24 **RP during Backsiphonage Conditions**
The supply pressure falls to 96 psi (perhaps because the lines are being flushed). Because the assembly is under static conditions, the pressure in the zone will remain at 95 psi (since the first check prevents backflow of water out of the zone and water in the zone will not move downstream until there is a demand for water from the customer). The pressure of 95 psi within the zone combined with the pressure exerted by the spring, 2.0 psi, will cause the relief valve to open and dump the water in the zone.

If the supply line pressure, instead of dropping toward 0 psi, quickly fluctuates between 100 and 96 psi, the RP will **"spit"** instead of dumping. The relief valve will open and close very quickly, so fast that only a little water will "spit" out of the relief port. Air in the water supply line sometimes causes minor line pressure fluctuations like this.

As water dumps from the relief port, the zone pressure drops. Once the zone pressure is reduced below the supply pressure, water from the supply side will flow past the first check valve into the zone. Water continues to flow from the relief port until pressure equilibrium is re-established at the lower water-line pressure. If the supply pressure increases back to 100 psi, the pressure in the zone will increase to 95 psi and the RP will function normally.

Recall that under backsiphonage conditions when the water is static, the relief valve will open and dump the water contained in the zone. If the second check valve is fouled under backsiphonage conditions, water will continually flow from the relief port if atmospheric pressure or some other source of backpressure (for instance, the weight of water in the consumer's system) causes a reverse movement of water. Under backsiphonage conditions, where the pressure at the backflow preventer has fallen below atmospheric pressure, any pressure on the customer's side of the RP greater than 14.7 creates backpressure. Since water always flows in the direction of lowest pressure, water will flow from the RP's relief port. The volume of the flow from the relief port (drip or gusher) depends on the extent to which check valve #2 is fouled and also on the amount of backpressure. Water will continue to flow from the RP as long as water is in the line. If, however, the customer uses all the water contained in the line downstream of the assembly, no water is available to drip from the relief port.

This concept is shown in Figure 5-25. The relief valve will open and dump water from the relief port as soon as the potable water supply pressure falls below the pressure applied by combined pressure of the relief valve spring and the backpressure from the building. The pressure at point B, resulting from the weight of water, exceeds the atmospheric pressure acting at point A. Therefore, the water within the apartment building will drain from the apartment plumbing system out through the relief port. Water will drip, run steady, or run intermittently depending on the water usage inside the building.

Of course, if the second check valve was fouled and the pressure on the consumer's side was greater than the supply pressure, the relief valve would have opened and drained even before the backsiphonage condition occurred. Figure 5-26 shows how this condition could develop. If a pump connected to the consumer's potable water line creates a pressure of 99 psi and the second check valve is leaking, the pressure acting to open the relief valve is 101 psi (99 psi + 2.0 psi), while the pressure acting to keep it closed is only 100 psi. Therefore, the relief valve will open. The assumption is made in this example that the backpressure was less than normal supply pressure. Therefore, even though the check valve was fouled, the hydraulic gradient was in the normal direction of flow, and there was no visible indication of a problem prior to the backsiphonage condition.

METHODS AND MECHANISMS FOR PREVENTING BACKFLOW

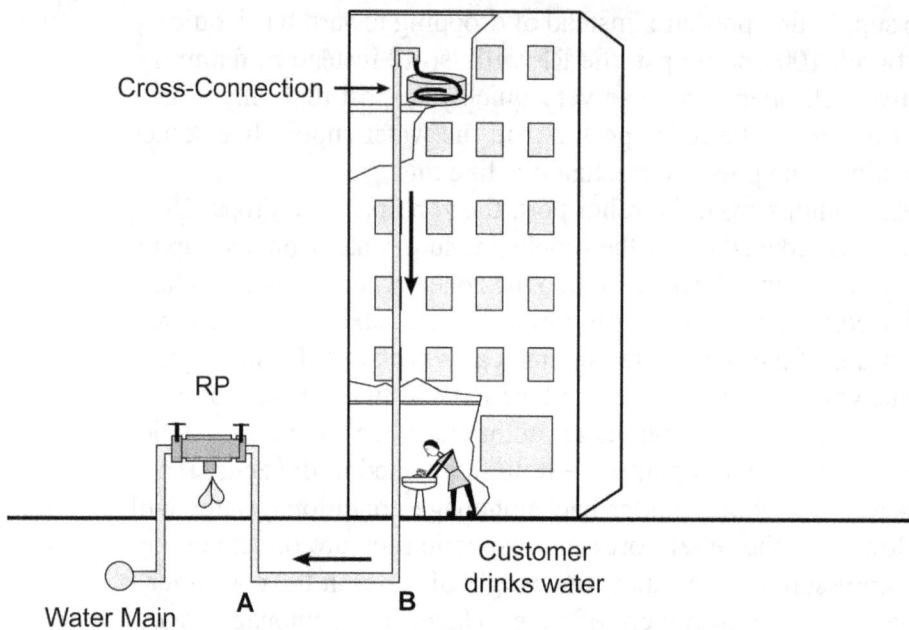

Figure 5-25 **RP: Failing Check Valve #2 under Backpressure and Backsiphonage Conditions**
The relief valve will open and dump water from the relief port as soon as the potable water supply pressure falls below the pressure applied by combined pressure of the relief valve spring and the backpressure from the building. The pressure at point B, resulting from the weight of water, exceeds the atmospheric pressure acting at point A. Therefore, the water within the apartment building will drain from the apartment plumbing system out through the relief port. Water will drip, run steady, or run intermittently depending on the water usage inside the building.

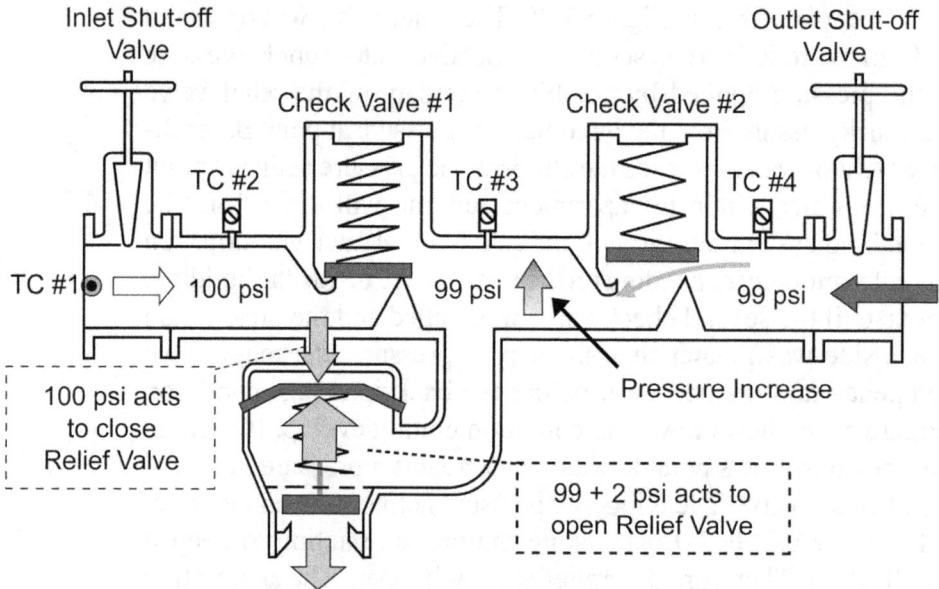

Figure 5-26 **Backpressure with a Leaking Check Valve #2**
If the second check valve was fouled and the pressure on the consumer's side was greater than the supply pressure, the relief valve would have opened and drained even before the backsiphonage condition occurred.

102

Like the DCVA, the RP is effective against both backpressure and backsiphonage. Because the relief valve provides added protection, the RP is considered adequate for high as well as low hazards. It also can be subjected to continuous pressure. Like the DCVA, the RP is traditionally used for containment, and it also is frequently installed in parallel at the point of service. It is the most versatile backflow preventer. The only other form of backflow prevention that can be used in all these different situations is the air gap. Recall that the air gap has one major problem associated with it—total loss of pressure. The RP does not cause a significant reduction in pressure; however, being a mechanical assembly, it does have some disadvantages—it is considerably more expensive than an air gap, it requires maintenance, and it should be tested at least annually.

The RP should be installed so that the vent of the differential relief valve is at least 12 inches and no more than 36 inches above grade or high water mark, with a minimum of 24 inches on all sides. It should be situated so the discharge of water from the relief port does not create an aesthetic problem. Moreover, discharges of water from the assembly should not create flooding of the port. Just as with the atmospheric and pressure vacuum breaker, it is essential that the air inlet port not be blocked or flooded. Because the air inlet port must not be flooded, the RP must not be installed in pits or vaults.

In general, installation should be horizontal with the relief port pointed downward so that it can drain easily. Vertical installations are only allowed if approved by the FCCC & HR. As with the DCVA, vertical installations should be considered carefully, because some assemblies have not been field-tested under these conditions.

If a drain line is installed under the relief port, a manufactured air gap should be provided to avoid creating a cross-connection between the assembly and the drain line (Figure 5-27). Furthermore, the relief port should not be reduced in size.

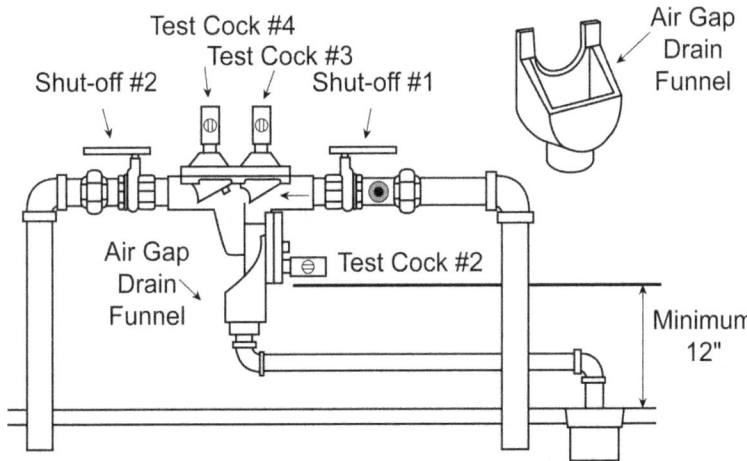

Figure 5-27   Drain Line Attached to an RP
In order for the relief port to function properly, it should not be blocked or reduced in size. If a drainline is provided, the appropriate air gap should be maintained.

METHODS AND MECHANISMS FOR PREVENTING BACKFLOW

Recall from earlier that a reduced pressure assembly can "spit," "dump," or "run steady" in terms of water flow from the relief port.

Spitting is caused by air in the system or fluctuations in line pressure. Figure 5-29 assumes a supply pressure of 100 psi, a relief valve with a spring pressure of 2.0 psi and spring pressure of 5.0 psi for the first check valve. This

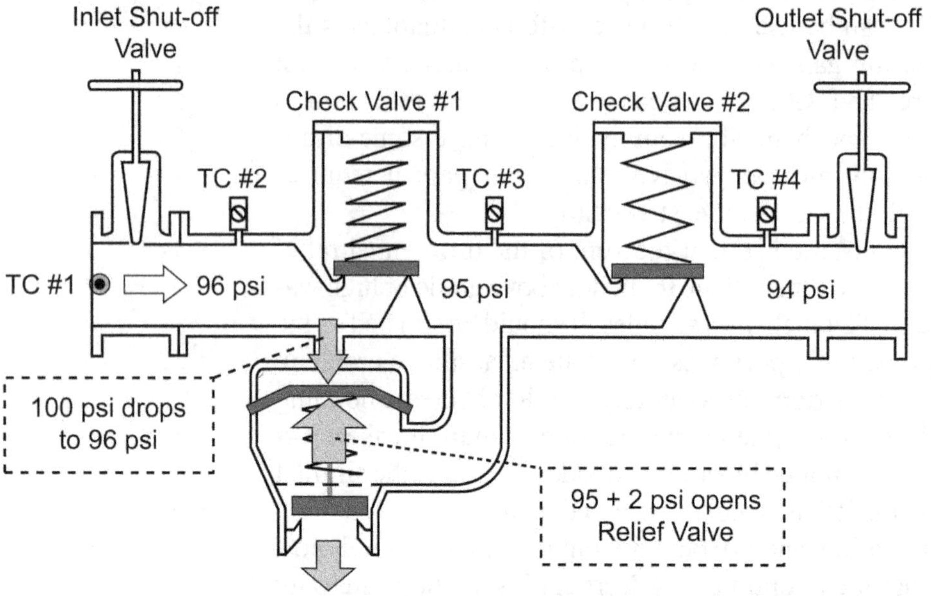

> Spitting is usually caused by line pressure fluctuations greater than the buffer value.

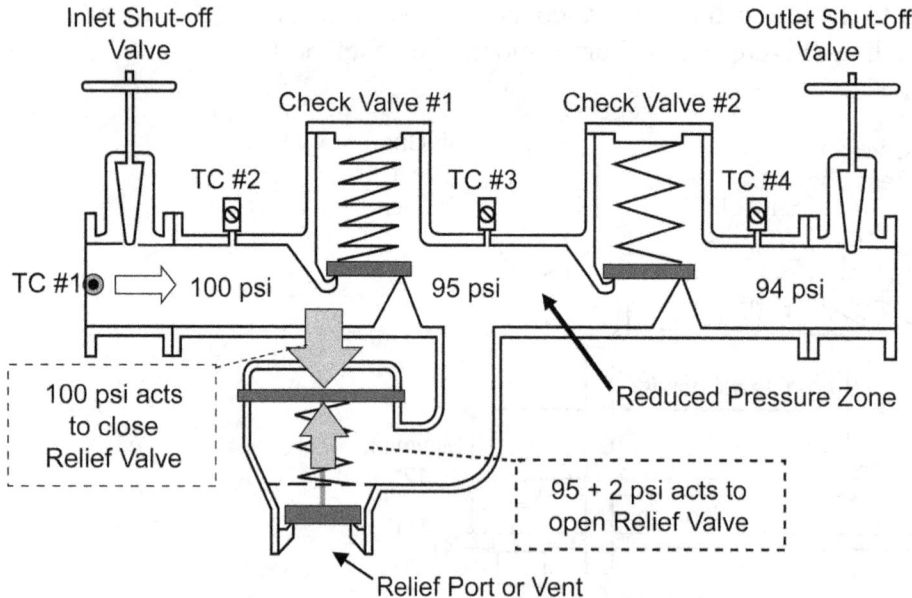

Figure 5-28   **Spitting: The Relief Valve Opens When Supply Pressure Drops**
If the supply line pressure quickly fluctuates between 100 psi and 96 psi, the RP will "spit." The relief valve will open when the supply pressure drops.

Figure 5-29   **The Relief Valve Closes When the Supply Pressure Increases**
If the supply line pressure quickly fluctuates between 100 psi and 96 psi, the RP will "spit." The relief valve will open and close very quickly, so fast that only a little water will "spit" out of the relief port.

will create a zone pressure of 95 psi and a pressure of 97 psi working to open the relief valve. In this case, the RP will not spit unless supply pressure fluctuates more than 3.0 psi. This 3.0 psi is referred to as the **buffer**.

Dumping occurs when a backsiphonage-backflow condition occurs or when a rapid reduction in supply line pressure happens. Figure 5-30 assumes a normal supply pressure of 100, a relief spring pressure of 2.0 psi and a spring pressure of 5.0 psi for the first check valve. If the supply pressure falls below 97 psi, the pressure of 95 psi within the zone plus the 2.0 psi exerted by the spring will cause the relief valve to open and dump the water in the zone and replace it with air. Once enough water pressure is released to lower the pressure in the zone below 93 psi, the relief valve will close and the pressure across check valve #1 will re-stabilize as illustrated in Figure 5-31.

The zone dumps or discharges at one time if there is a loss of pressure in the supply line.

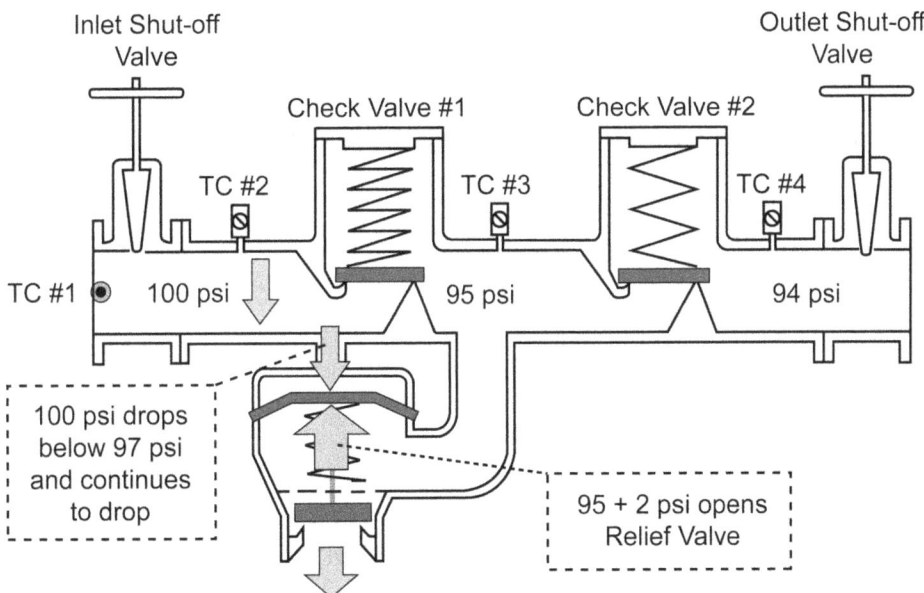

Figure 5-30   Dumping
If the supply pressure falls below 97 psi, the pressure of 95 psi within the zone plus the 2.0 psi exerted by the spring will cause the relief valve to open and dump the water in the zone and replace it with air.

Spitting and dumping are normal, expected responses. A continuous flow of water from the reduced pressure assembly, however, indicates that there is some sort of problem. Recall that water could continuously run from the assembly if the relief valve is stuck open, if check valve #1 is fouled or if check valve #2 is fouled during a backpressure-backflow condition or if the RP has been taken apart but the sensing line is still blocked when it is reassembled. Most frequently, check valve #1 is the culprit, since it is where most debris or contaminants in the water line will lodge.

A steady drip or continuous flow of water from the vent indicates that the RP is not working properly.

METHODS AND MECHANISMS FOR PREVENTING BACKFLOW

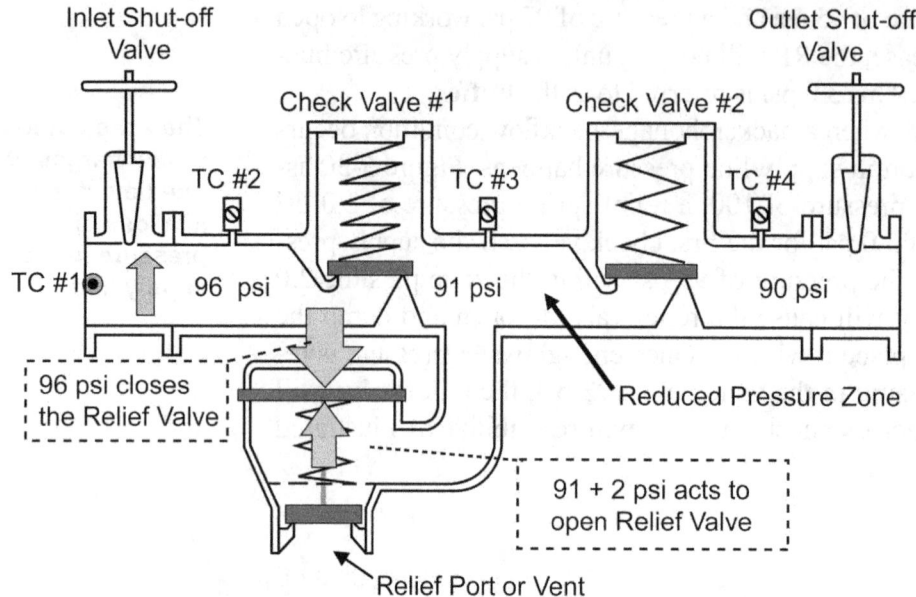

Figure 5-31 **The RP Will Adjust to the Drop in Supply Pressure**
Once enough water pressure is released to lower the pressure in the zone below 91 psi, the relief valve will close and the pressure across the check valve #1 will restabilize.

Air Gap: Approved Non-mechanical Backflow Prevention

The provision of an **air gap** provides the only absolute method for preventing backflows, because an air gap eliminates the cross-connection. An air gap is simply a physical separation of the potable water supply from all sources of contamination or pollution. An air gap is the simplest, and therefore, the least expensive method for preventing backflows. Unlike mechanical backflow preventers, an air gap does not require any mechanical parts to function. However, for an air gap to function dependably, certain specifications must be met.

There must be a minimum separation of two times the effected area or the inside diameter of the water supply outlet between the water supply outlet and the flood rim level of any receptacle or non-pressurized receiving vessel that could contain a hazardous substance. If the inside diameter of the water supply in Figure 5-32 is 1 inch, a 2-inch air gap must be provided between the water supply outlet and the flood rim of the non-pressurized receiving vessel or container.

The accepted method for measuring an air gap is at the flood rim of the container. Overflow outlets can easily become blocked, and therefore, cannot be relied upon. An absolute minimum of 1 inch, regardless of inside diameter of the supply outlet, must always be maintained between the water supply outlet and the flood rim, not the overflow. The 1/4-inch inside diameter water supply outlet in Figure 5-33 would require the absolute minimum 1-inch separation between the water supply outlet and the flood rim of the container. If the end of a pipe is cut on a diagonal, the air gap is meas-

There must be a minimum separation of two times the effected area (or the inside diameter) of the water supply outlet between the outlet and the flood rim level.

An absolute minimum of one-inch must always be maintained between the water supply outlet and the flood rim.

ured from the bottom of the diagonal cut. A one-inch water supply outlet that is cut on a diagonal must have a minimum 2-inch air gap, measured from the bottom of the cut to the top of the flood rim.

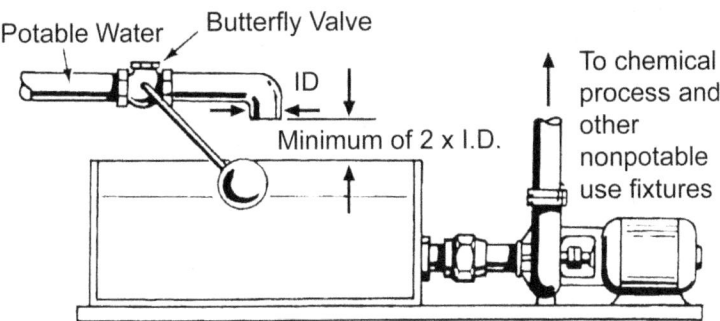

Figure 5-32   **Air Gap at a Booster Pump**
If the inside diameter (ID) of the water supply is 1 inch, a 2-inch air gap must be provided between the water supply outlet and the flood rim of the container.

The minimum separation is required because water will actually rise higher than the flood rim level of the container. This is due to the surface tension of water. Carefully overfill a glass just a little; the water level can actually get higher than the glass itself. Also, if a very strong vacuum is created in a water line, suction can draw water up through an air gap.

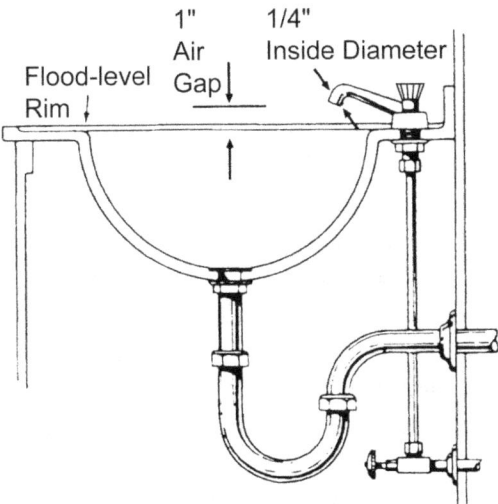

Figure 5-33   **Absolute Minimum Air Gap Separation**
The 1/4-inch inside diameter water supply outlet would require the absolute minimum 1-inch separation between the water supply outlet and the flood rim of the container.

METHODS AND MECHANISMS FOR PREVENTING BACKFLOW

If the water supply outlet is not freestanding, but instead is placed adjacent to a wall or is enclosed by walls on two or more sides, the minimum physical separation must be increased. This is necessary because if there are walls adjacent to the container, the water can move up even higher due to the effects of capillarity and surface tension. Capillarity is a combination of adhesive and cohesive forces where **adhesion** is the attraction of unlike particles and **cohesion** is the attraction of like particles. The attraction of water for itself and for the sides of the container causes it to move upward, against the force of gravity. The minimum air gap separation must be increased to prevent the water from rising high enough to eliminate the air gap and create an indirect cross-connection. If the water supply outlet is adjacent to a side wall (within three diameters of the water supply outlet), the minimum air gap must be three times the inside diameter of the outlet of the pipe. (Check local plumbing codes.)

Because air gaps are foolproof methods of backflow prevention *if* these minimum separations are provided, they can be used for any backflow situation: high hazard or low hazard, backsiphonage or backpressure. An air gap is considered the best method of backflow prevention, because it does not have any mechanical parts that could potentially malfunction. However, one disadvantage of the air gap is that it can easily be circumvented. For instance, a hose or extension of piping can easily eliminate an air gap. Therefore, where a water outlet is provided with threads for the attachment of a hose, steps should be taken to either remove the threads or provide another form of backflow protection at the outlet, because the chance that a hose will be connected and the air gap eliminated is very high.

An air gap can also be destroyed by raising the flood rim of the receiving vessel so that the minimum physical separation is no longer provided. However, the risk of an air gap being reduced or circumvented can be greatly diminished if consumers are educated about backflow. In addition, inspection programs can ensure that air gaps are being maintained. Another major disadvantage of an air gap is the loss of water pressure. Water pressure can be regenerated through the use of a surge tank and a booster pump.

On the other hand, the initial cost of an air gap is negligible compared to the initial cost of mechanical backflow preventers, especially where a high level of protection is needed. Also, air gaps do not have any mechanical parts that wear out and need replacement like mechanical assemblies. In addition, the cost for inspecting is much less than the testing and maintenance programs that must be established for mechanical backflow preventers as only a visual check is needed to ensure that the air gap is working properly.

A number of "manufactured" air gaps are on the market today. Air gaps should be provided using manufactured air gaps or simply maintaining the proper clearance between the supply pipe and flood rim. These devices help prevent water from splashing out around a drain. Often, consumers create a cross-connection in order to avoid splashing.

Adhesion is the attraction of unlike particles and cohesion is the attraction of like particles.

A major disadvantage of an air gap is the loss of water pressure.

An air gap is considered the best method of backflow prevention.

Summary of Approved Backflow Prevention Methods

Table 5-3 provides a summary of proper applications for various methods of backflow prevention and the advantages and disadvantages of each.

Table 5-3   Summary of Backflow Prevention Methods

	AVB	PVB	SVB	DCVA	RP	AIR GAP
High Hazard/Low Hazard	HIGH	HIGH	HIGH	LOW	HIGH	HIGH
Backpressure/Backsiphonage	Bs	Bs	Bs	Bp/Bs	Bp/Bs	Bp/Bs
Pressure Loss Significant	NO	NO	NO	NO**	NO**	YES
Continuous pressure allowed	NO	YES	YES	YES	YES	N/A
Vents to atmosphere	YES	YES	YES	NO	YES	YES
Pit/vault installation acceptable	NO	NO	NO	YES	NO	NO
Vertical installations are allowed	YES	YES	YES	***	***	N/A
Isolation/containment (commonly)	I	I	I	C	C	I
Parallel installation common	NO	NO	NO	YES	YES	N/A

N/A = Not Applicable   Bs = backsiphonage   Bp = backpressure

*   An AVB can be used for health hazards according to some plumbing regulations.
**   The maximum allowable pressure loss through a DCVA is 10 psi. The pressure loss through a RP assembly is approximately 7.0 to 14.0 psi.
***   Check manufacturer's literature.

Other Backflow Preventers

There are only four approved[2,3] methods of preventing backflow: vacuum breakers, double check valve assemblies, reduced pressure principle assemblies, and air gaps. There are several types of vacuum breakers, including: atmospheric, pressure, spill resistant, and hose-bibb.

Other backflow preventers, such as barometric loops and single checks, are not approved by FDEP regulation F.A.C. 62-555.360 for drinking water protection. These mechanisms might be used to protect substances other than potable water from backflow.

METHODS AND MECHANISMS FOR PREVENTING BACKFLOW

Barometric Loop

A barometric loop is created by installing water supply pipes so that the top of a loop extends at least 34 feet higher than the highest point of water used downstream (Figure 5-34). If a total vacuum were created by a large water demand, atmospheric pressure would move the contaminant up the potable water-line no more than 33.9 feet at sea level. Friction would actually reduce this theoretical height significantly. A loop height of 34 feet, then, provides a small margin of safety.

A barometric loop is only effective against backsiphonage.

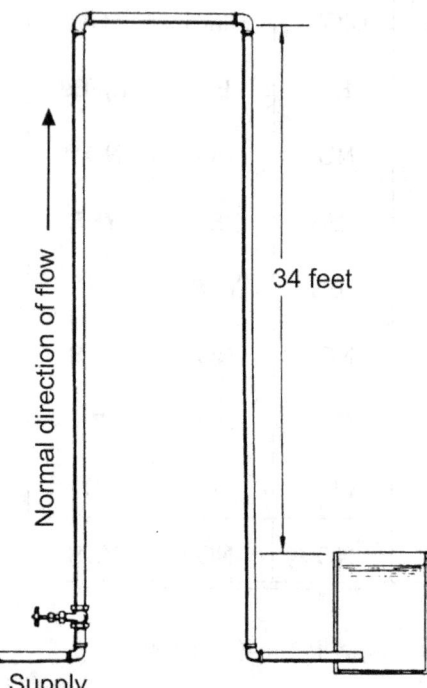

Figure 5-34   Barometric Loop
This 34-foot loop prevents backsiphonage-backflow because the force of atmospheric pressure can theoretically move water a maximum of 33.9 vertical feet in a total vacuum. In actuality, the effects of friction will reduce this height.

A barometric loop, however, is not effective against backpressure. If the non-potable supply pressure is greater than the potable supply pressure, the contaminant will simply flow through the loop into the potable water supply. This pressure can be provided by pumping or elevation. Therefore, a barometric loop is only effective against backsiphonage.

Barometric loops are not commonly used because the extra amount of piping needed to create the loop is usually more expensive than an equivalent form of backflow protection such as an atmospheric vacuum breaker. In addition, the loop creates a visual spectacle not commonly enjoyed by water consumers.

While the barometric loop does provide protection against backsiphonage-backflow, it does not prevent the contaminant from entering the loop itself. Therefore, any time a backsiphonage situation occurs, approximately 30 feet of the pipeline could become contaminated. If the barometric loop was part of the potable water line, 30 feet of pipe would then need to be thoroughly cleaned, or replaced, to ensure that the water passing through the pipe is potable when it reaches the consumer.

A barometric loop can be used to prevent backflows in systems where potable water supplies are not involved. The specific gravity of the material must be known, however, in order to calculate the necessary height of the loop. Specific gravity is the weight of the substance relative to water. If the specific gravity of a substance is greater than water's gravity, atmospheric pressure will not move the substance as high in the barometric loop as it does water. However, if the specific weight is less than that of water, which is the case with many organic solvents and gases, atmospheric pressure will move the substance much higher in the barometric loop. In this case, the height of the loop must be increased in order to prevent backsiphonage-backflow.

Single Check Valve

The **single check valve** is not actually considered a backflow preventer, since it provides little protection. A single check simply allows water or other substances to flow through it in the normal direction of flow but acts as a barrier to flow in the reverse direction (Figure 5-35). The single check may or may not contain a spring; in fact, the check itself might simply be a gate or a clapper, a hinged device that relies on gravity to move it into place as the flow rate decreases.

The single check valve is not considered a backflow preventer.

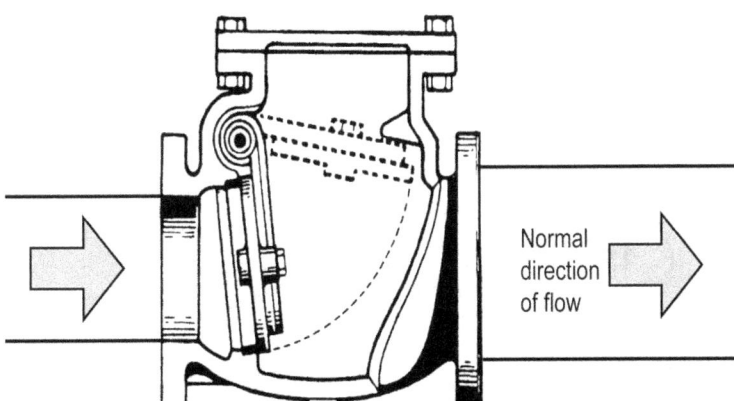

Figure 5-35   **Single Check Valve**
\tThe single check may or may not contain a spring; in fact, the check itself might simply be a gate or a clapper (a hinged device that relies on gravity to move it into place as the flow rate decreases).

METHODS AND MECHANISMS FOR PREVENTING BACKFLOW

The main problem with a single check valve is that it can easily become fouled. Once it is fouled, the device provides no protection against backflow. Another problem with the single check is that the device cannot be tested in-line. It cannot be visually inspected to determine that it is working properly.

The single check valve cannot be tested in-line.

The single check does provide slightly more protection against backpressure than the barometric loop, and, therefore, it could be used to deter (not prevent) both types of backflow. However, like the barometric loop, a single check should only be used where the need for backflow prevention is not essential. A single check device should never be used for protecting potable water.

This discussion, of course, does not include single checks that are part of an "approved assembly." When a single check is part of an approved assembly, it can be tested in-line. The single checks contained in some dual checks can be removed from the device and tested to ensure that they meet specifications. The single check contained in a PVB, a DCVA, the second check of a RP, or a dual check must meet AWWA standards. AWWA standards require that a single check be spring-loaded, soft-seated, and drip-tight in the normal direction of flow when the pressure on the supply side is 1.0 psi and the pressure on the customer side is atmospheric. These specifications exclude hinged and clapper-type devices.

The spring-loaded, soft-seated check valve must hold 1.0 psi in the direction of flow.

Dual Check Valve

A **dual check valve** (or residential dual check) (Figure 5-36) consists of two single checks in series. It is similar to a double check valve assembly and operates on the same basic principles. The major difference is that the dual check is a device, not an assembly. It lacks the shut-off valves and the test cocks provided on the DCVA (Figure 5-17). The dual check valve with test cocks is not designed to the same rigid specifications of the DCVA.

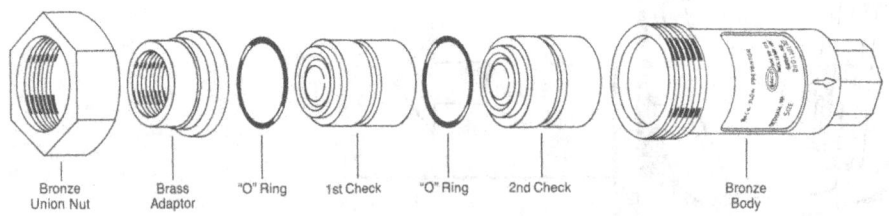

Bronze Union Nut | Brass Adaptor | "O" Ring | 1st Check | "O" Ring | 2nd Check | Bronze Body

Figure 5-36 Dual Check Valve
The single check valves contained within the dual check can be tested if they are removed from the device. A sight tube is used to determine whether the check valve will hold back 1.0 psi in the direction of flow.

Because the dual check cannot be tested in-line, it is not deemed adequate protection against backflow in potable water systems. The single checks within the device, however, can be taken out and replaced, and the old single checks can then be cleaned and tested at a later date. A sight tube is used to determine if the check valve will hold against 1.0 psi in the direction of flow. The dual check does provide slightly more protection than either the single check or the barometric loop. It provides limited protection against backsiphonage and backpressure.

Many local ordinances now require the installation of a dual check at the point of service to private residences. The logic is that some protection is better than none. More often than not, no form of backflow protection is provided at private residences, simply because the financial burden of buying, installing, and testing the backflow preventer is felt to outweigh the potential benefit. However, with the increase in domestic use of hazardous chemicals, such as photography chemicals, insecticides, fertilizers, and household cleaners, the risk of hazardous backflow from residential locations is increasing. Many communities compromise by using a dual check, because it provides some protection for limited expense. The device itself is less expensive than the DCVA. Also, dual checks are adapted so they can easily fit into meter boxes. For these reasons, they are more popular for residential use.

The major problem with this device is the false sense of security that it creates. The average life of the device is relatively short, with a small percentage of failures for one of the two checks within the first year and a 50% failure rate for both checks within 5 years. Moreover, reports of fairly high failure rates for devices tested prior to installation have been published.[4] Some communities are trying to cope with this problem by completely replacing the "guts" of the devices periodically (e.g., every 5 years). Periodic parts replacement increases the chance that a device will be functional when a backflow occurs.

The major problem with this device is the false sense of security that it creates.

The water purveyor has the primary responsibility for preventing backflow, and ultimately, must determine what level of protection should be provided at private residences. The water purveyor must weigh responsibilities and liabilities against the costs associated with providing backflow protection at these locations.

The water purveyor has the primary responsibility for preventing backflow.

Auxiliary Methods for Preventing Backflow

Auxiliary methods for preventing backflow include color-coding of pipes to differentiate potable from non-potable sources, the use of different types of pipes, meters and connections for non-potable systems, and the use of warning signs and tags on non-potable systems. Just as air gaps prevent backflows by eliminating cross-connections, these auxiliary methods help prevent backflows by eliminating or reducing the installation of cross-connections.

These auxiliary methods help simplify the process of identifying potable water supplies and non-potable sources. This is becoming

METHODS AND MECHANISMS FOR PREVENTING BACKFLOW

increasingly important as reuse/reclaimed water systems are installed. While these systems provide the essential service of water conservation, the potential for cross-connections between these systems and the potable system is very high. This potential can be significantly reduced if pipes are properly color-coded and if those who install and alter plumbing are educated about backflow and the color-coding system. A universal system of color-coding could prevent confusion and could also be helpful to inspectors as they review plans and conduct on-site inspections.

The use of different types of pipes and meters as well as warning signs and tags would all help to differentiate the potable water system from non-potable systems. The use of different-sized connections on non-potable systems would also help prevent accidental cross-connections.

A universal system of color-coding could prevent confusion.

Detector Check

The detector check is designed to detect small leaks or illegal water use that cannot be detected by the less-sensitive meter installed on the large line upstream from the detector check. In order for the small, more sensitive meter to measure low flows, the detector check must be designed so that low flows are routed through it. Therefore, the branch containing the smaller meter and check valve must present less resistance to flow than the more direct route straight through the larger assembly. The specifications for the larger assembly's check valve #1 spring have been modified so that it presents a higher resistance to flow than the smaller check valve. Consequently, low flows will pass through the small branch and be measured by the more sensitive meter.

Double Check Detector Assembly

A **double check detector assembly** (DCDA) consists of two approved double check valve assemblies installed as one unit. One assembly is smaller than the other and is installed as a protected by-pass around the main backflow preventer (Figure 5-37). The bypass branches out of the body of the larger assembly just after the inlet shut-off valve and terminates in the body of the large backflow preventer just prior to its outlet shut-off valve. It contains a specific water meter used to detect small flows. The maximum pressure loss through the assembly cannot exceed 10 psi under normal flow conditions.

Two off-the-shelf backflow preventers cannot be used as a "homemade" DCDA, because the resistance to flow would be the same in both backflow preventers even though they are of different sizes. The specification for the first check valve is the same no matter what the size, e.g., for a double check the first check valve must be drip-tight in the direction of flow against 1.0 psi regardless of the size of the assembly. Therefore, the water would pass though the most direct route and the "detector check" would not give accurate results.

The maximum pressure loss through the assembly cannot exceed 10 psi under normal flow conditions.

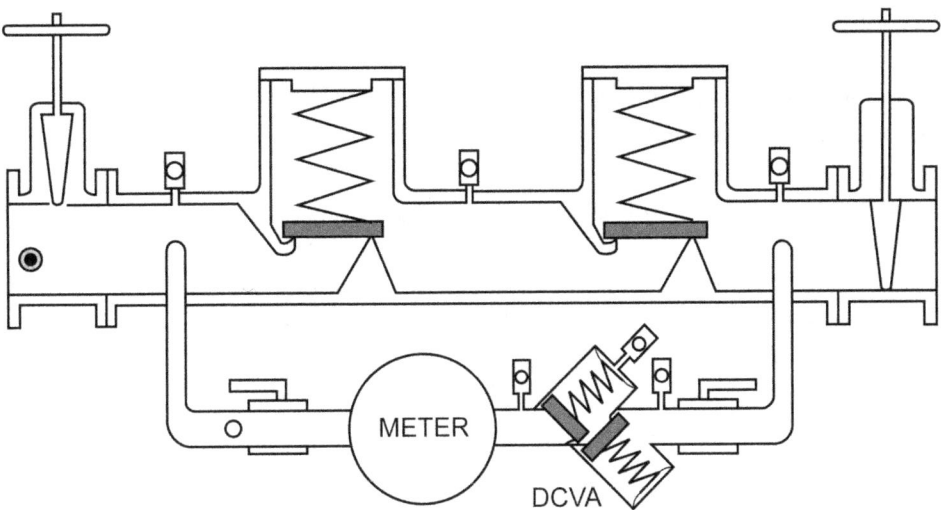

Figure 5-37 Double Check Detector Assembly
A double check detector assembly consists of a meter, a small backflow preventer, and two shut-off valves (one located prior to the meter and one located after the backflow preventer) installed on a large backflow prevention assembly.

Reduced Pressure Detector Assembly

A **reduced pressure detector assembly** (RPDA) consists of two approved reduced pressure backflow prevention assemblies installed as one unit. One assembly is smaller and is installed as a protected bypass around the main backflow preventer. The bypass branches out of the body of the main assembly just after the inlet shut-off valve and terminates in the body of the main backflow preventer just before its outlet shut-off valve. It contains a specific water meter used to detect small flows. The maximum size of these assemblies is 10 inches in diameter.

The type of backflow preventer in the by-pass assembly is always the same type as the main backflow preventer, thus providing the same level of protection in both branches. The RPDA is often used on fire sprinkler systems.

Commercial Fire Sprinkler Systems

The 1990 version of *American Water Works Association Manual M-14* listed several types of fire sprinkler systems that are connected to the public water system. The fire sprinkler systems were divided into six classes. This made it easier to determine which systems were high hazard and which systems were low hazard. With the new version of M-14, AWWA made an effort to clarify and simplify those different classes and to look at the potential hazards to the public water supply.

The water purveyor must also be mindful of state or provincial regulations pertaining to fire sprinkler systems that may limit the water purveyor's options for required backflow preventers on fire service lines. The water purveyor must also understand that installing a backflow preventer on an existing fire sprinkler system changes the hydraulics of the system. A fire protection engineer should recalculate the system before the backflow preventer is installed.

The anti-freeze additives must be non-toxic if the fire protection system is connected to the public water supply. The two acceptable anti-freeze additives are propylene glycol and pure glycerin. (See National Fire Protection Association Manual, Chapter 13, (NFPA-13) for more information on acceptable additives.)[5] Other additives that are toxic can be added only if the fire protection system is not connected to the public water supply.

If the hazard level is high, such as the addition of a non-toxic anti-freeze or a connection from the auxiliary water supply to a fire pumper truck, then the reduced pressure principle backflow preventer is required. If a fire sprinkler system does not have any additives and no chance for an auxiliary water supply to be connected exists, then the system is considered low hazard and the appropriate backflow preventer is the double check valve assembly.

A detector bypass on a backflow prevention assembly is not necessary if the fire protection system has an operating alarm switch that will indicate if water is flowing.

The maximum pressure loss through a double check valve assembly cannot exceed 10 psi. On average, a double check valve assembly may lose 3 or 4 pounds where a double check detector assembly may lose 7 or 8 pounds of pressure. Reduced pressure principle assemblies average about 10 psi pressure loss. Reduced pressure detector assemblies average 13 or 14 psi pressure loss.

Wet Pipe Fire Sprinkler System

A Wet Pipe Fire Sprinkler System is a "sprinkler system employing automatic sprinklers attached to a piping system containing water and connected to a water supply so that water discharges immediately from sprinklers opened by heat from a fire."[5]

Water standing in a non-flowing fire protection system may become stagnate and contaminated beyond acceptable drinking water standards. Some of the contaminants found in fire protection systems are anti-freeze additives, chemicals for corrosion control, wetting agents, oil, lead, cadmium, zinc and iron. The latest model building and plumbing codes and Occupational Safety and Health Administration (OSHA) regulations require the installation of an approved backflow prevention assembly on all new wet pipe fire sprinkler systems. For existing wet pipe fire sprinkler systems that have been determined to be of low hazard only, i.e., no chemical additives, the water purveyor should consider installing a testable backflow prevention assembly or require the use of an Underwriter's Laboratory listed alarm check valve with a water flow pressure switch as the alarm device. The water purveyor may elect to allow a modern Underwriter's Laboratory listed alarm check valve, which has a rubber disc, and a pressure switch to be installed instead of a double check valve assembly.

If a toxic chemical such as ethylene glycol, which is automotive antifreeze, is being utilized then the fire protection system is not allowed to be connected to the public water supply.

If a non-toxic additive is used in an existing system, then the reduced pressure principle backflow assembly is required. For low-hazard fire protection system sprinklers and when no chance of lead being leaked into the water exists, it is recommended that the check valve be maintained in accordance with NFPA 25.[6] (NFPA 25 is standard for the maintenance of water-based fire protection systems.) When an existing sprinkler system is significantly expanded or modified requiring a new hydraulic analysis, a double check valve assembly should be considered. For existing systems presenting a low hazard or having a lead containing alarm check valve, it is recommended that a Factory Mutual/Underwriter's Laboratory (FM/UL) rated double check valve assembly be installed.

On an existing wet charge system that is viewed to be a low hazard system and no chemical additives are involved, it is recommended that the water purveyor consider installing a double check valve assembly. If the wet charge system has non-toxic chemical additives, the appropriate protection is a reduced pressure principle assembly.

Deluge Sprinkler System

A Deluge Sprinkler System is a "sprinkler system employing open sprinklers that are attached to a piping system that is connected to a water supply through a valve that is opened by the operation of a detection system installed in the same areas as the sprinklers. When this valve opens, water flows into the piping system and discharges from all sprinklers attached thereto."[5]

A deluge system is a non-pressurized fire suppression system that is open to the atmosphere, with sprinklers and other outlets open and ready to flow water at all times. Generally these types of systems do not present a health hazard to the public water supply unless there is a long run of pipe with stagnate water between the tap on the main and the backflow preventer or check valve. No additional chemicals should be added to these systems. If no additional chemicals are added on an existing system, AWWA recommends a check valve with an alarm check. This is a directional control device and not a backflow preventer.

Combined Dry Pipe-Preaction Sprinkler System

A Combined Dry Pipe-Preaction Sprinkler System is a "sprinkler system employing automatic sprinklers attached to a piping system containing air under pressure with a supplemental detection system installed in the same areas as the sprinklers. Operation of the detection system actuates tripping devices that open dry pipe valves simultaneously and without loss of air pressure in the system. Operation of the detection system also opens listed air exhaust valves at the end of the feed main, which usually precedes the opening of sprinklers. The detection system also serves as an automatic fire alarm system."[5]

Dry Pipe Sprinkler System

A Dry Pipe Sprinkler System is a sprinkler system "employing automatic sprinklers that are attached to a piping system containing air or nitrogen under pressure, the release of which (as from the opening of a sprinkler) permits the water pressure to open a valve known as a dry pipe valve, and the water then flows into the piping system and out the opened sprinklers."[5]

Preaction Sprinkler System

A Preaction Sprinkler System is a "sprinkler system employing automatic sprinklers that are attached to a piping system that contains air that might or might not be under pressure, with a supplemental detection system installed in the same areas as the sprinklers."[5]

Dry pipe and preaction fire suppression systems are generally charged with air or nitrogen. The sprinklers and other outlets are closed until a rise in temperature is high enough to open a sprinkler, air is then released. The drop in air pressure is sensed by a valve, the valve opens, and water flows through the system and controls the fire. These systems contain air that is under pressure. A preaction system will have a supplemental fire detection system installed in the same areas as the sprinklers. These systems generally present a low hazard.

Residential Fire Sprinkler System (Single Family)

A residential fire sprinkler system typically receives water through a water main that is 1-and-½-inch or less in diameter. A double check valve assembly is recommended when a low hazard exists. Some parts of the country allow a flow through system, which is part of the main piping of the home. Because water is moving through these on a regular basis, installation of a backflow preventer is not recommended.

Assembly Installation

In general, all backflow preventers must be installed so that they can be easily tested and repaired. If double check valve assemblies must be installed in a pit, chamber, or vault, provisions must be made to ensure that the assembly does not become flooded. If an assembly is installed in a deep chamber, the chamber should be self-venting. The assemblies must be protected from vandalism and freezing, and water lines should not be used for electrical grounding purposes. Prior to actual installation, the water lines

Prior to actual installation, the water lines should be thoroughly flushed to remove loose materials that could foul the backflow preventer.

should be thoroughly flushed to remove loose materials that could foul the backflow preventer. In addition, it is sometimes recommended that a strainer be located prior to the assembly to trap any loose materials that could otherwise foul the assembly. Providing a blow-off valve after the backflow preventer is also a good idea. The blow-off valve can be used to remove grease and foreign materials that are produced when repairing the assembly, or to flush the customer's water line of any contaminants after a backflow incident without contaminating the backflow preventer. When installing an assembly or device, safety precautions must be observed. Additional safety precautions are covered in Chapter 6, Field Testing.

Thermal Expansion

When the temperature increases in a closed system the pressure also increases. If there is not a method to relieve the pressure then damage may occur. When the water heater or boiler heats the water, the pressure will increase. The excess pressure is relieved back through the water meter. The piping inside of a customer's facility or house is considered a closed system whenever a working backflow prevention device or assembly is installed on the main service line. If the temperature of the water increases to a dangerous level, the temperature and pressure (T & P) valve on the water heater should discharge a small amount of water to relieve the excess pressure. If the T & P valve fails to function, the excess pressure may cause a pipe to burst or cause damage to the water heater. Many communities require some additional protection from thermal expansion. One common method is to install an expansion tank (Figure 5-38).

Thermal expansion is the increase in pressure in a closed water system due to heating and expanding water.

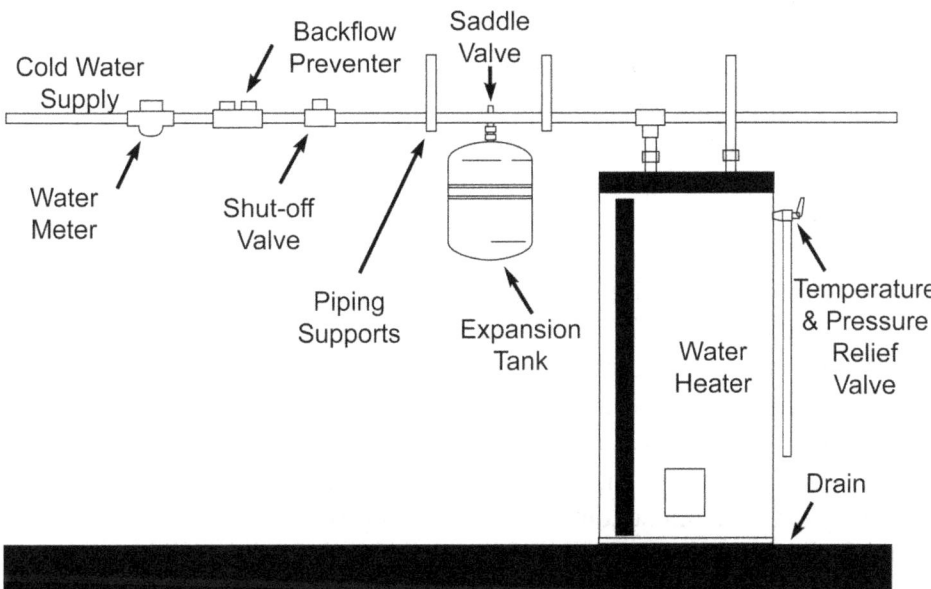

Figure 5-38 **Typical Piping System for a Customer**
An expansion tank is installed on an existing piping system to supply additional protection from thermal expansion.

METHODS AND MECHANISMS FOR PREVENTING BACKFLOW

Water purveyors should make an effort to educate their customers on thermal expansion. The purveyor should explain how the T & P valve works and that it should be exercised at least once a year. The purveyor may suggest that the customers contact their plumber to perform an inspection of their water heater.

Temperature and Pressure Valve

While the protection provided by backflow preventers clearly outweighs any drawbacks associated with them, backflow preventers can create hazardous conditions by preventing the backflow of water from water heaters. According to plumbing regulations, all water heaters are required to have T & P valves. These valves are designed to open and discharge water from the water heater when the temperature or pressure reaches a critical level. They function as a safety mechanism. However, these valves have a small percentage of failures, commonly attributed to improper installation and inappropriate usage or improper maintenance. For proper function, T & P valves need to be exercised periodically.

If the T & P valve fails and a backflow preventer (e.g., DCVA or RP) is installed on the potable water line, the pressure can build up to excessive levels since there is no place for the increased pressure to vent. Most commonly, the customer experiences a ruptured water pipe due to thermal expansion.

In Oklahoma, seven people were killed when a water heater exploded because the temperature probe of the T & P valve had been removed prior to installation.[7] In addition, other factors contributed to this explosion - the thermostat was broken and the heating element created super-heated steam.

When backflow preventers are installed, the customer should be informed about the problems created by non-functional T & P valves. This can be communicated through bill stuffers, newsletters, or flyers.

When backflow preventers are installed, the customer should be informed about the problems created by non-functional T & P valves.

Freeze Protection

In northern states, the problems that freezing temperatures create for backflow preventers are avoided by installing them inside buildings or enclosures. However, in much of the South, freezing temperatures are not carefully considered when plumbing installations are designed. While most backflow preventers may withstand freezing temperatures for short periods of time, extended freezes (i.e., over 10 hours) can burst the body (casting) of the assembly, rendering it useless. The most common method of protection is to wrap the assembly with insulating material. However, this provides very little protection during an extended freeze. Another common method is to encase the assembly within a fiberglass, wooden, or metal enclosure and supply a heat source. Many individuals build their own shelters, and also manufactured enclosures are available (Figure 5-39). One problem

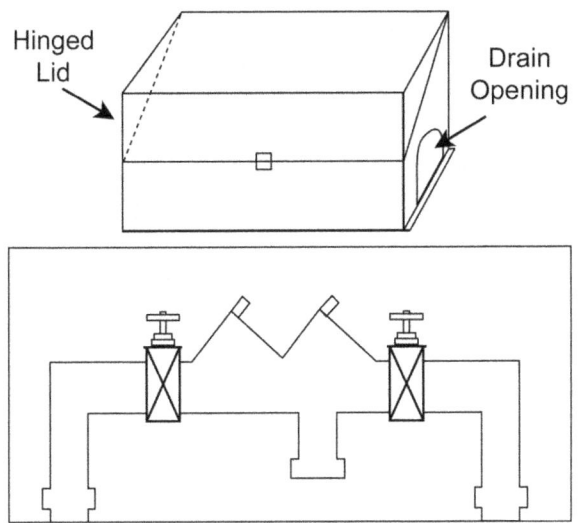

Figure 5-39 **Freeze Protection for Backflow Preventers**
This shelter includes a heat source as an added level of protection. It also has drain openings at each end to prevent flooding and a hinged lid so the backflow preventer can be accessed for testing purposes.

with these forms of protection is that they can block relief ports and hamper access to test cocks if they are not designed to properly fit the assembly. An additional problem with enclosures is that they prevent quick detection of backflow preventer malfunction.

Customers should be notified each year before the winter months about freezing related damage and how to provide protection for their backflow prevention assemblies. The assembly can be drained for lawn irrigation systems. For other outside installations, the customer should either let the water run continuously or provide a heated enclosure. A list of available enclosures can be found in Appendix Z.

Summary

Preventing the backflow of contaminants into the potable water system is absolutely essential to protect health and property. The best means to prevent backflow is to eliminate cross-connections through the use of a proper air gap as reviewed in this chapter. This chapter also introduced the backflow prevention assemblies and devices that can be employed to prevent backflows when cross-connections are necessary. In these cases, backflow preventers can be installed to reduce the chance of a backflow occurring. The choice of method depends on both the perceived level of hazard and on the type of backflow (backpressure or backsiphonage) that may occur.

High hazard situations require an atmospheric vacuum breaker, a spill-resistant vacuum breaker, a pressure vacuum breaker, or a reduced pressure principle assembly. However, only the RP is suitable for a high hazard situation that may involve backpressure, thus, AVB, SVB and PVB are generally only used for isolation purposes. Any of the previous methods, as well as by a double check valve assembly, can be used in low hazard situations. If there is the potential for backpressure, only the RP or the DCVA can be utilized. Other less-reliable methods are available for non-potable systems, but these can never be used in a situation where public health would be jeopardized by a backflow incident.

Cost can be a concern when choosing a backflow preventer. AVBs, SVBs, and PVBs are inexpensive, while RPs and DCVAs are a larger investment. However, the hazard level and anticipated hydraulic conditions at each cross-connection must take priority over cost when selecting a method of backflow prevention.

CHAPTER FIVE

Chapter Five Review

5-1 What single item at home causes the most incidents of backflow?
garden hose

5-2 What is the appropriate height of an Air Gap?
2 x ID - min 1"

5-3 What is the correct height to install an AVB?
6 in

5-4 What is the biggest disadvantage to the customer of using an AIR GAP?
loss of pressure

5-5 What protection does a PVB provide?
high, low hazard
Backsiphonge

5-6 What is the correct height to install an RP?
12 in above grade

5-7 What is the correct height to install a PVB?
12 in above piping

5-8 What type of protection does a DCVA provide?
low hazard
Backsiphonge + Back pressure

5-9 What is the maximum pressure drop allowed across a DCVA?
10 psi

5-10 What is the minimum standard for the spring in Check Valve #1 in an RP?
5 psi

5-11 What is the minimum spring tension on the Relief Valve in an RP?
2 psi

5-12 Name the two major requirements of an FCCC & HR Approved Backflow Prevention Assembly?
Testable + repairable in line

5-13 Why do we install backflow preventers in parallel?
maintane service

5-14 What special conditions are required when you install an AVB?
No value downstream, backsiphonage only, 6 in above

METHODS AND MECHANISMS FOR PREVENTING BACKFLOW

5-15 Under what circumstances would you only allow the use of an air gap?

Water to sewer connection

5-16 What is the minimum pressure drop allowed across a spring-loaded, soft-seated check valve? (AWWA standard)

1 psi

5-17 Which check valve in a PVB, a DCVA and an RP has a minimum 1.0-psi loss specification?

PVB-CV, DC-CV1 + CV2, RP-CV2

5-18 What is the visual difference between a DCVA and a DCDA?

has meter + bypass + second backflow

5-19 What is the difference between a PVB and an AVB?

PVB: Assembly, Testable
AVB: device, no valve

CHAPTER FIVE

Summary of Backflow Prevention Applications

Check in the appropriate locations to indicate the proper application for each of the backflow preventers listed. When the chart is complete it will serve as a handy reference.

	Atmospheric Vacuum Breaker	Pressure Vacuum Breaker Assembly	Spill-resistant Vacuum Breaker Assembly	Double Check Valve Assembly	Reduced Pressure Assembly	Air Gap	
Backsiphonage							High Hazard
							Low Hazard
Backpressure							High Hazard
							Low Hazard

125

REFERENCES

1. *Water Distribution Operator Training Handbook* (2nd Edition). 1976, 1999. American Water Works Association, Denver, CO.

2. *Manual of Cross-Connection Control*, (9th Edition). 1993. Foundation for Cross-Connection Control and Hydraulic Research, University of Southern California, University Park, CA.

3. *Recommended Practice for Backflow Prevention and Cross-Connection Control (M14)*, 1990, American Water Works Association, Denver, CO.

4. Jolley, R. Personal communication, March 29, 1990.

5. "Installation of Sprinkler Systems," *NFPA 13*, 2002. National Fire Protection Association, Quincy, MA.

6. *Standard for the Inspection, Testing, and Maintenance of Water-Based Fire Protection Systems (NFPA 25)*. 2002. National Fire Protection Association, Quincy, MA.

7. *Water Heater Explosion in School*, (F-SX 826), Watts Regulator Company, North Andover, MA.

CHAPTER SIX

FIELD TESTING

The previous chapter discussed air gaps and their use in eliminating cross-connections and, thus, prevent backflow. Mechanical backflow preventers were also discussed as a method for preventing backflow. Both air gaps and mechanical assemblies can easily be circumvented and their effectiveness destroyed through jumper connections and bypasses. To prevent this, it is essential that backflow preventers be inspected periodically. When an air gap is utilized, a visual check is sufficient. Mechanical backflow preventers, on the other hand, require testing to determine that they are functioning properly. They also need periodic maintenance and repair. The focus of this chapter will be on how to test mechanical backflow preventers, with additional information on different types of test gauges and their uses.

> **Mechanical backflow prevention assemblies must be field tested at least once a year (annually).**

The primary reason for periodic testing of mechanical assemblies is to ensure that they continue to provide adequate protection against backflow (Appendix I). The movement of water through these assemblies, whether corrosive or depositing, causes deterioration that will affect their functioning. Testing the assembly can help detect some of these problems, but only if the tests are done accurately. Testing is usually performed using commercially available test gauges.

Test Gauges

Many manufacturers supply kits for testing backflow preventers. A list of manufacturers is provided in Appendix G. Manufacturer's instructions should be followed on the care, calibration, and repair of the test gauges. The gauges should have their accuracy verified at least once a year (Appendix G). Many manufacturers require that the kit be returned for repair and calibration.

> **The gauges should have their accuracy verified at least once a year.**

There are three basic types of test gauges: the differential pressure gauge, which can be analog or digital, the duplex pressure gauge, and the sight tube. In most areas of the United States, the duplex gauge is no longer used.

Differential Pressure Gauge

A **differential pressure gauge** can be used to test reduced pressure principle assemblies, double check valve assemblies, pressure vacuum breaker assemblies, and spill-resistant vacuum breakers.

FIELD TESTING

As its name implies, a differential pressure gauge measures pressure differentials. It measures the difference in pressure between the high-pressure side and the low-pressure side of a check valve. The general design of the differential pressure gauge must include three pressure hoses and two, three, or five needle valves. These needle valves must allow the separate movement of water from the high-pressure hose to both the low-pressure hose and the by-pass or vent hose. The gauge itself measures the difference in pressure between the low-pressure hose and the high-pressure hose (Figure 6-1).

Schematic diagrams of two types of differential pressure gauges are provided below (see Figures 6-2 and 6-3). The major difference in the components of the two types of gauges is that the one depicted in Figure 6-3 does not have separate bleed valves. The bleed valves are used to remove air from the backflow preventer and the test gauge.

An electronic test gauge contains transducers and functions both as a duplex pressure gauge and a differential pressure gauge. A schematic of this is included in Figure 6-5. The hoses are attached to the test cocks located on the backflow prevention assembly. When the test cocks are opened, the hoses supply water pressure from different portions of the assembly to the gauge. As soon as the test cocks are opened, the gauge is on-line. In other words, water pressure is always supplied to both sides of the gauge, no matter which test gauge needle valves are opened or closed (Figure 6-2).

Although the gauge depicted above does not have separate bleed valves, air can still bleed from the assembly and the test gauge by opening the vent hose (Figure 6-3).

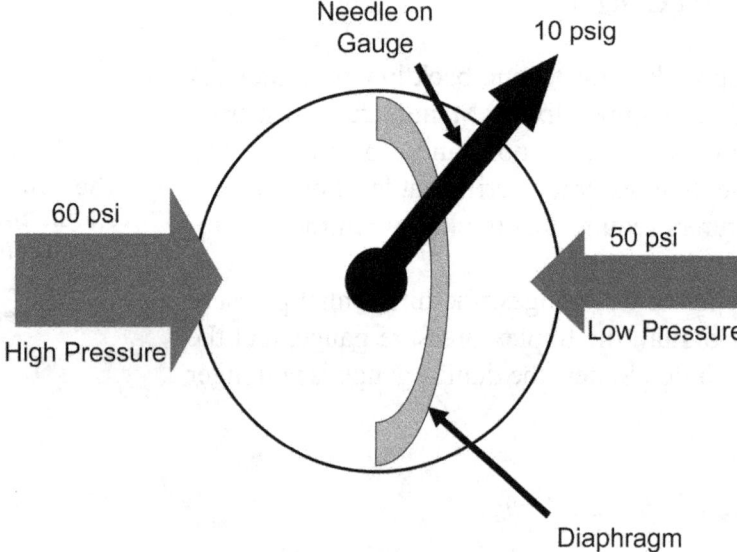

Figure 6-1 Differential Gauge: Diaphragm
Water pressure is channeled to a diaphragm within the kit. This diaphragm controls the needle on the face of the gauge.

CHAPTER SIX

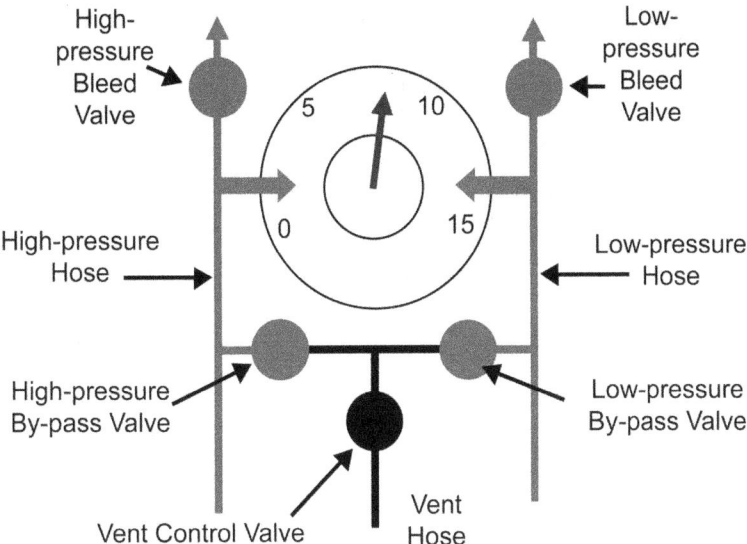

Figure 6-2 **Differential Test Kit**
The gauge will read the pressure differential between the pressures supplied to the high and low sides of the gauge. Note: The position of the high-pressure and low-pressure hoses may be reversed.

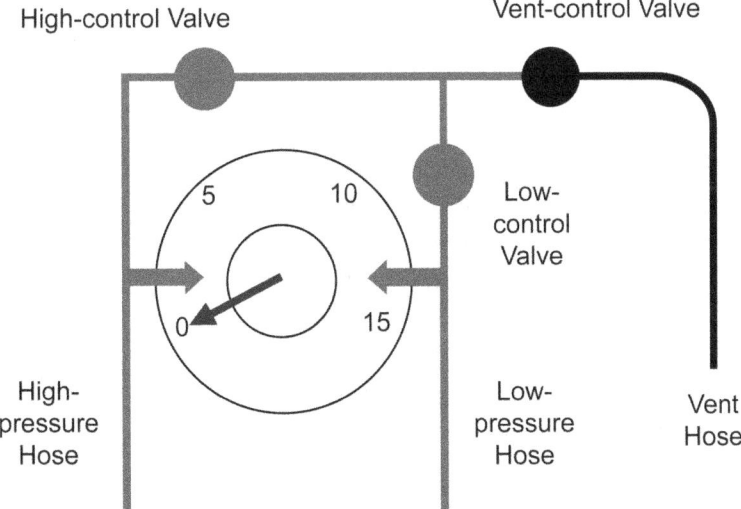

Figure 6-3 **Differential Test Kit: Without Separate Bleed Valves**
If the vent-control is open, air can be bled from the test kit and the backflow preventer through the vent hose.

When the bleed valves are opened, water pressure forces air out of the bleed line. The control valves are used during testing to control the by-pass of water from the high-pressure hose to the low-pressure hose. For instance, the high pressure contained on the upstream side of the #1 check in the RP assembly can be routed from the high-pressure line connected to

FIELD TESTING

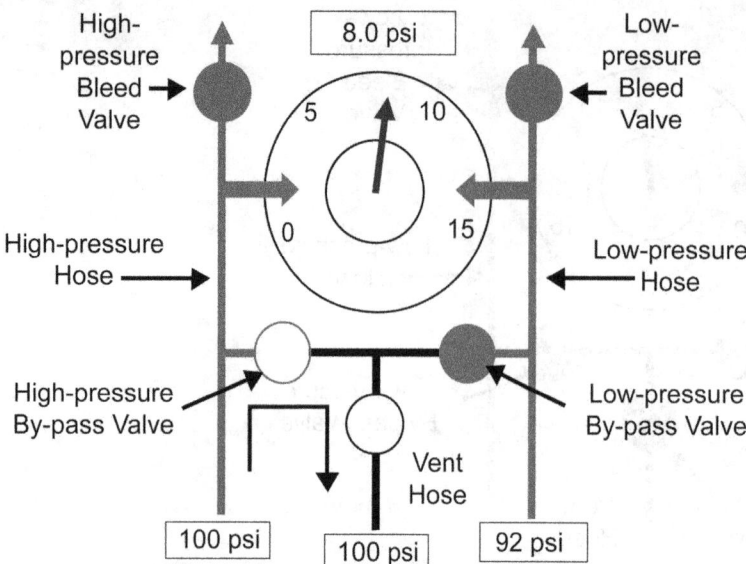

Figure 6-4 Differential Test Kit: With Pressure Differences Illustrated
If the high control valve and the vent-control valve are left open, the high pressure will pass from the assembly through the test kit and back to the assembly. Because the low-pressure, bypass valve is closed, the gauge will still measure the difference in pressure supplied by the high-pressure hose and the low-pressure hose.

test cock #2 through the vent hose (but not through the gauge) to the downstream side of check valve #2 (Figure 6-9). If the high control valve and the vent-control valve are left open, the high pressure will pass from the assembly through the by-pass (but not through the gauge) and back to the assembly. The low-pressure hose is connected to test cock #3. Because the low-pressure, by-pass valve is closed, the gauge will still measure the difference in pressure supplied by the high-pressure hose and the low-pressure hose.

When testing the pressure vacuum breaker assembly, the differential pressure gauge is used to ensure that the air inlet valve opens at 1.0 psi or above. It can also verify that the single check valve contained in the pressure vacuum breaker assembly will hold tight against 1.0 psi in the direction of flow. This is called a "direction of flow" test.

When testing the double check valve, the differential pressure gauge is used to test the assembly to be certain that each check valve closes when water is moving through the check assembly. This is also called a "direction of flow" test.

When testing a reduced pressure assembly, the differential pressure gauge is used to verify that check valve #1 will maintain a pressure differential of at least 5.0 psi. This is called a "differential pressure" test. It also determines the opening point of the relief valve and will confirm that check valve #2 holds tight against backpressure.

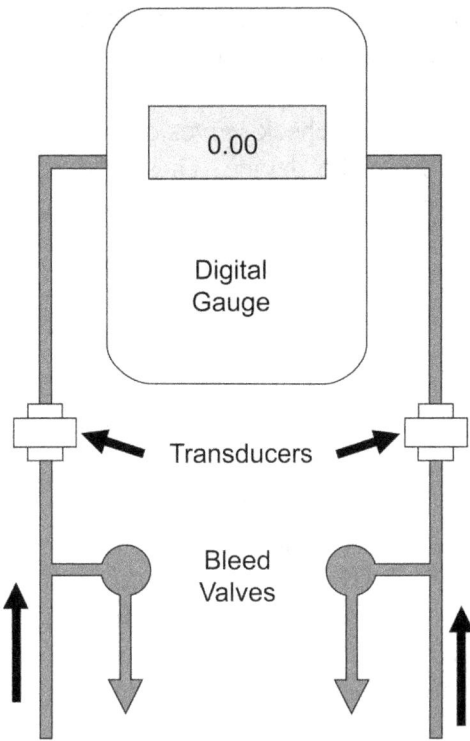

Figure 6-5   Digital Gauge
Usually the hoses and the gauge needles are color-coded for ease in determining which pressure reading corresponds to which hose.

Sight Tube

A **sight tube** is used to test a double check valve assembly and some dual check valves. A sight tube is merely a plastic tube with numbers and adaptor fittings that allow it to be attached to the test cocks of the backflow assembly. One of the advantages of this type of test gauge is that it has no moving parts, so it requires very little maintenance. The unit has o-ring seals that should be gently tightened to the backflow preventer to avoid damaging the threads.

Once the sight tube is attached to the test cock, it is filled with water to a minimum height of 28 inches above the point where water is discharged from the downstream test cock. The weight of this 28-inch water column is applied to the check valve to determine that the check will hold back 1.0 psi in the direction of flow. A disadvantage of the sight tube is that it cannot be utilized unless adequate space (at least 28 inches) is provided above the backflow preventer. Therefore, this method of testing a DCVA cannot be conducted when assemblies are installed within 28 inches of a ceiling. Test results from a sight tube only indicate pass or fail. The site tube does not give you a value. Note: When testing with a sight tube, the tester must wait for 2 minutes after the water level has stabilized before determining results.

FIELD TESTING

The sight tube can test the check valves contained in a DCVA. AWWA specifies that these check valves must be able to hold drip tight against 1.0 psi in the direction of flow.

A sight tube can also be utilized to test the check valves of a dual check if they can be isolated or removed from the device. These check valves must also be able to hold drip tight against 1.0 psi in the direction of flow.

Testing

This section provides instruction on the appropriate methods of testing backflow prevention assemblies. Chapter 7 will cover what to do should the backflow preventer fail the test. While manufacturers provide specific testing instructions for their test gauges, it is recommended that the test instructions provided in this manual also be followed. The test methods that follow are modified procedures of the FCCC & HR and represent the best method for testing backflow prevention assemblies. Many other useful test procedures are often followed, but they have shortcomings that restrict their use to specific conditions. The test procedures described here will work on any size or brand of backflow prevention assembly.

Spill-resistant vacuum breakers, pressure vacuum breakers, double check valve assemblies, double check detector assemblies, reduced pressure assemblies, and reduced pressure detector assemblies must be tested at least once a year. Record keeping is a crucial part of the testing process, and the results of these tests should be recorded on a standard form. Appendix E contains a sample test form. The test form should be completed at the time of the test. This helps to ensure that the proper assembly is being tested and that accurate information is recorded. The test form should include the following information:

1) Customer's name and address
2) Location of the assembly
3) Date of installation
4) File number, tap number or some other type of identification number for the assembly
5) Manufacturer
6) Type of assembly
7) Size of assembly
8) Model number
9) Serial number
10) Meter number or account number
11) Serial number of test gauge
12) Date gauge was last calibrated
13) Whether assembly passed or failed field test
14) Date and time field test was performed
15) Name and certificate number of tester
16) Remarks

> Sample Remark: Tester should note if the assembly was not installed correctly.

Any problems with the assembly due to vandalism or weather also should be noted on the form in a "comments" or "remarks" section, along with notes on the conditions of the area surrounding the assembly. The area around pressure vacuum breakers and reduced pressure assemblies should be examined for evidence of frequent dumping or water leaks, which are indicated by tall grass, moss, or algae growth, soil erosion, or puddled water around the assembly. The reason for the test should be indicated (i.e., routine inspection or response to a complaint), as well as what the backflow prevention assembly is preventing from entering public water. All of this information is essential, and it will be discussed further in Chapter 8, Developing Cross-Connection Control Programs.

Prior to testing any backflow preventer the customer should be notified that testing will to be conducted and also apprised of how long the water will be turned off. Contacting the customer well ahead of time can prevent this short disruption of service from causing unnecessary hardship. When possible, testing should be scheduled for a time that is convenient for the customer. This is less important if the customer has parallel backflow preventers, since **parallel installations** allow one backflow preventer to be tested while the other is still in service.

Testing the Reduced Pressure Principle Assembly

The reduced pressure principle assembly is tested using a differential pressure gauge. The differential pressure gauge will measure the difference in pressure between the high-pressure hose and the low-pressure hose. As soon as the test gauge is connected to the assembly and the test cocks opened, the gauge is on-line. Water pressure is always supplied to both sides of the gauge whether the control valves and bleed valves are open or closed (Figure 6-6).

Objectives of the Tests: The objectives of the tests are to ensure that check valve #1 is holding tight; to record the relief valve opening point; to ensure that check valve #2 will hold tight against backpressure; to record the differential pressure across check valve #1; to confirm that the outlet shut-off valve is holding tight; and to determine and record the differential pressure drop across check valve #2.

Test 1: Ensure that check valve #1 is holding tight.

Check valve #1 must be holding tight in order to continue testing the RP. The differential pressure across check valve #1 is monitored during tests one through four. The high-pressure hose is connected to test cock #2, which is located on the supply side or high-pressure side of check valve #1. The low-pressure hose is connected to test cock #3, which is located on the downstream or low-pressure side of

---

Parallel installations allow one backflow preventer to be tested while the other is still in service.

The differential pressure gauge will measure the difference in pressure between the high-pressure hose and the low-pressure hose.

Accurate readings on the differential pressure gauge can only be accomplished during a no-flow static condition.

check valve #1. The value on the gauge should be 5.0 psi or higher. This is an apparent pressure differential. If the outlet shut-off valve has a slight leak that has not been detected and the customer is using water, a flow through the assembly will occur. This flow will alter the pressure differential across check valve #1. The apparent reading on the gauge will be a bit higher than the correct reading.

Test 2: Record the relief valve opening point. This value must be 2.0 psi or above.

The relief valve is allowed to open to determine at what pressure the RV will open and vent water to atmosphere. A cross-connection is created across check valve #1 by using the high and low-pressure hoses combined with the high and low by-pass control needle valves on the test kit. Water is allowed to trickle into the zone between the check valves very slowly. The low by-pass control needle valve cannot be opened more than one-quarter turn.

Test 3: Ensure that check valve #2 will hold tight against backpressure.

A cross-connection is created between test cock #2 and test cock #4. Supply pressure travels up the high-pressure hose and is bypassed into the vent hose, which is connected to test cock #4. Supply pressure is applied downstream to the area between check valve #2 and the closed outlet shut-off valve. If check valve #2 is leaking, water will pass into the zone, the zone pressure will increase and the needle on the gauge will drop until the relief valve opens again. If the outlet shut-off valve is leaking and backpressure is present, the needle on the gauge should rise and indicate that higher pressure is present.

Test 4: Record the differential pressure across check valve #1. This value should be at least 5.0 psi.

Simply reset the gauge with the low-bleed needle valve and record the value shown on the gauge as the differential pressure across check valve #1. This value should be at least 5.0 psi and greater than the relief valve.

Test 5: Confirm that the outlet shut-off valve is holding tight.

Turning off test cock #2 will terminate the supply pressure to the gauge through the high-pressure hose. The high side of the gauge is now getting pressure from the downstream side of check valve #2 through the vent hose. If the shut-off valve is leaking and water is flowing to the customer, the gauge will indicate a loss of pressure by the needle dropping. If backpressure is present, the needle on the gauge should rise. If the needle remains steady, record that the shut-off valve as closed tight.

Test 6: Determine and record the differential pressure drop across check valve #2.

(This is an optional test.) If the outlet shut-off valve was determined to be leaking in the previous step, this test *cannot* be performed. However if the outlet shut-off valve was holding tight, connect the

low-pressure hose to test cock #4 and the high-pressure hose to test cock #3. The reading on the gauge will indicate the differential pressure across check valve #2 and should be 1.0 psi or higher. Recall in test #1 that if the outlet shut-off valve has a slight leak and the customer is using water, a flow through the assembly will occur. This flow will alter the pressure differential across check valve #2. The reading on the gauge will be too high. A leaking check valve #2 should show a 0 psi reading on the gauge. With water flowing, the needle may show above 1.0 psi, which would be a false positive reading.

Step-by-step Testing Procedure for the Reduced Pressure Principle Assembly

Test 1: Ensure that check valve #1 is holding tight.

1) Before conducting field testing, the customer should be contacted and a testing date scheduled. Immediately prior to testing, the customer should again be notified that water service will be discontinued temporarily.
2) Verify that the appropriate backflow preventer is being tested, and note the general conditions of the backflow preventer and the surrounding area, look for signs of water spillage around the assembly.
3) Flush the test cocks. Open test cock #4 and leave it running, open test cock #3, open test cock #2 slowly, open test cock #1. When all test cocks are running clean water, close test cock #1, close test cock #2, close test cock #3, and finally, close test cock #4.
4) Install brass fittings and flair adapters.

> Proper test cock flushing moves foreign matter that may foul the check valves or damage the test gauge. Opening test cocks #4 and #3 first and leaving them running simulates flow through the assembly and prevents premature opening of the relief valve. Opening test cock #4 relieves any backpressure.

5) Close all needle valves on gauge.
6) Attach the high-pressure hose to test cock #2.
7) Attach the low-pressure hose to test cock #3.
8) Bleed air from the test gauge and assembly by opening test cock #3, then the low-bleed valve. Water will discharge through the clear hose. Next, *SLOWLY* open test cock #2. Open high-bleed needle valve.

> If test cock #2 is opened too quickly the relief valve will open prematurely because insufficient pressure is being supplied to the high-pressure side of the diaphragm. As a result, an accurate test of the condition of the assembly cannot be acquired. Exercising the relief valve prior to testing must be avoided.

Exercising the relief valve prior to testing must be avoided.

FIELD TESTING

9) Close the outlet shut-off valve.
10) When all the air is expelled from the gauge, close the high-bleed valve. After all the air has been purged from the gauge and hoses, close the low-bleed valve. The low-bleed valve must be closed last.
11) Observe the pressure differential across check valve #1. It should be holding steady and the relief valve should not be dripping.

> This ends the initial test. This is the apparent pressure differential across check valve #1. This value is not recorded.

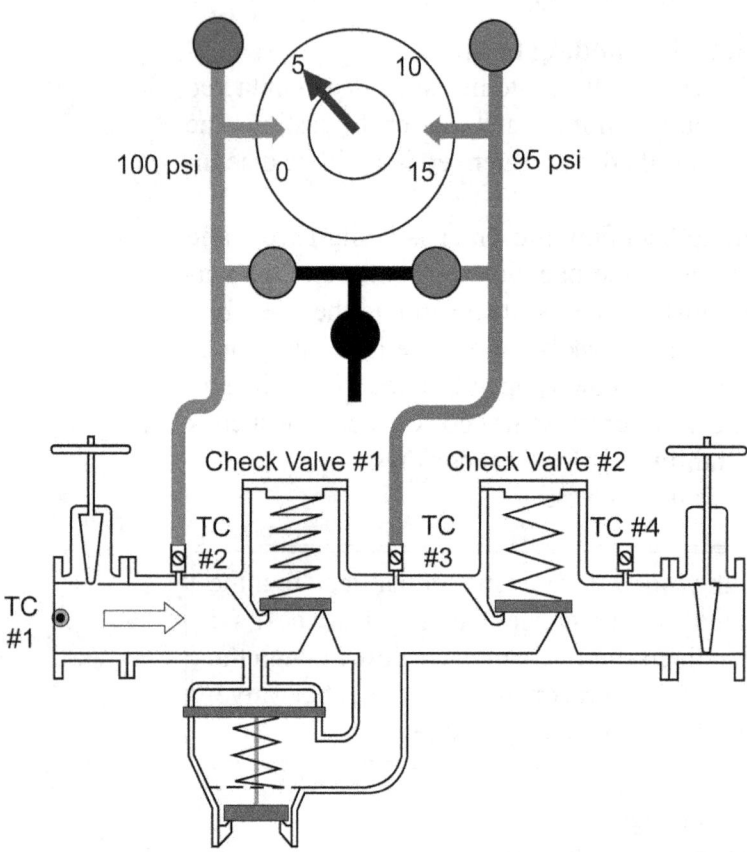

Figure 6-6 Determining the Pressure Differential Across Check Valve #1 on an RP
The high-pressure hose is connected to test cock #2 and the low-pressure hose is connected to test cock #3.

Test 2: Record the relief valve opening point. This value must be 2.0 psi or above.

1) Open high-control, by-pass needle valve one full turn. Open the low-control, by-pass needle valve slightly until the needle on the gauge starts to drop. DO NOT OPEN VALVE MORE THAN ONE QUARTER TURN.

> This allows water pressure from the supply side to pass through the test gauge into the zone, simulating a leaking check valve #1.

2) The first drops of water coming from the relief valve vent indicate when the relief valve opens. Record the opening point of the relief valve. The value should be 2.0 psi or greater (Figure 6-7).

> The opening point can be determined by placing a hand under the assembly and noting when water begins to drip from the assembly.

3) Close the low-control valve. Leave the high-control valve open.

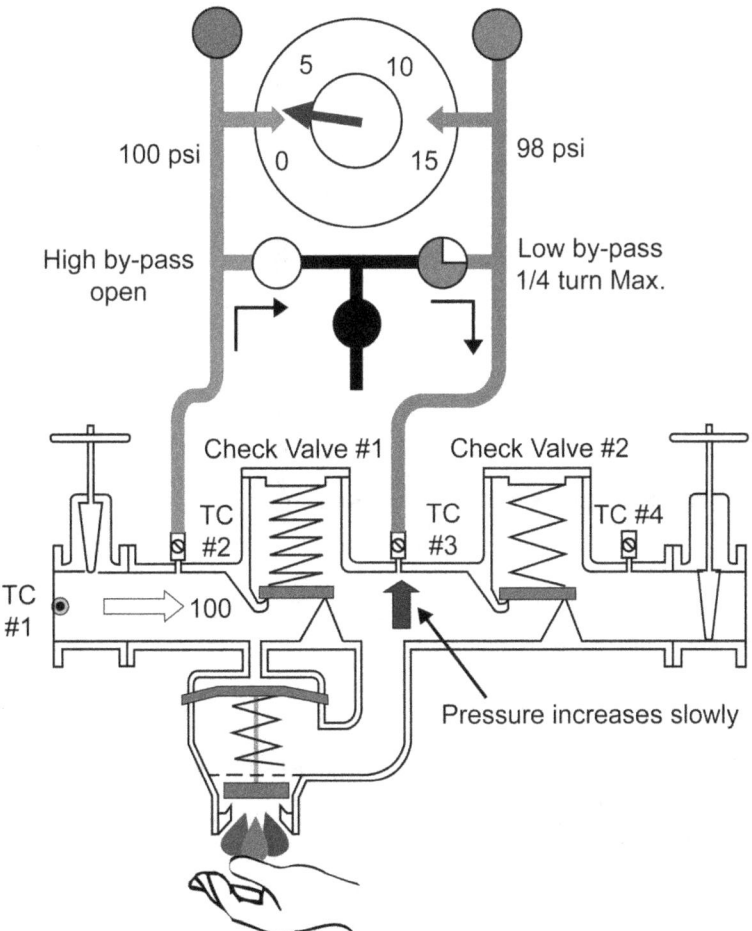

Figure 6-7 Determining the Opening Point of the Relief Valve on an RP
Essentially, the test kit is acting as a bypass around the check valve. The equalization in pressure across the check valve causes the zone pressure to increase to the point where the zone pressure plus the relief valve spring open the relief valve and discharge water.

FIELD TESTING

> Note: If the needle on the gauge starts to drop but stops before the relief valve opens, the outlet shut-off valve may be leaking with flow through the assembly. Connect an extra bypass hose from test cock #1 to test cock #4 (Figure 6-8).

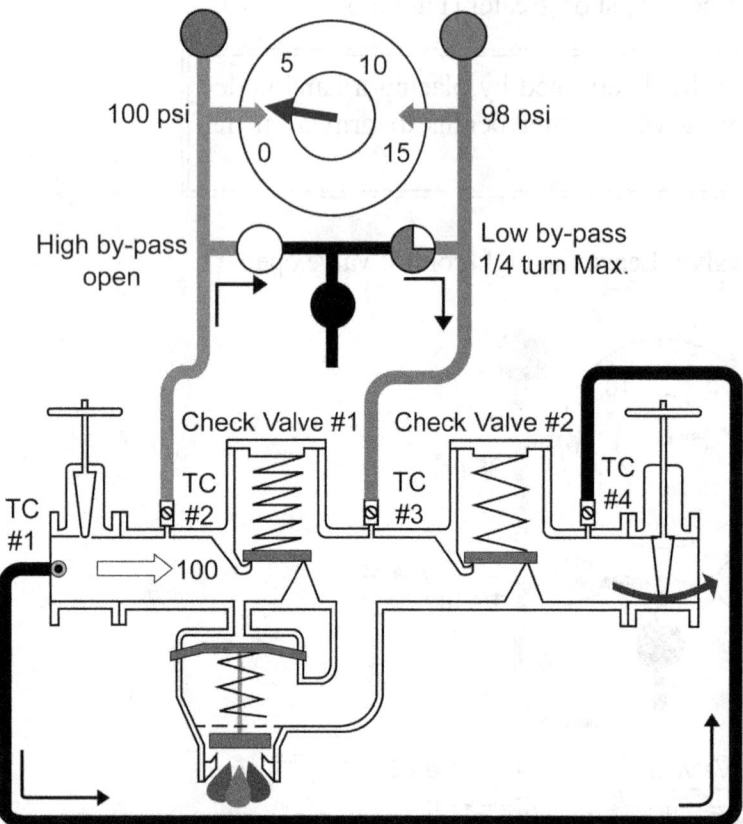

Figure 6-8   Testing an RP with a Leaking Outlet Shut-off Valve
If the needle on the gauge starts to drop but stops before the relief valve opens, the outlet shut-off valve may be leaking with flow through the assembly. Connect an extra by-pass hose from test cock #1 to test cock #4.

Test 3: Ensure that check valve #2 will hold tight against backpressure.

1) Open the vent-control valve and bleed air from the vent hose.
2) While the water is running, attach the vent hose to test cock #4. Close the vent-control valve.
3) Open test cock #4.
4) Open the low-bleed valve to re-establish line pressure throughout the assembly. This will reset the gauge to the original differential across check valve #1. Close the low-bleed valve.
5) Open the vent-control valve. If the indicated pressure differential remains steady and water does not drip from the relief valve vent, then check valve #2 is reported as "closed tight." If the pressure

differential falls to the relief valve opening point then check valve #2 is noted as "leaking." Record the status of check valve #2.

> Opening the vent-control valve with the high-control valve open allows high pressure from the supply side to pass through the test gauge to the backside of check valve #2. Some movement of the needle may result from compression of the check valve disc. If the needle on the gauge rises, then backpressure is present. Immediately close the vent-control valve. If the vent-control valve is left open, the gauge acts as a by-pass around the RP.

> Recall that the pressure differential across check valve #1 is still being monitored as long as check valve #2 is holding tight. On some assemblies, the relief valve may discharge a small amount of water. If this happens, reset the gauge with the low-bleed valve. If the relief valve still drips, then check valve #2 is noted as "leaking." Note: This is a pass/fail test. Do not record any values.

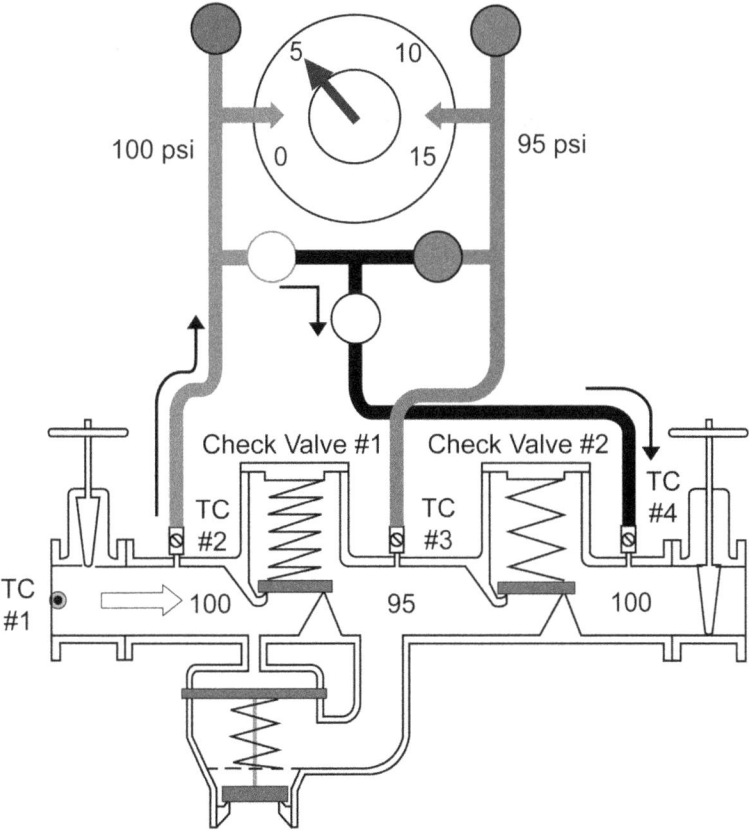

Figure 6-9   Determining That Check Valve #2 of an RP Will Hold Tight Against Backpressure
Opening the by-pass control valve allows water from the supply side to be routed through the test kit to the back side of check valve #2.

FIELD TESTING

Test 4: Record the differential pressure across check valve #1. This value should be at least 5.0 psi.
1) If check valve #2 was leaking, close the vent-control by-pass valve, otherwise leave open. The vent-control and the high-pressure by-pass valve must be open to perform tests 3, 4 & 5. Open the low-bleed valve to re-establish line pressure throughout the assembly, and to reset the gauge at the differential across check valve #1. Close the low-bleed valve.
2) Record the gauge reading. This is the pressure differential across check valve #1. The reading should be at least 5.0 psi. If a minimum buffer of 3.0 psi is required, this reading should be 3.0 psi greater than the relief valve opening (Test 2).

Test 5: Confirm that the outlet shut-off valve is holding tight.
1) Close test cock #2. The high-pressure side of the gauge is now receiving high pressure from test cock #4. If the needle on the gauge drops, the outlet valve is leaking and the customer is using water. If the needle on the gauge rises, the outlet valve is leaking and back-pressure is present. Note: If the customer is not using any water, you may not be able to determine whether the outlet valve is leaking (Figure 6-10).

> Caution: If the outlet shut-off valve is leaking, you cannot perform the next test, Test 6.

CHAPTER SIX

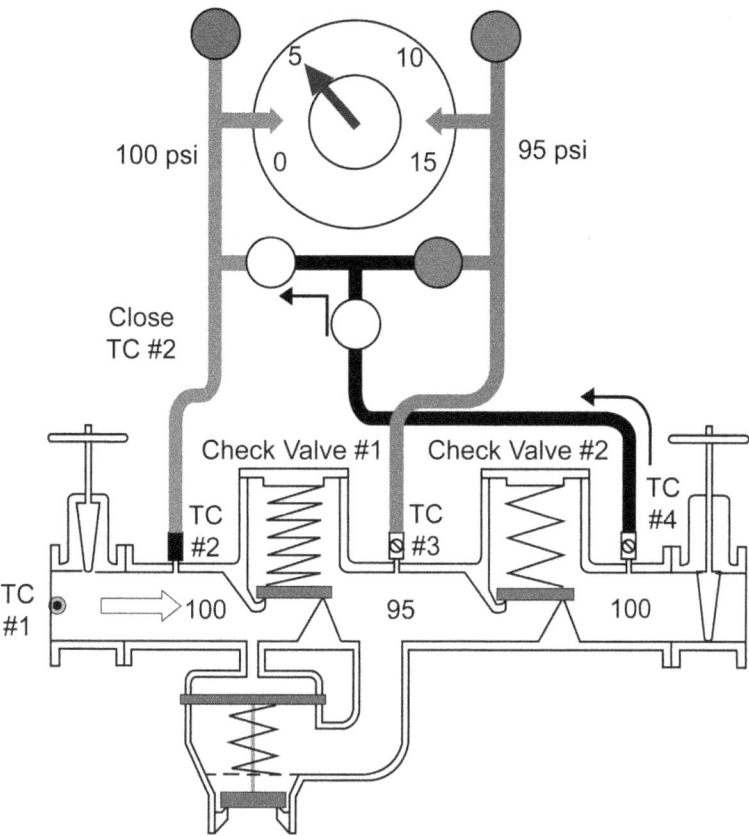

Figure 6-10   Confirm That the Outlet Shut-off Valve Is Not Leaking
Close test cock #2. The high-pressure side of the gauge is now receiving high pressure from test cock #4. If the needle on the gauge drops, the outlet valve is leaking and the customer is using water.

Test 6: Determine and record the differential pressure drop across check valve #2.

1) Close the vent-control valve.
2) Close all test cocks. Remove vent hose from test cock #4.
3) Move low-pressure hose from test cock #3 to test cock #4. Move the high-pressure hose from test cock #2 to test cock #3.
4) Close all needle valves on gauge.
5) Bleed air from the test gauge and assembly by opening test cock #4, and then open the low-bleed valve. Water will discharge through the clear hose. Next, open test cock #3. Open the high-bleed needle valve.
6) When all the air is expelled from the gauge, close the high-bleed valve. After the needle has climbed past the original reading, close the low-bleed valve. The low-bleed valve should be closed *slowly*.

FIELD TESTING

7) Record the pressure differential across check valve #2. It should be 1.0 psi or greater.

> Assemblies installed before 1993 may have springs in check valve #2 that hold tight at values less than 1.0 psi. If the gauge is indicating a value less than 1.0 psi (0.7, 0.8, 0.9) but holding steady, check valve #2 may be determined to be holding tight.

> Note: If the outlet shut-off valve is determined to be leaking in Test 5, this test will not give accurate results. A failing check valve #2 may test as tight if there is flow through the assembly.

A failing check valve #2 may test as tight if there is flow through the assembly.

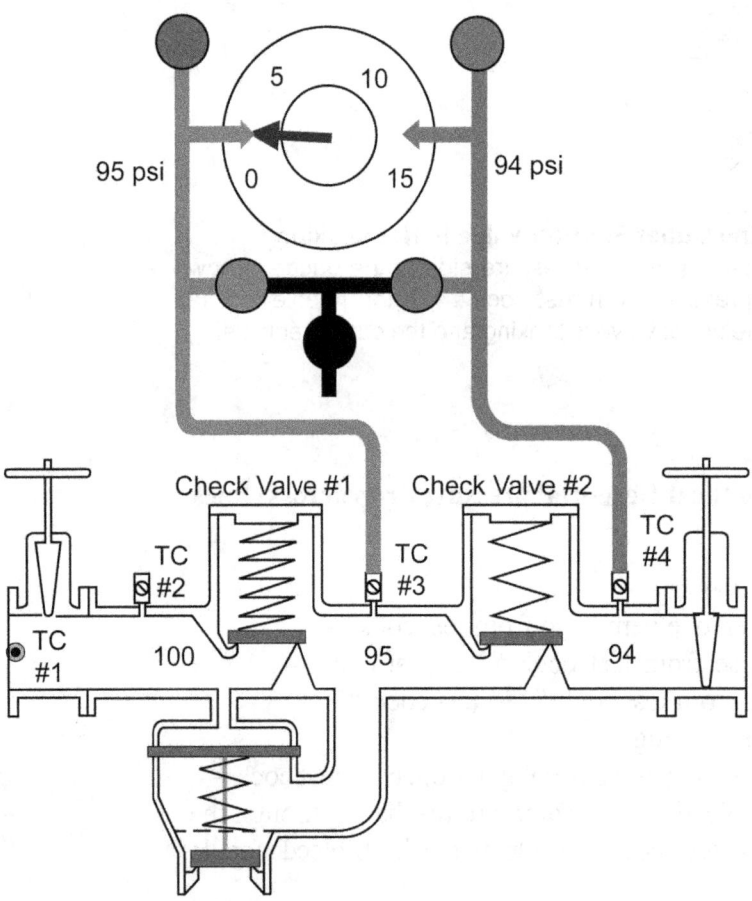

Figure 6-11 **Testing the Differential Pressure Across Check Valve #2**
If the outlet shut-off valve leaks and water is flowing through the assembly, then the reading will be a false positive.

Final Steps
1) Close all test cocks. Remove high and low hose from assembly.
2) Open all needle valves on test kit.
3) Open the outlet shut-off valve slowly to restore service to the customer.

Testing the Reduced Pressure Principle Assembly with a Leaking Outlet Shut-off Valve

When testing the relief valve opening of the RP, if the relief valve will not open and drip water from the zone or if the needle on the gauge starts to drop but stops before the opening point, then a leaking outlet shut-off valve may be the problem. The relief valve opening point may often be determined, however, without replacing the outlet shut-off valve. To determine the relief valve opening point, attach an extra by-pass hose to test cock #1 and connect to test cock #4, then open test cock #1, open test cock #4 slowly with your hand under the RV vent. This allows water pressure from the supply side of check valve #1 to be routed to the backside of check valve #2 through the by-pass hose and compensates for the leak. If the volume of the leak exceeds the capacity of the by-pass hose, the outlet shut-off valve may need to be replaced (Figure 6-8).

> The relief valve opening point may often be determined, however, without replacing the outlet shut-off valve.

Testing the Reduced Pressure Detector Assembly

Test 1: Check the detector meter for function.
1) Before testing the assembly, close the outlet shut-off valve of the by-pass assembly. Open test cocks #3 and #4 on the by-pass RP to check the volume of water passing through the meter. If the flow is low then the meter may be clogged with debris. Clean the meter screen before testing the by-pass assembly.

Test 2: Field test the RPDA.
1) When finished with the by-pass RP test, leave the outlet shut-off valve of the by-pass RP closed and begin testing the main or larger RP.
2) Open the by-pass outlet shut-off valve and begin testing the by-pass RP using the field test procedure for the RP.
3) When finished with Test 6 on the main RP and before opening the main RP outlet shut-off valve, open the outlet shut-off valve on the by-pass RP.

Test 3: Check the detector meter again for low flows.
1) Open test cock #4 on the main RP and check the meter to see if the low flow indicator is turning.
2) Close test cock #4 and then slowly open the outlet shut-off valve of the main RP.

FIELD TESTING

Testing the Double Check Valve Assembly

Objectives of the Tests: To determine the tightness of the check valves and to record the pressure drop across each of the check valves.

Step-by-step Testing Procedures for the Double Check Valve Assembly, Differential Pressure Gauge Single-hose Method

Test 1: Test check valve #1 for a minimum of 1.0 psi in the direction of flow under normal, no backpressure conditions.

1) Before conducting field testing, the customer should be contacted and a testing date scheduled. Just prior to testing, the customer should again be notified that water service will be temporarily discontinued.
2) Verify that the appropriate backflow preventer is being tested and note the general conditions of the backflow preventer and the surrounding area.
3) Flush the test cocks.

> This is done to remove any lodged foreign materials that might interfere with the test.

4) Install a bleed-off tee on test cock #2 and a short, clear tube on test cock #3. Fill short tube with water. Note: a short tube is added to the downstream test cock to ensure that air does not enter the assembly. The top of the short tube must be higher than the check valve. This short tube also acts as a sight glass when testing check valve #2.
5) Position the test gauge centerline and the end of the low-pressure hose at same elevation as the discharge point from test cock #3. Close all needle valves on the gauge.
6) Attach the high-pressure hose from the test gauge to the bleed-off tee on test cock #2 (Figure 6-12).
7) Open test cock #2. Open the high-bleed needle valve and bleed all air from hose and test gauge. Close high-bleed needle valve.
8) Close the outlet shut-off valve.
9) Close the inlet shut-off valve.
10) Open test cock #3. Allow the water to drain from the downstream side of check valve #1.

> If the water does not stop flowing, the inlet shut-off valve is leaking. Follow the procedures for Testing the Double Check Valve Assembly with a Leaking Inlet Shut-off Valve (page 146).

11) Record the gauge reading. It must be 1.0 psi or greater.
12) Close test cocks #2 and #3.
13) Disconnect the high-pressure hose, and then open the inlet shut-off valve.

CHAPTER SIX

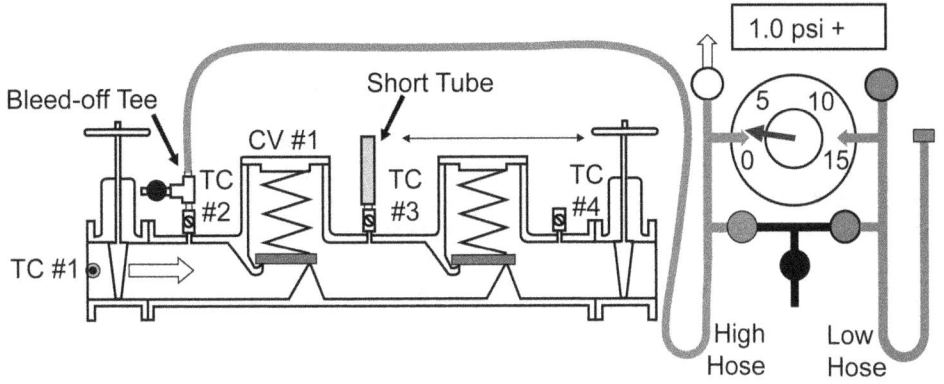

To prevent damage to your gauge, never pressurize the assembly when gauge hoses are connected to the test cocks.

Figure 6-12   Testing Check Valve #1 on a DCVA
The high-pressure hose is connected to the bleed-off tee on test cock #2. Center of gauge and end of low-pressure hose are at the same level that water discharges from the short tube connected to test cock #3.

Test 2: Test check valve #2 for a minimum of 1.0 psi in the direction of flow under normal, no backpressure conditions.

1) Install bleed-off tee on test cock #3 and a short, clear tube on test cock #4. Note: a short tube is added to the test cock #4 to serve as a sight glass. The top of the short tube must be higher than the check valve.

2) Position the test gauge centerline and the end of the low-pressure hose at same elevation as the discharge point from the short tube connected to test cock #4. Close all needle valves on the gauge.

3) Attach the high-pressure hose from the test gauge to the bleed-off tee on test cock #3 (Figure 6-13).

4) Open test cock #3. Open the high-bleed needle valve and bleed all air from hose and test gauge.

5) Close high-bleed needle valve.

6) Fill the short tube with water.

7) Close the inlet shut-off valve.

8) Open test cock #4 and allow the water to drain from the downstream side of check valve #2.

> If the water does not stop flowing, the outlet shut-off valve may be leaking with backpressure.

9) Record the gauge reading. It must be 1.0 psi or greater.

10) Close test cocks #3 and #4.

11) Open all needle valves on the gauge.

12) Disconnect the high-pressure hose, and open the inlet shut-off valve and then slowly open the outlet shut-off valve.

FIELD TESTING

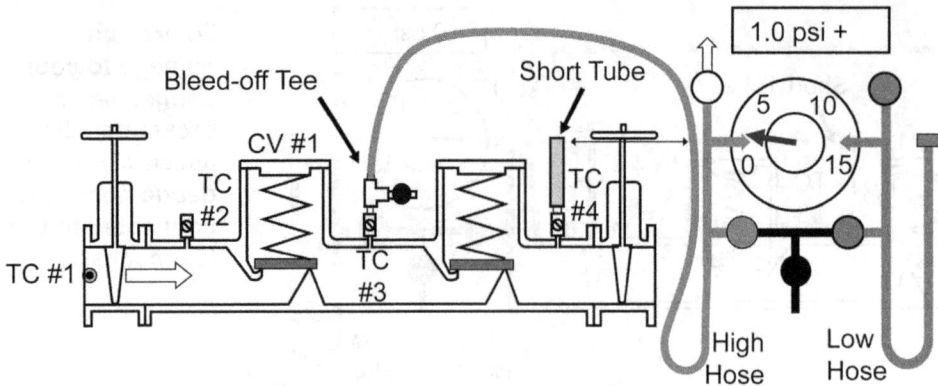

Figure 6-13 **Testing Check Valve #2 on a DCVA**
The high-pressure hose is connected to the bleed-off tee on test cock #3. Center of the gauge and the end of the low-pressure hose must be at same level that water discharges from the short tube connected to test cock #4.

Testing the Double Check Valve Assembly with a Leaking Inlet Shut-off Valve, Differential Pressure Gauge Method

Test 1: To test check valve #1 of a DCVA when the inlet shut-off valve is leaking, the excess flow of water must be compensated for by allowing that water to flow to atmosphere.

1) A bleed-off tee must be connected to test cock #2. Attach the high-pressure hose to the bleed-off tee on test cock #2. Then a short, clear tube is connected to test cock #3. Fill short tube with water. Note: a short tube is added to the downstream test cock to make sure that air does not enter the assembly. The top of the short tube must be higher than the check valve. This short tube also acts as a sight glass when testing check valve #2 (Figure 6-13).

2) Close outlet shut-off valve.

3) Open test cock #2 slowly to fill the high-pressure hose of the gauge. Bleed the gauge through the high-bleed needle valve.

4) Close the inlet shut-off valve.

5) Open test cock #3 to drain the downstream side of the check valve. If water continues to flow from the short tube on test cock #3 then the inlet shut-off valve may be leaking. The needle valve on the bleed-off tee is opened slightly; this allows the water added as a result of the leak to be vented to atmosphere and eliminates the effects of the leak. The needle valve should be adjusted so that only a small amount of water flows out of test cock #3. This will be a slight positive drip. Once this adjustment has been made, record the reading on the gauge as the static pressure drop across check valve #1.

6) To complete the test, close test cock #3 and test cock #2. Remove the hose, tee, and needle valve. Open the inlet shut-off valve.

> If water continues to flow from the downstream test cock, make sure the needle valve was adjusted properly by repeating the steps described above. If the water stops flowing from the needle valve on the compensating tee, but continues to flow from test cock #3, then check valve #2 and the outlet shut-off valve are leaking allowing water to backflow. This usually does not affect the test unless the size of the leak and the amount of backpressure on the assembly is so great that the excess pressure cannot be discharged though the open test cock.

Test 2: To test check valve #2 of a DCVA when the inlet shut-off valve is leaking, the excess flow of water must be compensated for by allowing that water to flow to atmosphere.

1) A bleed-off tee must be connected to test cock #3. Attach high-pressure hose to bleed-off tee on test cock #3. The short tube is attached to test cock #4. Fill the short tube with water (Figure 6-13).

2) Open test cock #3 slowly to fill the high-pressure hose of the gauge. Bleed the gauge through the high-bleed needle valve.

3) Close the inlet shut-off valve.

4) Open test cock #4 to drain the downstream side of check valve #2. The needle valve on the bleed-off tee is again opened slightly; this allows the water added as a result of the leak to be vented to atmosphere and eliminates the effects of the leak. The needle valve should be adjusted so that only a small amount of water flows out of test cock #4. This will be a slight positive drip. Once this adjustment has been made, record the reading on the gauge as the static pressure drop across check valve #2.

5) To complete the test, close test cock #4 and test cock #3. Remove the hose, tee, and needle valve. Open the inlet then open the outlet shut-off valves.

Testing the Double Check Valve Assembly with a Leaking Outlet Shut-off Valve, Differential Pressure Gauge Method

Test 1: To test check valve #1 for a minimum of 1.0 psi in the direction of flow when the outlet shut-off valve is leaking and the customer is using water.
Same as Test 1 for Step-by-step Testing Procedures for the Double Check Valve Assembly with a Leaking Inlet Shut-off Valve, Differential Pressure Gauge Method (page 146).

Test 2: To test check valve #2 of a DCVA when the outlet shut-off valve is leaking and the customer is using water, the elevation of the test kit must be lowered to the water level in the assembly (Figure 6-14).

FIELD TESTING

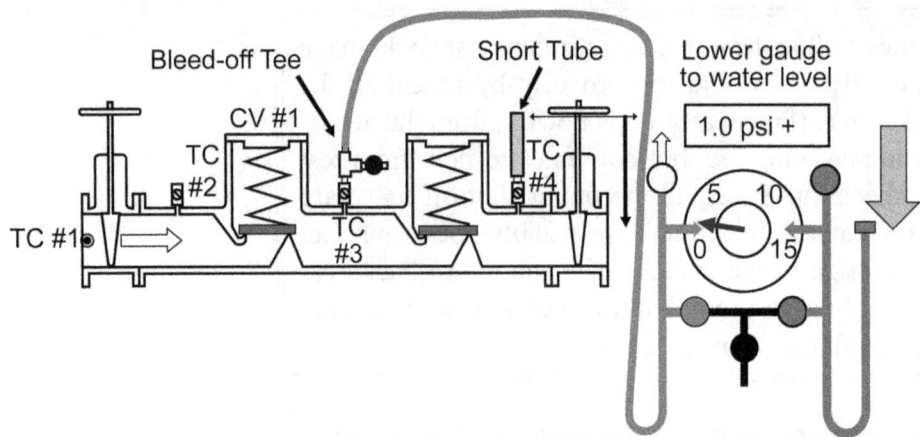

Figure 6-14 Testing Check Valve #2 with Water Flowing to Customer through a Leaking Outlet Check Valve
Lower center of gauge and end of low-pressure hose to the centerline of assembly.

1) A bleed-off tee must be connected to test cock #3. The short tube is attached to test cock #4. Fill the short tube with water.
2) Open test cock #3 slowly to fill the high-pressure hose of the gauge. Bleed the gauge.
3) Close the inlet shut-off valve.
4) Open test cock #4 to drain the downstream side of the check valve. If the water level drops in the clear tube, the outlet shut-off valve is leaking. Lower the center of the gauge and the end of the low-pressure hose to the approximate center of the assembly. Record the gauge reading. It must be 1.0 psi or greater.
5) To complete the test, close test cock #4 and test cock #3. Remove the hose, tee, and needle valve. Open the inlet then open the outlet shut-off valves.

If the water level drops in the short tube while testing check valve #2, lower the centerline of the gauge and end of low-pressure hose to center of the assembly.

Table 6-1 Testing the DCVA - Single Hose Differential Pressure Gauge Method

	Test 1 - CV1	Test 2 - CV2
NORMAL	Record CV1 differential ≥1.0 psi.	Record CV2 differential ≥1.0 psi.
CV1 LEAKS	STOP and repair.	Record CV2 differential ≥1.0 psi.
CV2 LEAKS	Record CV1 differential ≥1.0 psi.	STOP and repair.
Inlet shut-off valve LEAKS	Add compensating or bleed-off tee and needle valve to TC2. Add short tube to TC3.	Add compensating tee and needle valve to TC3. Add short tube to TC4.
Outlet shut-off valve LEAKS	Should not affect test.	Water will continue to run out of TC4 if backpressure is present. If the water drops in short tube, lower centerline of test gauge to center of assembly.

Sight Tube Method

These instructions may be more up-to-date than the instructions supplied by the manufacturer since the latter may give false results. The sight tube method uses the weight of water to determine whether the check valve is working properly. The sight tube is simply a plastic tube designed to hold more than a 28-inch (1 psi) column of water. It has a special fitting for connecting to the test cocks of the assembly.

Step-by-step Testing Procedures for the Double Check Valve Assembly, Sight Tube Method

Objectives of the Tests: This test determines if the check valves will hold 1.0 psi in the direction of flow. Recall that this is the AWWA specification for a single check valve.

Test 1: To check that check valve #1 will hold 1.0 psi in the direction of flow under normal no backpressure conditions (Figure 6-15).

1) Before conducting field testing, the customer should be contacted and a testing date scheduled. Just prior to testing, the customer should again be notified that water service will be temporarily discontinued.
2) Verify that the appropriate backflow preventer is being tested and note the general conditions of the backflow preventer and the surrounding area.

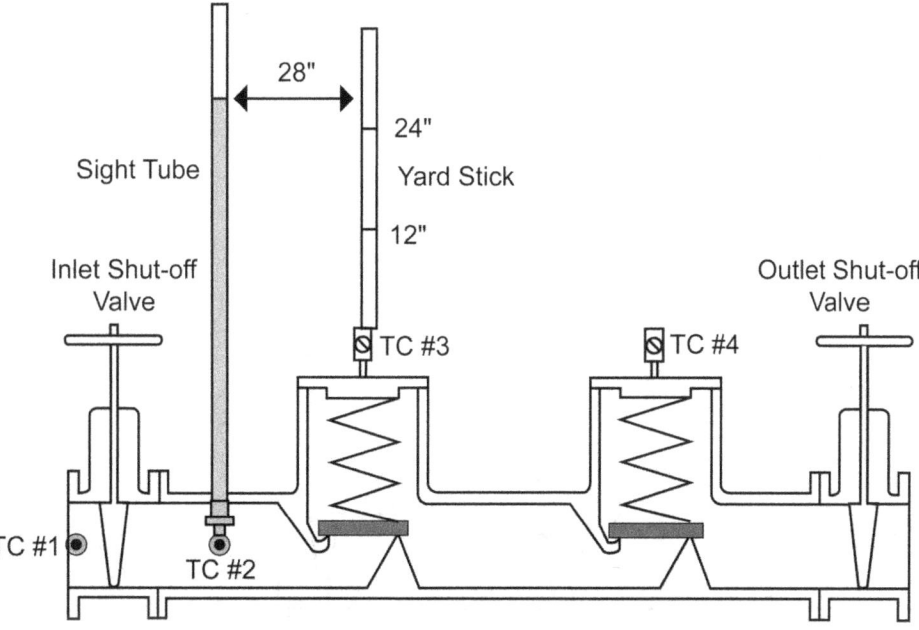

Figure 6-15   Testing Check Valve #1 of the DCVA with Sight Tube
The sight tube is connected to test cock #2 of the assembly for testing check valve #1.

FIELD TESTING

3) Flush the test cocks. This is done to remove any lodged foreign materials that might interfere with the test.
4) Install bleed-off tee on test cock #2.
5) Attach the sight tube to test cock #2.
6) Attach a short, clear tube and ell (if necessary) to test cock #3.

> The short tube is used to prevent air from entering test cocks located on the side of an assembly. If the base of the sight tube is at a different elevation than the top of the short tube, a yardstick is used to determine where 1.0 psi is located above the short tube on test cock #3.

7) Close the outlet shut-off valve.
8) Open test cock #2 and fill the sight tube so that the water level in it will be at least 28 inches above the water level in the short, clear tube or ell attached to test cock #3. Close test cock #2.
9) Close the inlet shut-off valve.
10) Open test cock #3 to relieve backpressure and then open test cock #2.
11) Observe whether the level in the sight tube is maintained at least 28 inches above the water level at the top of the short tube on test cock #3. Wait two minutes after the water level stabilizes.
12) Record the results (pass or fail).
13) Close test cocks #2 and #3, disconnect the sight tube.

Test 2: To check that check valve #2 will hold back 1.0 psi in the direction of flow under normal no backpressure conditions (Figure 6-16).

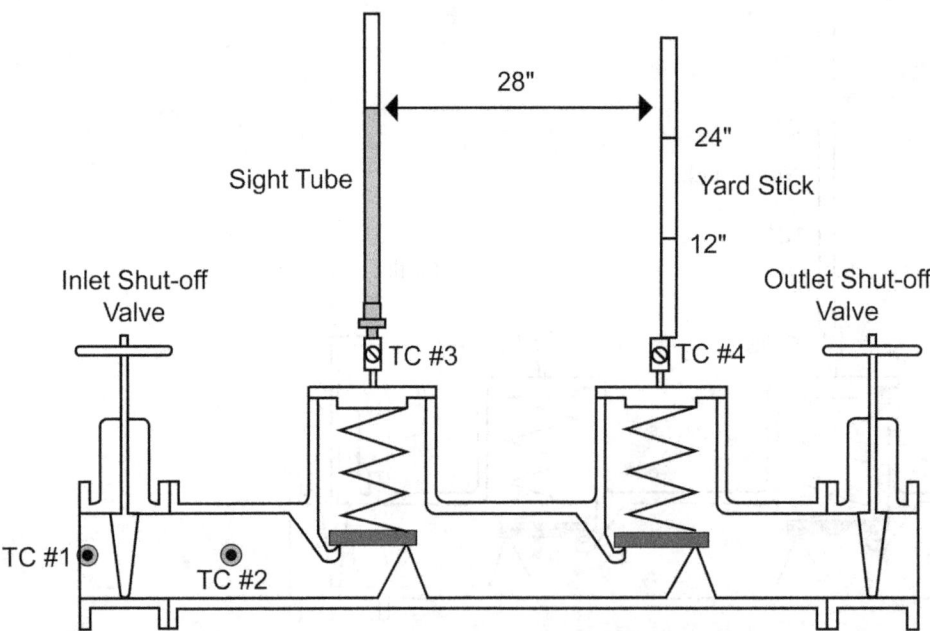

Figure 6-16   Testing Check Valve #2 of the DCVA with Sight Tube
The sight tube is connected to test cock #3 for testing check valve #2.

1) Open the inlet shut-off valve.
2) Install bleed-off tee on test cock #3.
3) Attach the sight tube to test cock #3.
4) Attach a short, clear tube and ell (if necessary) to test cock #4.

> The short sight tube is used to prevent air from entering test cocks located on the side of an assembly. If the base of the sight tube is at a different elevation than the top of the short tube, a yardstick is used to determine where 1.0 psi is located above the short tube on test cock #4. The short tube also acts as a sight glass to determine if the outlet shut-off valve is leaking and the customer is using water.

5) Open test cock #3 and fill the sight tube so that the water level in it will be at least 28 inches above the water level in the short, clear tube or ell attached to test cock #4. Close test cock #3.
6) Close the inlet shut-off valve.
7) Open test cock #4 and then open test cock #3.
8) Observe whether the level in the sight tube is maintained at least 28 inches above the water level at test cock #4. Wait 2 minutes after the water level stabilizes.
9) Record the results. (Pass or Fail)
10) Close test cocks #3 and #4, disconnect the sight tube, and open the inlet shut-off valve and then slowly open the outlet shut-off valve.

Testing the Double Check Valve Assembly with a Leaking Inlet Shut-off Valve, Sight Tube Method

Test 1: To test check valve #1 of a DCVA when the inlet shut-off valve is leaking, a bleed-off tee with a needle valve must be connected to test cock #2.

1) Flush the test cocks. This is done to remove any lodged foreign materials that might interfere with the test.
2) Install bleed-off tee on test cock #2.
3) Attach the sight tube to the bleed-off tee at test cock #2.
4) Attach a short tube and ell (if necessary) to test cock #3.
5) Close the outlet shut-off valve.
6) Open test cock #2 slowly to fill the sight tube.
7) Close test cock #2 when the tube is filled.
8) Close the inlet shut-off valve.
9) Open test cock #3 and allow the water to drain. If water continues to flow from the short tube on test cock #3 then the inlet shut-off valve may be leaking. Adjust the needle valve on the bleed-off tee until there is a slight positive drip at test cock #3.

FIELD TESTING

10) Open test cock #2.
11) Observe the level in the sight tube. If it remains stable for two minutes at least 28 inches above the water level at test cock #3 and water stops flowing from test cock #3, then check valve #1 is holding tight to a minimum of 1.0 psi in the direction of flow. Record the results (pass or fail).
12) Close test cock #3 and test cock #2. Remove the sight tube, tee and needle valve. Open the inlet shut-off valve.

> If water continues to flow from the downstream test cock, check and make sure the needle valve was adjusted properly by repeating the steps described above. If the water stops flowing from the needle valve on the compensating tee, but continues to flow from test cock #3, then check valve #2 and the outlet shut-off valve are leaking allowing water to backflow. This usually does not affect the test unless the size of the leak and the amount of backpressure on the assembly is so great that the excess pressure cannot be discharged through the open test cock.

Test 2: To test check valve #2 of a DCVA when the inlet shut-off valve is leaking, a bleed-off tee with a needle valve must be connected to test cock #3.

1) Install bleed-off tee on test cock #3.
2) Attach the sight tube to the bleed-off tee at test cock #3.
3) Attach a short tube and ell (if necessary) to test cock #4.
4) Open test cock #3 slowly to fill the sight tube.
5) Close test cock #3 when the tube is filled.
6) Close the inlet shut-off valve.
7) Open test cock #4 and allow the water to drain. If water continues to flow from the short tube on test cock #4 then the inlet shut-off valve may be leaking. Adjust the needle valve on the bleed-off tee until there is a slight positive drip at test cock #4.
8) Open test cock #3.
9) Observe the level in the sight tube. If it remains stable for 2 minutes at least 28 inches above the water level at test cock #4 and water stops flowing from test cock #4, then check valve #2 is holding tight to a minimum of 1.0 psi in the direction of flow.
10) Close test cock #4 and test cock #3. Remove the sight tube, tee, and needle valve.
11) Open the inlet and outlet shut-off valves.

> If water continues to flow from the downstream test cock, make sure the needle valve was adjusted properly by repeating the steps described above. If the water stops flowing from the needle valve on the compensating tee but continues to flow from test cock #4, then the outlet shut-off valve is leaking and allowing backflow. This usually does not affect the test unless the size of the leak and the amount of backpressure on the assembly is so great that the excess pressure cannot be discharged though the open test cock.

Testing the Double Check Detector Assembly

Test 1: Check the detector meter for function.
1) Before testing the assembly, close the outlet shut-off valve of the by-pass assembly. Open test cocks #3 and #4 on the by-pass DC to check the volume of water passing through the meter. If the flow is low then the meter may be clogged with debris. Clean the meter screen before testing the by-pass assembly.

Test 2: Field test the DCDA.
1) Open the by-pass outlet shut-off valve and begin testing the by-pass DC using the field test procedure for the DCVA.
2) When finished with the by-pass DC test, leave the outlet shut-off valve of the by-pass DC closed and begin testing the main or larger DC.
3) When finished with Test #2 on the main DC and before opening the main DC outlet shut-off valve, open the outlet shut-off valve on the by-pass DC.

Test 3: Check the detector meter again for low flows.
1) Open test cock #4 on the main DC and check the meter to see if the low flow indicator is turning.
2) Close test cock #4 and then slowly open the outlet shut-off valve of the main DC.

Testing the Pressure Vacuum Breaker

The pressure vacuum breaker is tested by using a pressure differential pressure gauge. When conducting this test, it is essential that the differential pressure gauge be held level with the assembly. If the low-pressure hose is permanently attached to the gauge, then the end of the low-pressure hose must be at the same level as the center of the gauge. Any difference in elevation will affect the readings. This is because of the pressure head created by the weight of the water between the centerlines of the test gauge and

FIELD TESTING

pressure vacuum breaker. Incorrect conclusions may be drawn from the test results if the test gauge is not setup properly.

Objectives of the Test: These tests will determine if the air inlet port will open at 1.0 psi or above and if the single check valve will hold back at least 1.0 psi in the direction of flow.

Step-by-step Testing Procedures for the Pressure Vacuum Breaker

Test 1: To determine the opening point of the air inlet valve (Figures 6-17 and 6-18).

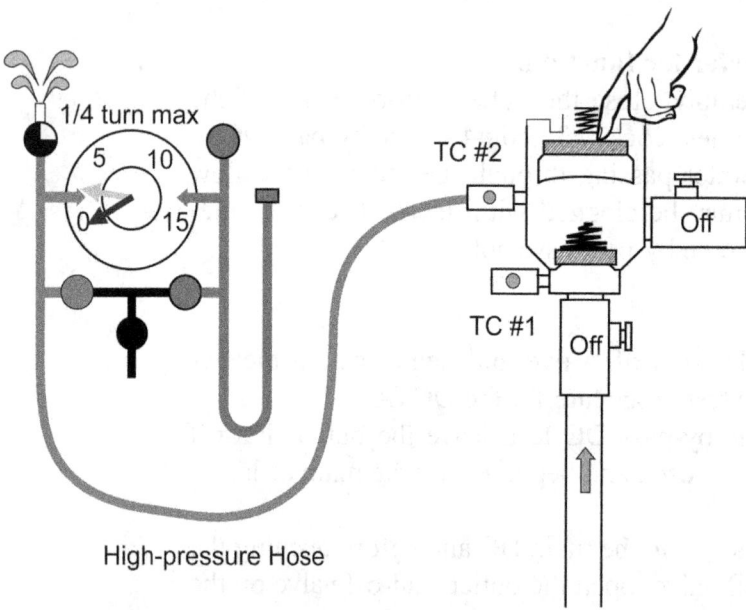

Figure 6-17   Testing the Opening Point of the Air Inlet Valve in a PVB with a Five-valve Differential Pressure Gauge
Open the high-bleed no more than quarter turn to lower the pressure under the air inlet valve.

1) Before conducting field testing, the customer should be contacted and a testing date scheduled. Before testing is actually conducted, the customer should again be notified that water service will be temporarily discontinued.
2) Verify that the appropriate backflow preventer is being tested, and note the general conditions of the backflow preventer and the surrounding area, looking for signs of water spillage around the assembly.

3) Remove the canopy.

> This should be done carefully as bees and other biting and stinging insects commonly make this area their home.

4) Flush the test cocks. This is done to remove any lodged foreign materials that might interfere with the test. Install brass fittings.
5) Close all needle valves on gauge.
6) Lift gauge and end of low-pressure hose to same elevation as test cock #2.
7) Attach the high-pressure hose to test cock #2. Slowly open test cock #2.
8) Bleed air from the test gauge and assembly by opening the high-bleed valve.
9) Close the bleed valve when all the air is removed.
10) Close the inlet shut-off valve; close the outlet shut-off valve.
11) Place a finger on top of the air inlet port to determine its opening point. Do not press down.
12) Open the high-bleed valve slowly to the atmosphere (one quarter turn maximum).

> This reduces the pressure trapped between the check valve and the air inlet valve. Water contained within the assembly will flow through test cock #2 and out of the bleed line of the test gauge, because the pressure inside the assembly (line pressure) is higher than atmospheric pressure. Water trapped between the inlet shut-off valve and the check valve during static conditions will also flow past the check valve, because the single check is only designed to hold tight against 1.0 psi in the direction of flow. If the high side bleed valve needs to be opened more than one quarter turn, the inlet shut-off valve may be leaking. If the inlet shut-off valve is leaking, the assembly can still be tested. See Testing a Pressure Vacuum Breaker with a Leaking Inlet Shut-off Valve (page 157).

13) Record the pressure at which the air inlet valve opens. The air inlet should open while there is at least 1.0 psi still in the chamber.
14) Close test cock #2, disconnect the high-pressure hose, and open the inlet shut-off valve.

To prevent damage to your test gauge, never pressurize the assembly when gauge hoses are connected to the test cocks.

FIELD TESTING

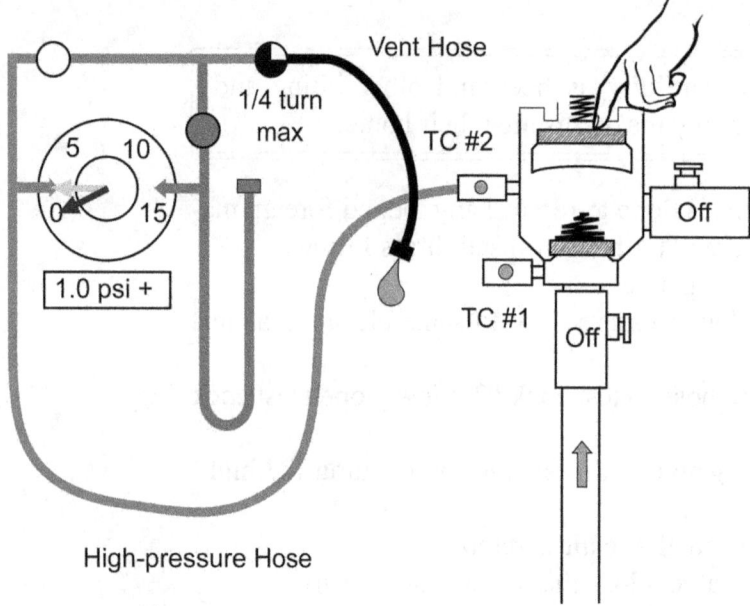

Figure 6-18   **Testing the Opening Point of the Air Inlet Valve in a PVB with a Three-valve Differential Pressure Gauge**
Open the high control one full turn. Open the vent-control valve no more than a quarter turn to lower the pressure under the air inlet valve.

Test 2: To determine if the single check will hold tight against 1.0 psi in the direction of flow (Figure 6-19).

1) Attach the high-pressure hose to test cock #1.
2) Bleed air from the test gauge by using the high-bleed valve.
3) Close the bleed valve when all the air is removed.
4) Close the inlet shut-off valve. The outlet shut-off valve should still be closed from Test #1.
5) Open test cock #2.
6) When the water stops running out of test cock #2, record the pressure indicated on the gauge. If the check valve is meeting specifications, the pressure indicated on the differential pressure gauge will not fall below 1.0 psi.
7) Replace the canopy.
8) Close both test cocks, disconnect the high-pressure hose from test cock #1, and open the inlet shut-off valve then the outlet shut-off valve.

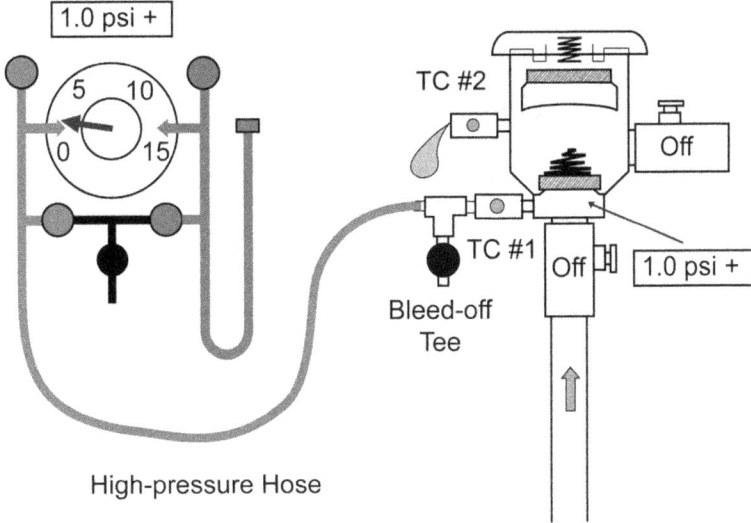

Figure 6-19 **Verifying That the Single Check of a PVB Will Hold Back 1.0 psi in the Direction of Flow**
Water will flow past the check valve and out test cock #2. If the check valve is working properly, it should hold back 1.0 psi in the direction of flow.

Testing the Pressure Vacuum Breaker with a Leaking Inlet Shut-off Valve

A leaking inlet shut-off valve is indicated during Test #1 (testing the air inlet opening point) if the high-bleed valve must be opened more then one quarter turn or water continues to flow out of the test gauge. To test the assembly under these conditions, just prior to opening the high-bleed valve, test cock #1 should be open slightly to divert the leak but maintain the reading on the gauge. The check valve should maintain the pressure trapped between the check valve and the outlet shut-off valve and, thus, prevent the gauge needle from dropping and the air inlet from opening prematurely.

A leaking inlet shut-off valve is also indicated during Test #2 (testing the check valve in the direction of flow) by water continuously flowing out of test cock #2. Under these conditions the check valve cannot be tested in the direction of flow. To test the check valve, install a 1/4-inch street ell or elbow in test cock #2 so that the open end faces up. Install a bleed-off tee with needle valve on test cock #1. Attach the high-pressure hose to the bleed-off tee. Fill the street ell with water by slowly opening test cock #2. Open test cock #1. Bleed the high side of the gauge. Close the inlet shut-off valve. Open test cock #2. The leak at the inlet shut-off valve can flow out of the bleed-off tee. Adjust the needle valve on the bleed-off tee until there is a slight positive drip coming from test cock #2. If the water level in the street ell remains constant and the gauge reads 1.0 or greater then the check valve can be recorded as "tight" (Figure 6-20).

FIELD TESTING

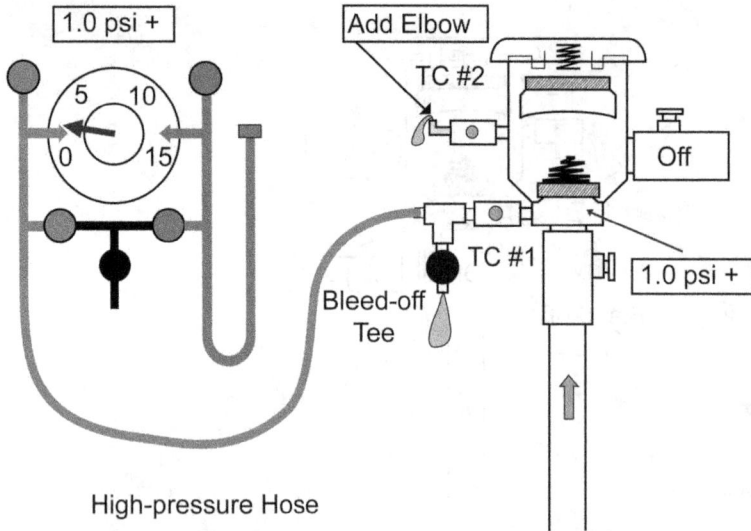

Figure 6-20 Testing the Check Valve in a PVB with a Leaking Inlet Shut-off Valve
Add an elbow on test cock #2. Adjust the bleed-off tee until there is a slight positive drip at the elbow.

Step-by-step Testing Procedures for the Spill-resistant Vacuum Breaker

The spill-resistant vacuum breaker is tested by using a pressure differential pressure gauge. When conducting this test, the differential pressure gauge must be held level with the assembly. If the low-pressure hose is permanently attached to the gauge, then the end of the low-pressure hose must be at the same level as the center of the gauge. Any difference in elevation will affect the readings. This is because of the head created by the weight of the water between the centerlines of the test gauge and pressure vacuum breaker. Incorrect conclusions may be drawn from the test results if the test gauge is not setup properly.

Objectives of the Test: These tests will determine if the air inlet port will open at 1.0 psi or above, and if the single check valve will hold back at least 1.0 psi in the direction of flow

Test 1: To determine if the check valve will hold tight against 1.0 psi in the direction of flow.
1) Close inlet shut-off valve.
2) Open vent (air bleed) to lower pressure in body.
3) When the needle on the gauge stops dropping, record the pressure indicated on the gauge. If the check valve is meeting specifications, the pressure indicated on the differential pressure gauge will not fall below 1.0 psi. Record this value.
4) Replace the canopy.
5) Open all needle valves on the gauge.
6) Close the test cock, close air bleed screw, disconnect the high-pressure hose, and open the inlet shut-off valve, then the outlet shut-off valve.

158

CHAPTER SIX

Test 2: To determine the opening point of the air inlet valve.
1) Before conducting field testing, the customer should be contacted and a testing date scheduled. Before testing is actually conducted, the customer should again be notified that water service will be temporarily discontinued.
2) Verify that the appropriate backflow preventer is being tested, and note the general conditions of the backflow preventer and the surrounding area, looking for signs of water spillage around the assembly.
3) Remove the canopy.

> This should be done carefully since bees and other biting and stinging insects commonly make this area their home.

4) Flush the test cock. This is done to remove any lodged foreign materials that might interfere with the test. Open the air bleed (vent) to discharge any air. Install brass fitting on test cock. Close all needle valves on gauge.
5) Lift gauge and end of low-pressure hose to same elevation as the center of the assembly.
6) Attach the high-pressure hose to test cock.
7) Open test cock slowly.
8) Bleed air from the test gauge and assembly by using the high-bleed needle valve.
9) Close the high-bleed needle valve when all the air is removed.
10) Close the outlet shut-off valve; close the inlet shut-off valve.
11) Place a finger or small object on top of the air inlet port to determine its opening point. Do not press down.
12) Open the air bleed valve slowly to atmospheric pressure.

> This reduces the pressure trapped between the check valve and the air inlet valve. Water contained within the assembly will flow out of the air bleed and out of the high-bleed line of the test gauge, because the pressure inside the assembly (line pressure) is higher than atmospheric pressure. If the high side bleed valve needs to be opened more than one quarter turn, the inlet shut-off valve may be leaking. If the inlet shut-off valve is leaking, the assembly can still be tested. See Testing a Spill-resistant Vacuum Breaker with a Leaking Inlet Shut-off Valve (page 160).

13) Open the high-bleed needle valve on the gauge no more than a quarter turn.
14) Record the pressure at which the air inlet valve opens. The air inlet should open while there is at least 1.0 psi still in the chamber.
15) Close the air bleed (vent), close high bleed needle valve on gauge.
16) Open inlet shut-off valve slowly.

FIELD TESTING

Testing a Spill-resistant Vacuum Breaker with a Leaking Inlet Shut-off Valve (Figure 6-21)

A leaking inlet shut-off valve is indicated during Test #1 (testing the air inlet opening point). If water continues to flow out of the air bleed valve and the needle does not drop on the gauge the inlet shut-off valve may be leaking. Install a bleed-off tee on the test cock. Attach the high-pressure hose to the bleed-off tee. Slowly open the bleed-off tee needle valve until there is a slight drop in pressure indicated on the gauge. Record the pressure at which the air inlet valve opens. With a slight drip coming from the air bleed screw, the needle on the gauge should stop at a value at 1.0 psi or greater. Record this value for the check valve.

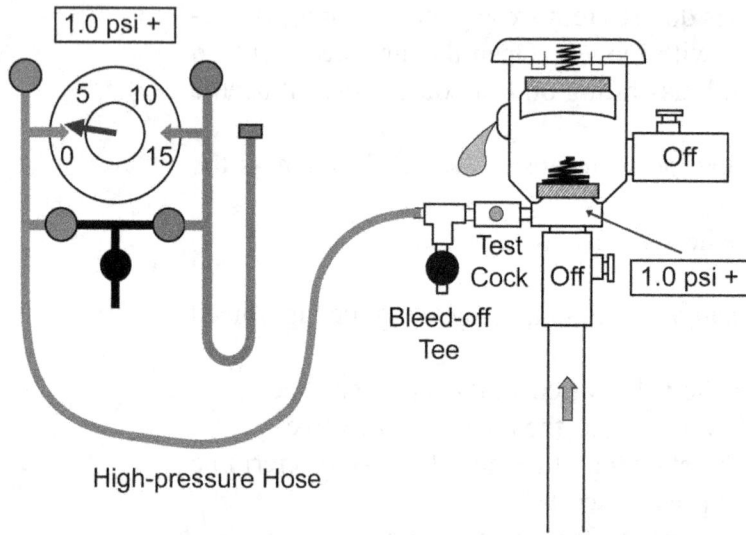

Figure 6-21 **Testing a SVB With a Leaking Inlet Shut-off Valve**
A bleed-off tee is attached to the test cock to test when the inlet shut-off valve leaks.

Testing Follow-up

Once testing is completed, the test form should be reviewed to ensure that all the essential information is provided. The owner of the backflow preventer must be provided with a copy of the test results, and a copy of the test results should be retained by the tester. Most water purveyors also require a copy of the test results. In addition, some water purveyors may require that the backflow preventer be tagged. An **approval tag** (Figure 6-22) provides a visual cue that the backflow preventer has been tested and found to be working properly. This allows the water purveyor's field personnel to easily determine whether water service should be provided to a new customer or restored to an existing customer. Some utility companies will require a different tag color each year.

If the mechanical backflow preventer fails the test, the owner must be notified that it requires repair and/or replacing. In addition, if the backflow

> An approval tag provides a visual cue that the backflow preventer has been tested and found to be working properly.

```
      ANY
    COMPANY
    UTILITY
PHONE: (000) 555-5555
    CERTIFIED
BACKFLOW ASSEMBLY
ASSEMBLY #_____

EXPIRES:
JAN FEB MAR APR MAY JUN
JUL AUG SEP OCT NOV DEC
```

Figure 6-22   Approval Tag

An approval tag provides a visual cue that the backflow preventer has been tested and found to be working properly.

preventer is not owned by the water purveyor, the water purveyor should also be informed that it is no longer operating properly. Some water purveyors also require that the backflow preventers be "red-tagged" when they fail field testing (Figure 6-23). The **red tag** provides the same basic function as the approval tag. It provides a visual indication of the internal conditions of the backflow preventer. The red tag also provides a psychological impetus to repair the backflow preventer.

The red tag provides the same basic function as the approval tag. It provides a visual indication of the internal conditions of the backflow preventer.

The red tag provides a visual indication of the internal conditions of the backflow preventer.

```
    FAILED
    ANY
    COMPANY
    UTILITY

PHONE: (000) 555-5555
DATE:_____
ASSEMBLY #_____
```

Figure 6-23   Red Tag

FIELD TESTING

The next step is for the customer to determine who will repair the backflow preventer if repairs are not handled by the water purveyor.

Summary

In order to prevent backflow of contaminants into the public water supply, it is essential that backflow preventers be properly tested and maintained. Manufacturers state that backflow prevention assemblies should be tested at least annually. The International Plumbing Code requires that assemblies be tested according to manufacturers' installation instructions. According to the Florida Building Code, "where the manufacturer of the assembly does not specify the frequency of testing, the assembly shall be tested at least annually."[1] (Appendix I) If tests are performed correctly, they give an accurate indication of the working condition of the backflow preventer. A review of present and past test results can also indicate needed repairs and maintenance. Moreover, routine maintenance is essential to ensure that the backflow preventer will continue to work properly between testing periods.

Chapter Six Review

6-1 What are three basic methods for testing a check valve?
differential pressure, direction of flow, backpressure

6-2 In Test #3, how do you test check valve #2 in an RP? (Based on answer to question 6-1)
back pressure

6-3 What type of test do you use on check valve #2 in a DCVA?
(Based on answer to question 6-1)
direction of flow

6-4 What type of test do you use on the check valve in a PVB?
(Based on answer to question 6-1)
direction of flow

6-5 What type of test do you use on check valve #1 in an RP?
differential pressure

6-6 Where is test cock #1 located on an RP? What is its purpose?
before SO inlet, make up for leaking SO outlet

6-7 Where is test cock #2 located on an RP?
between SO inlet and CV#1

6-8 What is maximum drop allowed across a DCVA or a DCDA?
10.0 psi

6-9 Which shut-off valve is closed when testing a DCVA?
both

6-10 Which gauge(s) is (are) used to test a DCVA or DCDA?
DP, duplex, sight tube

6-11 With backpressure on check valve #2, what does the relief valve do? If check valve #2 leaks?
nothing, drips if CV#2 leaks

(Please use answers below for questions 6-12 through 6-14.)

DUMP   SPIT   DRIP

6-12 Water pressure fluctuation will cause the relief valve of the RP to *spit*

FIELD TESTING

6-13 Water will steadily __drip__ out of the vent of the relief valve if check valve #1 leaks or check valve #2 leaks with backpressure from the customer.

6-14 Water will __dump__ from the vent of the relief valve of the RP if there is a loss of system pressure. (backsiphonage)

6-15 What is the minimum value for the PVB air inlet valve opening? The check valve?
__1.0 psi__

6-16 When testing the air inlet opening of the PVB, you connect which hose where?

6-17 What is the minimum value for check valve #1 of the DCVA? Check valve #2? __1.0 psi__

__high hose to TC #2__

6-18 When testing check valve #2 of the DCVA, you connect which hose where?
__high hose to TC #3__

6-19 What is the last step when testing the RP?
__Turn water back on__

6-20 When testing check valve #1 of the RP, you connect which hoses where?
__high hose → TC #2__
__low hose → TC #3__

CHAPTER SIX

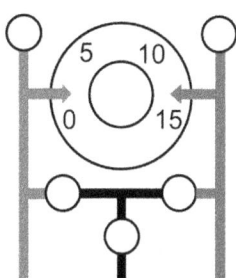

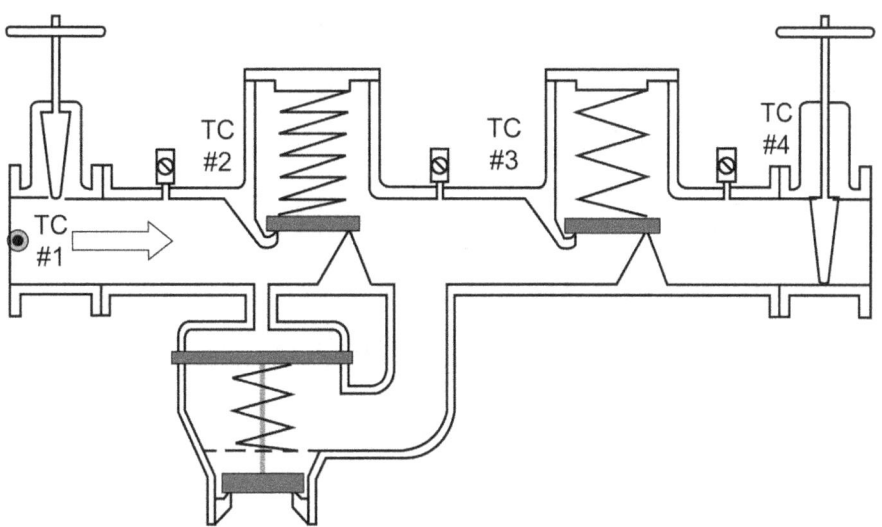

FIELD TESTING

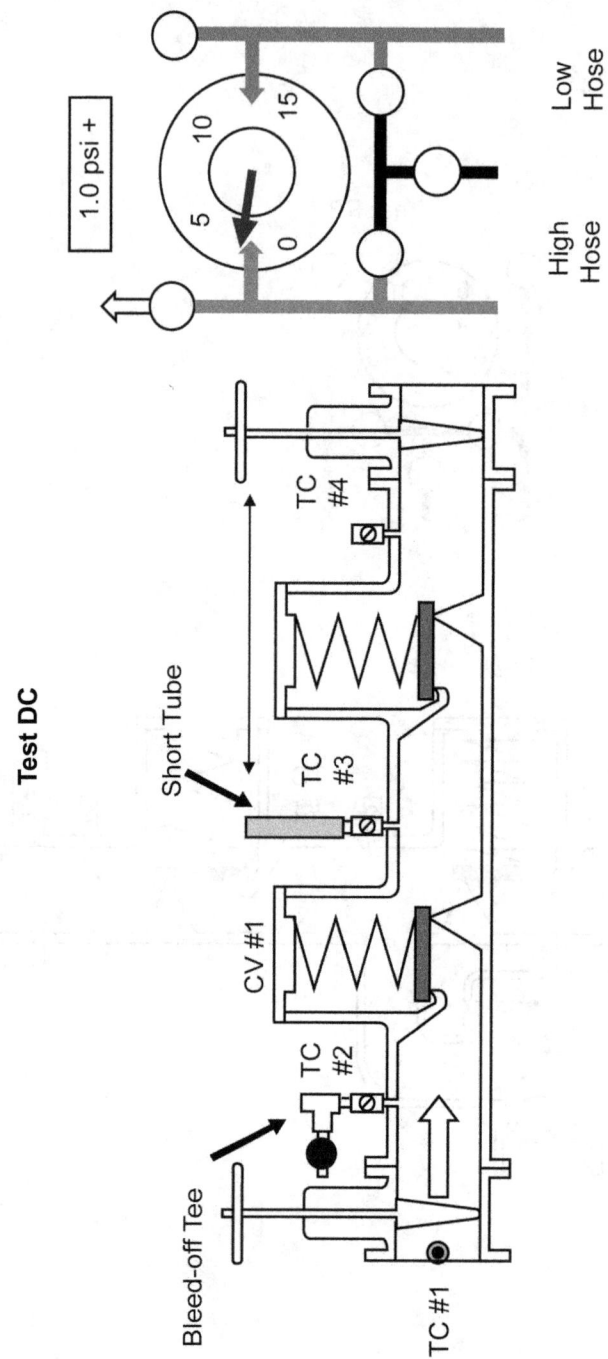

CHAPTER SIX

Test PVB

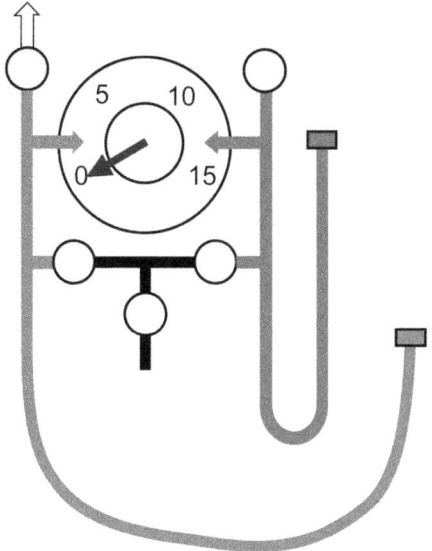
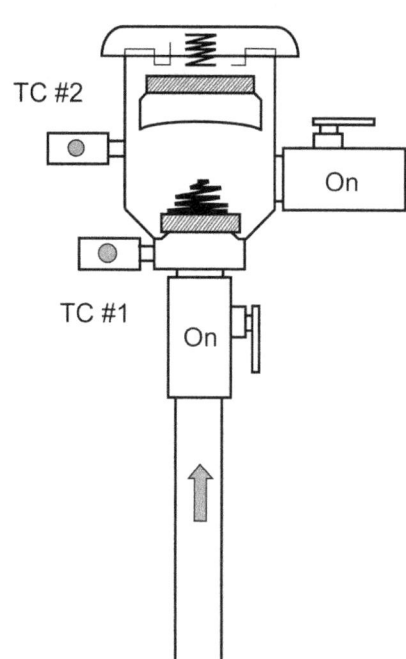

High-pressure Hose

Test SVB

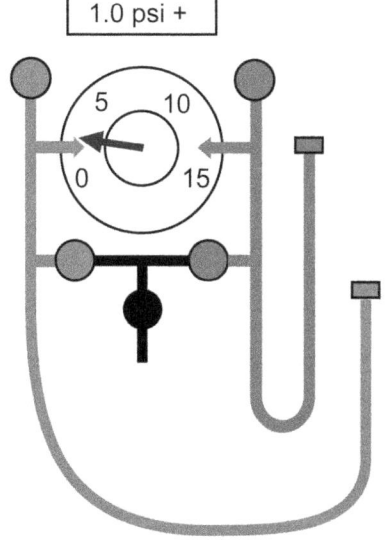
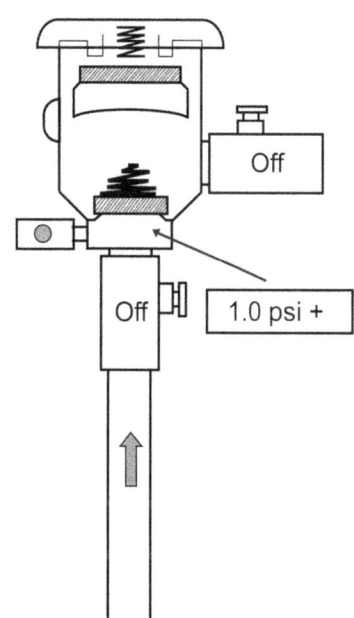

High-pressure Hose

FIELD TESTING

REFERENCES

1. *Florida Building Code - Plumbing*, effective March 2002. Florida Department of Community Affairs, Tallahassee, FL.

CHAPTER SEVEN
TROUBLESHOOTING, MAINTENANCE AND REPAIR

Troubleshooting Reduced Pressure Principle Assemblies

Keeping in mind how and why backflow preventers work to prevent backflow is important when troubleshooting. The first step when approaching a malfunctioning assembly is to note exactly what is happening. Check the troubleshooting guides in Tables 7-2 and 7-3 for potential causes of the problem. The information obtained from the test should give additional clues to the cause of the problem. Again, the troubleshooting guide can provide potential causes. Often, one symptom can have many different causes. All of this information is then used to determine what repairs need to be made. While experience is often the best teacher, the troubleshooting guide will serve as a valuable tool until that experience can be gained. This guide does not cover every situation that might occur, but it will significantly reduce the time spent troubleshooting problems. For each problem, the guide lists the most likely cause first, then other possible causes.

Failing First Check Valve: If the first check valve fails, the zone pressure could theoretically increase until it equals the supply line pressure, attempting to create equilibrium across the check valve. If the relief valve is working properly, this equilibrium is never reached, since the combined zone pressure and the relief valve spring pressure force the relief valve open, dripping water from the zone before pressure equilibrium is reached. The example below (Figure 7-1) assumes the same operating conditions before the leak occurs.

Under normal flow conditions, the pressure inside the zone is 95 psi and the pressure acting to open the relief valve is 97 psi. If the first check fails, the pressure inside the zone will increase toward 100 psi. When the pressure inside the zone exceeds 98 psi, the total pressure acting to open the diaphragm is more than 100 psi. This is sufficient pressure to open the relief valve and continuously flow water from the zone (Figure 7-2).

Testing the Reduced Pressure Principle Assembly with a Leaking Outlet Shut-off Valve (Figure 7-3)

If the relief valve will not open and drip water from the zone or if the needle on the gauge starts to drop but stops before the opening point when testing the RP's relief valve opening, then an outlet shut-off valve may be the problem. If there is no flow to the customer, the leak will not be

TROUBLESHOOTING, MAINTENANCE AND REPAIR

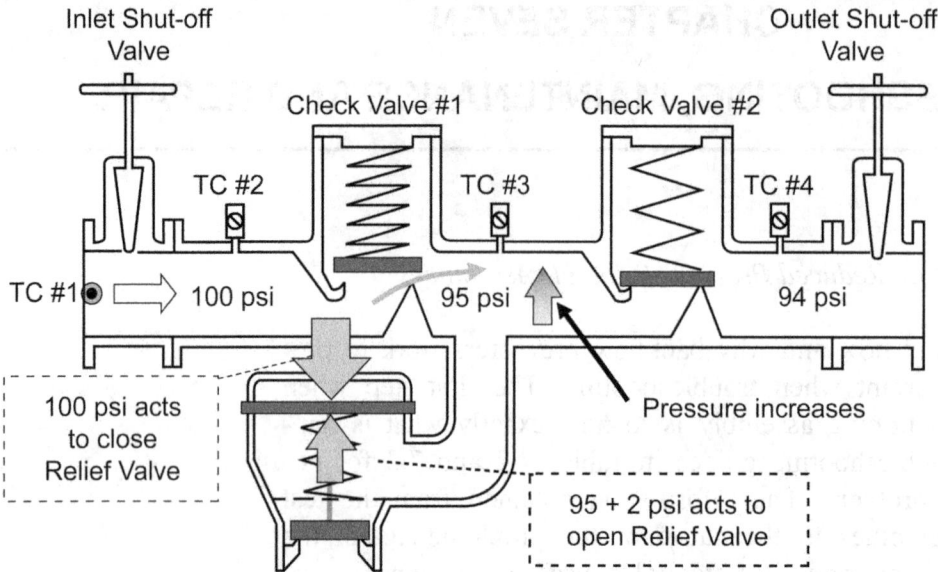

Figure 7-1   **Failing Check Valve #1**
Under normal flow conditions, the pressure inside the zone is 95 psi and the pressure acting to open the relief valve is 97 psi. If the first check fails, the pressure inside the zone will increase toward 100 psi.

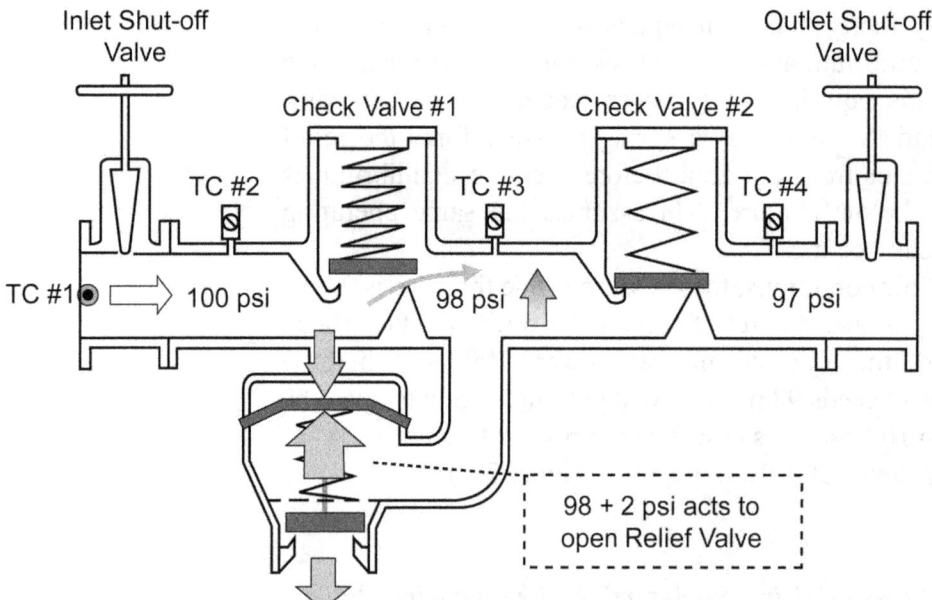

Figure 7-2   **Failing Check Valve #1**
When the pressure inside the zone reaches 98 plus psi, the total pressure acting to open the diaphragm is 100 plus psi. This is sufficient pressure to open the relief valve and continuously flow water from the zone.

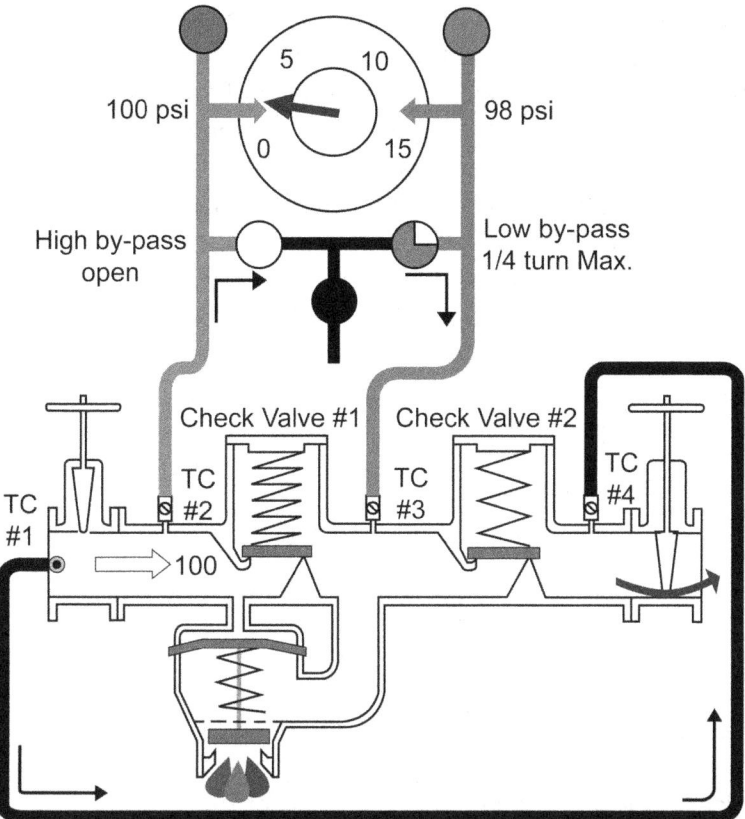

Figure 7-3 Testing an RP with a Leaking Outlet Shut-off Valve
If the needle on the gauge starts to drop but stops before the relief valve opens, the outlet shut-off valve may be leaking with flow through the assembly. Connect an extra by-pass hose from test cock #1 to test cock #4.

noticed and the relief valve will test as normal. If water is flowing through a leaking outlet shut-off valve, the relief valve opening point may often be determined without replacing the outlet shut-off valve. To determine the relief valve opening point, an extra by-pass hose is attached to test cock #1 and connected to test cock #4 then open test cock #1, open test cock #4 slowly with your hand under the RV vent. This allows water pressure from the supply side of check valve #1 to be routed to the backside of check valve #2 through the by-pass hose and compensates for the leak. If the volume of the leak exceeds the capacity of the by-pass hose, the outlet shut-off valve may need replacing. The low by-pass control valve must not be opened more than ¼ turn (Figure 7-1).

Failing Second Check Valve: Recall that if the second check valve fails under normal operating conditions, the assembly continues to function and no visible indication of failure is given. Testing, however, would detect that the check is not working properly.

The low by-pass control valve must not be opened more than ¼ turn.

TROUBLESHOOTING, MAINTENANCE AND REPAIR

In a backpressure-backflow condition, however, failure of the second check valve causes the relief valve to open and drain the contaminant from the zone. The port opens because the pressure in the zone increases until the combined zone pressure and relief valve spring pressure exceeds the supply line pressure. The relief valve will continue to open and drain water from the zone as long as the backpressure (along with the relief valve spring) is greater than the pressure on the supply side of the diaphragm. The example depicted in Figure 7-4 assumes a check valve #1 spring pressure of 5.0 psi, a relief valve spring pressure of 2.0 psi, and a supply pressure of 100 psi; therefore, normally the pressure working to open the relief valve is 97 psi. The relief valve opens even though the backpressure, created by the pump, is less than the supply line pressure. It will remain open as long as the backpressure is greater than 98 psi.

Table 7-1 Testing an RP with a Leaking Outlet Shut-off Valve

RP	NORMAL	CV1 LEAKS	CV2 LEAKS	OUTLET SHUT-OFF VALVE LEAKS	OUTLET SHUT-OFF AND CV2 LEAKS
Test 1 CV1	Observe CV1, leaks or tight.	STOP, repair, then re-test.	Observe CV1, leaks or tight.	Observe CV1 leaks or closed tight.	Observe CV1, leaks or tight.
Test 2 RV	Record RV opening. Must be > than or = to 2.0 psi.		Record RV opening. Reset gauge.	RV will not open – add by-pass hose from TC1 to TC4.	RV will not open – STOP RV test. Add by-pass hose from TC1 to TC4.
Test 3 CV2	Backpressure test on CV2, leaks or tight. RV does not drip.		Backpressure test. RV drips. Record CV2 leaks.	Backpressure test on CV2. Observe gauge.	RV may drip.
Test 4 CV1	Reset and record CV1 > RV + 3.0 psi or 5.0 psi min.		Close vent control, reset and record CV1.	Reset and record CV1.	Stop testing. Repair CV2.
Test 5 Outlet Valve	Close TC2. Observe gauge.		No action! Must repair and retest.	Record that outlet shut-off valve leaks.	Stop testing. Repair CV2.
Test 6 CV2	Move hoses and record CV2 > than or = to 1.0 psi.		No action!	You cannot perform differential test on CV2.	You cannot perform differential test on CV2.

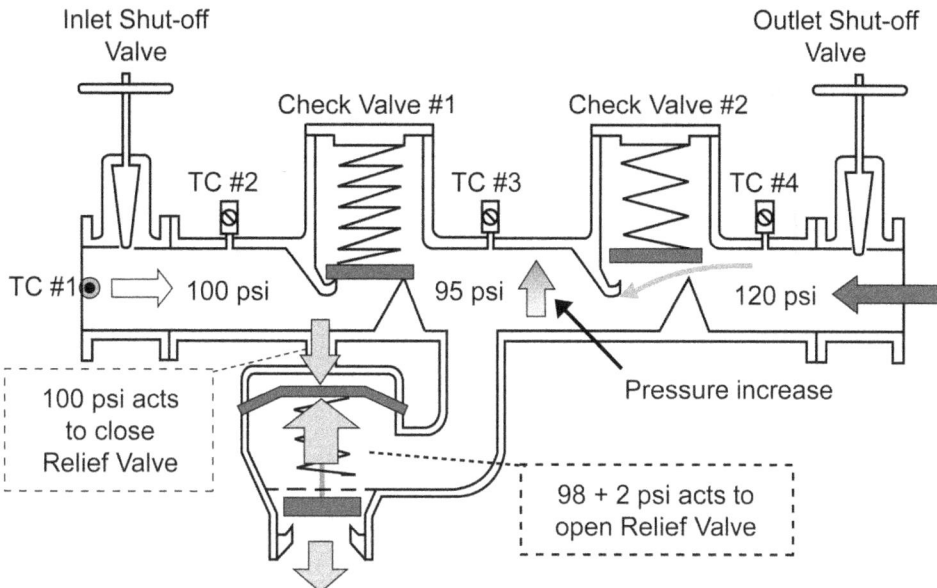

Figure 7-4 RP: Failing Check Valve #2 during Backpressure Conditions
If a pump connected to the consumer's potable water line creates a pressure of 120 psi and check valve #2 is leaking, the pressure acting to open the relief valve is 100 plus psi (98 plus psi + 2.0 psi), while the pressure acting to keep it closed is only 100 psi. Therefore, the relief valve will open.

Recall that under backsiphonage conditions when the water is static, the relief valve will open and dump the water contained in the zone. If the second check valve is fouled under backsiphonage conditions, water will continually flow from the relief port if atmospheric pressure or some other source of backpressure (for instance, the weight of water in the consumer's system) causes a reverse movement of water. Under backsiphonage conditions, where the pressure at the backflow preventer has fallen below atmospheric pressure, any pressure on the customer's side of the RP greater than 14.7 creates backpressure. Since water always flows in the direction of lowest pressure, water will flow from the RP's relief port. The volume of the flow from the relief port (drip or gusher) depends on the extent to which check valve #2 is fouled and also on the amount of backpressure. Water will continue to flow from the vent of the RP as long there is water in the line. If, however, the customer uses all the water contained in the line downstream of the assembly, no water is available to drip from the relief port.

This concept is shown in Figure 7-5. The relief valve will open and dump water from the relief port as soon as the potable water supply pressure falls below the pressure applied by combined pressure of the relief valve spring and the backpressure from the building. The pressure at point B, resulting from the weight of water, exceeds the pressure acting at point A. Therefore, the water within the apartment building will drain from the apartment plumbing system out through the relief port.

TROUBLESHOOTING, MAINTENANCE AND REPAIR

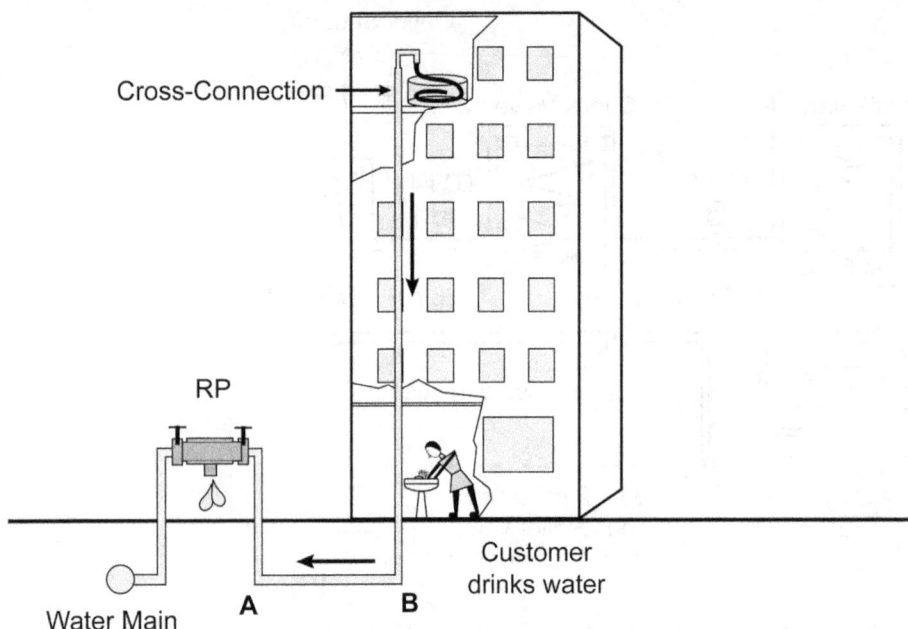

Figure 7-5 RP: Failing Check Valve #2 under Backpressure and Backsiphonage Conditions
The relief valve will open and dump water from the relief port as soon as the potable water supply pressure falls below the pressure applied by combined pressure of the relief valve spring and the backpressure from the building. The pressure at point B, resulting from the weight of water, exceeds the pressure acting at point A. Therefore, the water within the apartment building will drain from the apartment plumbing system out through the relief port. Water will drip, run steady, or run intermittently.

Water will drip, run steady, or run intermittently depending on the water usage inside the building. If the second check valve was fouled and the pressure on the consumer's side was greater than the supply pressure, then the relief valve would have opened and drained even before the backsiphonage condition occurred. Figure 7-6 shows how this condition could develop. If a pump connected to the consumer's potable water line creates a pressure of 99 psi and the second check valve is leaking, the pressure acting to open the relief valve is 101 psi (99 psi + 2.0 psi), while the pressure acting to keep it closed is only 100 psi. Therefore, the relief valve will open. The assumption is made in this example that the backpressure was less than normal supply pressure. Therefore, even though the check valve was fouled, the hydraulic gradient was in the normal direction of flow, and no indication of a problem prior to the backsiphonage condition was visible.

Clogged Sensing Line: If the high-pressure sensing line becomes clogged with supply pressure trapped on the high-pressure side of the diaphragm, the relief valve might stay closed even during backpressure or backsiphonage conditions. For instance, if the first check valve becomes fouled and the pressure inside the zone increases, the relief valve remains

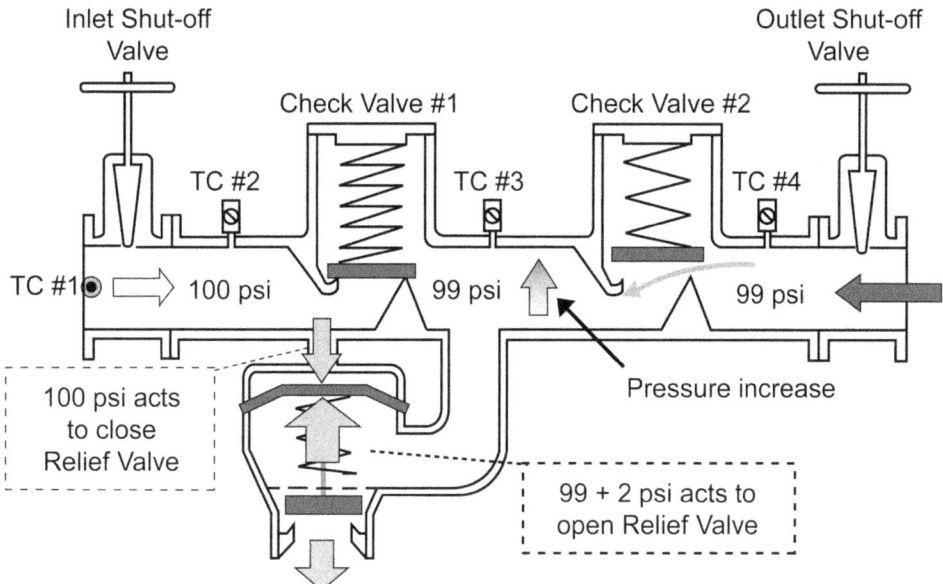

Figure 7-6   Backpressure with a Leaking Check Valve #2
If the second check valve was fouled and the pressure on the consumer's side was greater than the supply pressure, the relief valve would have opened and drained even before the backsiphonage condition occurred.

closed because of the high pressure trapped in the sensing line. The relief valve will not open even though the pressure acting to open the relief valve is greater than the pressure in the sensing line. The diaphragm controlling the relief valve cannot move because water contained inside the sensing line cannot be compressed.

The relief valve would also stay closed if a failed second check valve allowed backpressure into the zone (Figure 7-7). If the second check valve leaked during backpressure conditions and the pressure in the zone increased to 120 psi, the relief valve will not open even though the pressure working to open the relief valve (122 psi) is greater than the pressure supplied by the sensing line (100 psi), because water inside the sensing line cannot be compressed. The diaphragm cannot move to allow the relief valve to open.

On the other hand, if the high-pressure sensing line was clogged with no pressure applied to the diaphragm from the supply side, the relief valve will fully open and discharge water until the sensing line is cleared and pressure is once again provided to the supply side of the diaphragm. This situation commonly occurs when the assembly is taken apart for cleaning but the sensing line is not cleaned adequately. The example below (Figure 7-8) assumes a static condition, a first check spring pressure of 5.0 psi, a relief valve spring pressure of 2.0 psi, and a supply pressure of 100 psi. The pressure inside the zone is 95 psi, and the pressure acting to open the relief valve is 97 psi. If the sensing line is clogged and the pressure acting to

175

TROUBLESHOOTING, MAINTENANCE AND REPAIR

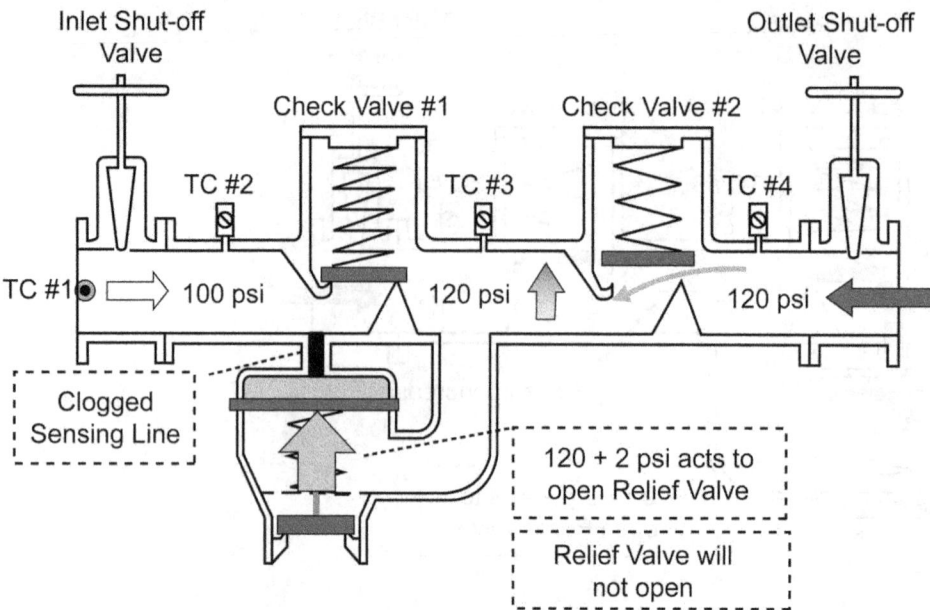

Figure 7-7 RP: Clogged Sensing Line under Backpressure Conditions
Assume a static condition, a check valve #1 spring pressure of 5.0 psi, a relief valve spring pressure of 2.0 psi, a supply pressure of 100 psi, and a pressure of 100 psi trapped in the sensing line. If the second check valve leaked during back pressure conditions and the pressure in the zone increased to 120 psi, the relief valve will not open even though the pressure working to open the relief valve (122 psi) is greater than the pressure supplied by the sensing line (100 psi), because water inside the sensing line cannot be compressed. The diaphragm cannot move to allow the relief valve to open.

keep the relief valve closed is 0 psi, the relief valve will open and continue to gush water from the zone until the sensing line is cleared allowing water pressure to the upstream side of the diaphragm.

Failing Relief Valve: Finally, the relief valve itself can become stuck open or closed due to mechanical wear or debris. If the relief valve is stuck closed, the RP is essentially functioning as a DCVA. Therefore, if a backpressure situation occurs, the RP will still provide protection as long as one of the check valves holds tight. If the relief valve is stuck open, obviously the customer will not receive any water until the problem has been corrected. Relief valve failure may be totally independent of other problems with the assembly and can make troubleshooting RP problems very difficult.

Example Problem 1: Customer complaint that a RP is dumping water on the ground.

1) Arrive at the site and observe that the water runs intermittently—a steady dribble for several minutes, then nothing, then a steady dribble, then nothing.

176

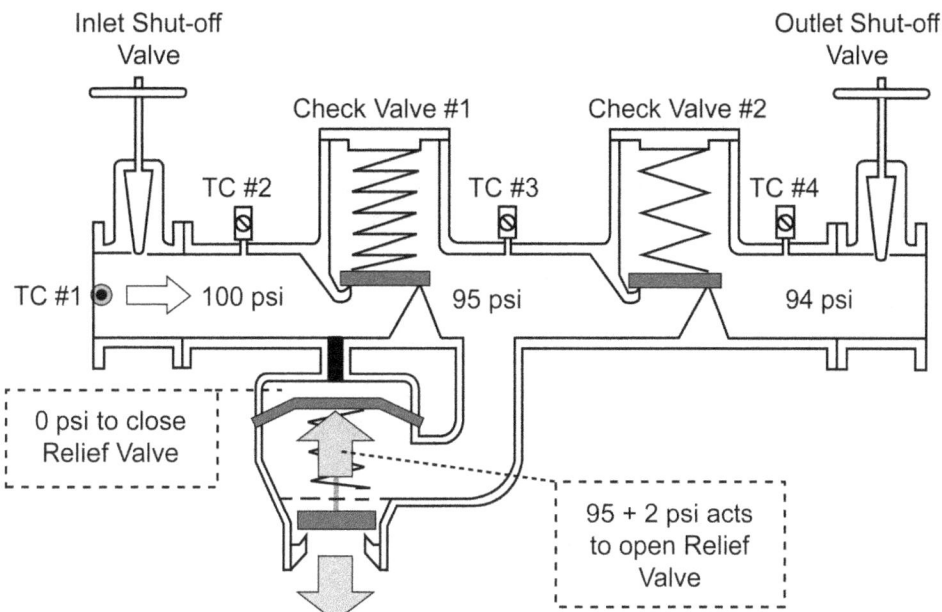

Figure 7-8 Clogged Sensing Line with 0 psi in the Sensing Line
The pressure inside the zone is 95 psi, and the pressure acting to open the relief valve is 97 psi. If the sensing line is clogged and the pressure acting to keep the relief valve closed is 0 psi, the relief valve will open and continue to discharge water from the zone until the sensing line is cleared allowing water pressure to the upstream side of the diaphragm.

2) Under field conditions, the most likely cause for periodic discharge of water from the relief valve is a fouled check valve #2 coupled with backpressure. The second most likely cause is fluctuation of inlet pressure.

3) Start to test the RP. The last step in Test #1 is to turn off the outlet shut-off valve. The water runs continuously from the relief port. Re-evaluate the problem; the problem must not be from backpressure coupled with a fouled check valve #2. Closing the outlet shut-off valve would eliminate backpressure (assuming the outlet shut-off valve is closing properly). The second alternative, fluctuation in line pressure, also does not seem likely since it would continue to produce the same symptoms seen prior to turning off the outlet shut-off valve. The most likely cause now would appear to be a leaking check valve #1 as indicated in the troubleshooting guide (Table 7-2).

4) Test the assembly. It will be impossible to test this assembly until check valve #1 has been repaired.

5) Determine possible causes. The most likely cause for this is debris or particles between the check and seat.

Why did water only *periodically* flow from the relief port? If check valve #1 is fouled, shouldn't water continuously flow from the relief port?

177

TROUBLESHOOTING, MAINTENANCE AND REPAIR

Table 7-2   Troubleshooting the RP

When you arrive	Action	Observation	Diagnostics	Action
Relief valve is not dripping.	Test #1-close outlet shut-off valve.	RV does not drip.	CV1 is tight.	Continue testing RP.
Relief valve is not dripping.	Test #1-close outlet shut-off valve.	RV begins to drip.	CV1 is leaking.	Repair CV1.
Relief valve is dripping.	Test #1-close outlet shut-off valve.	RV stops dripping.	CV2 leaks with backpressure.	Repair CV2.
Relief valve is dripping.	Test #1-close outlet shut-off valve.	RV continues to drip.	CV1 or RV leaks.	Open TC4. *

Opening test cock #4 simulates the customer using water. If the drip stops, then CV 1 is fouled. If the drip continues, most likely the RV needs to be repaired.

Table 7-3   Troubleshooting the DCVA

CONDITION	POSSIBLE PROBLEM	ACTION	RESULTS
CV1 fails test or differential pressure gauge reads less than 1.0 psi.	1. Debris trapped between disk and seat. 2. CV1 disk is damaged. 3. CV1 seat or seat o-ring is damaged.	Repair CV1	Retest DCVA
CV2 fails test or differential pressure gauge reads less than 1.0 psi.	1. Debris trapped between disk and seat. 2. CV2 disk is damaged. 3. CV2 seat or seat o-ring is damaged.	Repair CV2	Retest DCVA

Maintenance and Repair

Maintenance and repair of backflow prevention assemblies are absolutely essential to help ensure that the assembly functions properly under backflow conditions. In addition, routine maintenance and repair help to protect the original investment. A routine maintenance and repair program not only saves the consumer money but also saves time.

Maintenance and repair are necessary because of the problems created by water flowing through the assembly. Depositing water and tuberculation build-ups can foul check valves or clog-sensing lines in RPs.

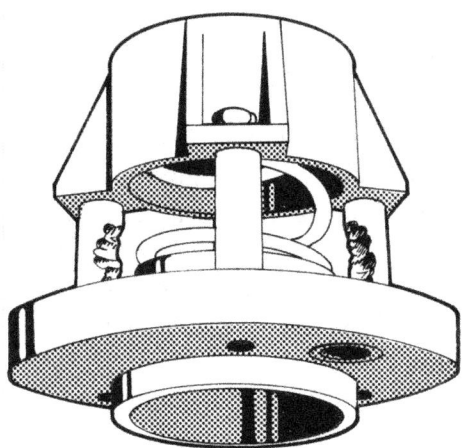

Figure 7-9 Wear on the Plastic Check Valve Guide
The plastic guides that hold the check valve in place can become worn simply by the movement of water through the check valve.

Corrosive water can disintegrate metal parts. The use of dissimilar metal pipes and the use of dissimilar metals in backflow preventers can result in the disintegration of metal parts (galvanic corrosion). Even if the water is chemically benign, the simple movement of water through the assembly causes wear. The plastic valve guides shown in Figure 7-9 have been worn by valve abrasion and the movement of water through the check valve. In addition, high temperatures and pressures also have an effect on the internal parts of backflow preventers. Finally, foreign materials such as sand grains, hairpins, and pebbles occasionally enter the potable water lines and foul check valves. Even though testing can detect many problems with the internal workings of the backflow preventer, it cannot detect everything. For example, the excessive wear on the check valve guides shown in Figure 7-9 would probably not be detected during testing. Reviewing the test results of a particular backflow preventer over a number of years can also provide an indication as to whether the assembly may require disassembly and cleaning. Any time the test results of a backflow preventer vary significantly from past results, this indicates a need for cleaning and inspection. Other indications of the need for maintenance and repair are supplied by evidence of dumping or dripping around RPs or PVBs. Every assembly should be disassembled, cleaned and inspected every 3 to 5 years to detect these special types of problems. Any maintenance should be recorded on the testing form. For example, cleaning and lubricating should be noted under the "comments" section of the form, including the type of lubricant used. Also, signs of wear should be recorded on the test form.

Some manufacturers recommend that certain parts be lubricated periodically. If parts are lubricated, the lubricant should be approved by the manufacturer, as some lubricants react with plastic and cause early aging. Lubricants must also be food grade quality as approved by the Food and

TROUBLESHOOTING, MAINTENANCE AND REPAIR

Drug Administration, since they are in contact with the potable water supply. Normally only o-rings are lubricated to help hold the o-ring in the proper position.

Routine maintenance should also include simple safety precautions such as plugging the test cocks of DCVAs that are installed in pits. This is done to prevent contaminants from entering the potable water supply through the test cocks in the event the pit floods. While pit and vault installations should be avoided whenever possible and the specifications for these installations should prevent flooding, plugging the test cocks provides an added margin of safety.

After routine maintenance is performed, the backflow preventers should be retested to ensure that they function properly.

Safety

Whenever backflow prevention assemblies are tested or repaired, adequate safety precautions must be taken to prevent accidents. Prior to testing or repairing assemblies, the water lines should be checked to determine if they are being used as an electrical ground. While the practice of using water lines as an electrical ground is highly discouraged (because it contributes to early deterioration of the pipes), a check is necessary prior to beginning work. If the pipe is being used as a ground, appropriate steps should be taken to remove this grounding wire and find a replacement grounding source.

If an assembly is installed in a deep pit or vault, the atmosphere should be checked before entering to ensure the pit does not contain any toxic gases. Conversely, when backflow preventers are installed near ceilings, above drop ceilings, or in other hard-to-reach locations, care should be used to prevent falls. (Consult Occupation Safety and Health Administration regulations for worker safety.)

While the testing procedure is the same for every assembly of a given type no matter what its size, extra care should be used when performing maintenance or repair work on large assemblies. Large assemblies are, generally, considered to be those over 2-inches in diameter. Obviously, the best course of action is to receive training prior to repairing these large assemblies so that accidents can be avoided. Always read the maintenance instructions for the specific backflow prevention assembly.

Repair

While this chapter provides some general information on repair, it is not a complete repair manual. Manufacturer's literature should be reviewed thoroughly before making any repairs. In order to make repairs quickly and correctly, it is important to be prepared. This chapter covers some materials and tools that will be useful for making repairs, as well as some precautions that are necessary when making repairs.

Any time a repair is made, the safety precautions discussed earlier should be followed. In addition, manufacturer's recommendations should be heeded. For instance, many assemblies must be taken apart in a specific manner for safety reasons, and some manufacturers require that certain parts be lubricated periodically, while other parts should not be lubricated. When making repairs, replacement parts should meet the same specifications that parts in the original assembly were required to meet. In general, swapping parts should be avoided. The springs in a DCVA may be interchangeable, but those in the RP are not. The disks of a check valve should only be flipped as an emergency measure until replacement parts are received.

It is recommended that the repairer wear some form of protective clothing, since tuberculation within the assembly causes staining on clothing and skin. Goggles or safety glasses also should be worn.

At a minimum, the repair kit should include the following items:

1) a white drop cloth for spreading beneath the assembly to catch stray parts.
2) a spring steel wire (or something similar), used for removing debris from test cocks or unclogging them.
3) a length of plastic tubing and a plastic bucket, for flushing test cocks and catching discharge water in areas that should not get wet.
4) a mirror, used to read serial numbers of assemblies when they are in hard-to-read places.
5) a flashlight, useful when assemblies are installed in dimly lit areas (e.g., most utility rooms).
6) a plastic tackle or tool box, for storing tools and equipment used for making repairs.

In addition, many assemblies require specialized tools for disassembly. Often the manufacturers' literature contains instructions on how these tools can be made or where they can be purchased. If possible, it is best to learn repair from an experienced field technician or a good repair course.

Use only factory authorized replacement parts.

Read the manufacturer's repair instructions before taking the assembly apart.

Items to Check When the Backflow Preventer Fails

Reduced Pressure Principle Assembly and
Reduced Pressure Detector Assembly

Check Valve #1

Check valve disk – Look for damage or cuts on either side of the disk.

Check valve guides – Remove any deposited material on the guides and guiding surfaces.

Check valve seat – Check for dings or nicks on the seating edge. Also check for wear and signs of corrosion.

TROUBLESHOOTING, MAINTENANCE AND REPAIR

Check seat o-ring – If the disk or seat are not damaged, then check the seat o-ring last.

Check valve spring – Inspect the spring to be sure it is not bent or misshaped. Only replace spring if it is bent or broken. (Check the manufacturer's instructions.)

Check Valve #2

Check valve disk – Look for damage or cuts on either side of the disk.

Check valve guides – Remove any deposited material on the guides and guiding surfaces.

Check valve seat – Check for dings or nicks on the seating edge. Also check for wear and signs of corrosion.

Check seat o-ring – If the disk or seat are not damaged, then check the seat o-ring last.

Check valve spring – Inspect the spring to be sure it is not bent or misshaped. Only replace spring if it is bent or broken. (Check the manufacturer's instructions.)

Relief Valve Assembly

Relief valve disk – Look for damage or cuts on either side of the disk.

Relief valve seat – Check for dings or nicks on the seating edge. Also check for wear and signs of corrosion.

Relief seat o-ring – If the disk or seat are not damaged, then check the seat o-ring last.

Relief valve diaphragm(s) – Check for ruptures in the diaphragm. Also check that sharp metal edges are not in contact with the diaphragm. Rolling diaphragms can become pinched or restricted in their ability to roll.

Stem o-rings – Check o-rings for damage.

Air vents – Some relief valve assemblies have air vents to prevent a vacuum lock. Make sure any air vents are open.

Sensing lines, tubes, or channels – Check high and low pressure sensing lines to make sure they are clear and not clogged or plugged.

Relief valve spring – Replace spring if it is bent or broken. (Check manufacturer's instructions)

Double Check Valve Assembly and
Double Check Detector Assembly

Check Valve #1

Check valve disk – Look for damage or cuts on either side of the disk.

Check valve guides – Remove any deposited material on the guides and guiding surfaces.

Check valve seat – Check for dings or nicks on the seating edge. Also check for wear and signs of corrosion.

Check seat o-ring – If the disk or seat are not damaged, then check the seat o-ring last.

Check valve spring – Inspect the spring to be sure it is not bent or misshaped. Only replace spring if it is bent or broken. (Check the manufacturer's instructions.)

Check Valve #2

Check valve disk – Look for damage or cuts on either side of the disk.

Check valve guides – Remove any deposited material on the guides and guiding surfaces.

Check valve seat – Check for dings or nicks on the seating edge. Also check for wear and signs of corrosion.

Check seat o-ring – If the disk or seat are not damaged, then check the seat o-ring last.

Check valve spring – Inspect the spring to be sure it is not bent or misshaped. Only replace spring if it is bent or broken. (Check the manufacturer's instructions.)

Pressure Vacuum Breaker Assembly and
Spill-Resistant Vacuum Breaker Assembly

Air Inlet Valve

Air inlet disk – Look for damage or cuts on either side of the disk.

Air inlet spring (if present) - Replace spring if it is bent or broken.

Air inlet seat – Check for dings or nicks on the seating edge. Check for signs of the disk bonding to the seat. Also check for wear and signs of corrosion.

Air inlet o-ring – Check for a damaged or broken o-ring.

Check Valve

Check valve disk – Look for damage or cuts on either side of the disk.

Check valve guides – Remove any deposited material on the guides and guiding surfaces.

Check valve seat – Check for dings or nicks on the seating edge. Also check for wear and signs of electrolysis.

Check seat o-ring – If the disk or seat are not damaged, then check the seat o-ring last.

Check valve spring – Inspect the spring to be sure it is not bent or misshaped. Only replace spring if it is bent or broken. (Check the manufacturer's instructions.)

Atmospheric Vacuum Breaker Device

Check Valve

Air inlet disk – Look for damage or cuts on the disk.

TROUBLESHOOTING, MAINTENANCE AND REPAIR

Air inlet seat – Check for dings or nicks on the seating edge. Check for signs of the disk bonding to the seat.

Residential Dual Check Device

Check Valve #1
Check valve disk – Look for damage or cuts on either side of the disk.
Check valve seat – Check for dings or nicks on the seating edge.
Check seat o-ring – If the disk or seat are not damaged, then check the seat o-ring last.
Check valve spring – Replace check assembly if spring is bent or broken.

Check Valve #2
Check valve disk – Look for damage or cuts on either side of the disk.
Check valve seat – Check for dings or nicks on the seating edge.
Check seat o-ring – If the disk or seat are not damaged, then check the seat o-ring last.
Check valve spring – Replace check assembly if spring is bent or broken.

General Items to Check When the Backflow Preventer Fails

- Read the factory maintenance instructions before opening the assembly.
- Use proper safety equipment when disassembling or re-assembling the backflow preventer.
- Use only factory authorized replacement parts. You cannot make or use any parts other than those produced by the manufacturer for that particular assembly. There are several companies that specialize in factory parts. These companies can get you the necessary parts next day delivery. Check around for availability and prices.
- Do not replace the springs unless they are bent or broken. The stainless steel springs should not weaken over time.
- Do not use abrasive materials when cleaning the seats. The seats must maintain their edge in order to work properly.
- Over time, plastic and rubber parts will react to conditions (temperature and/or chemicals). The most common reaction is swelling. Replace when necessary.
- Check for a buildup of depositing material on moving parts.
- It is usually less expensive to repair the assembly than replace it.
- Replace non-testable devices with testable assemblies whenever possible.

> Note #1: Check seat o-rings: "There are a few models where the seat uses a gasket seal (not an o-ring). More common problems causing a failure are stem o-rings on the larger assemblies, bent or cracked disk holders on the smaller assemblies, and improper placement and lubrication of o-rings on the small in-line check designs. On some larger models the removal of the seat has led to damage of the epoxy coating around the seat area. This may cause a larger problem than the one you are trying to fix."[1]
>
> Note #2: This is not an all-inclusive list and proper field test procedures can help diagnose the location of the failure.

Corrosion

Electrolytic corrosion is an electrical phenomenon: electrolysis. Aside from unusual types of corrosion such as bacterial or direct chemical attack, the corrosion process, as normally encountered in pipeline work, is basically electrochemical in nature.

The pipe is essentially a length of metal embedded in an electrolyte. For many different reasons, electrical potentials may vary from one spot on a pipe to another with the result that anodic areas and cathodic areas will exist. These different electrical potential areas are the basis for an electrolytic "corrosion cell."

There are certain conditions that must be present before an electrolytic corrosion cell can function:

1) There must be an anode and a cathode.
2) There must be an electrical potential between the anode and cathode. This potential can result from a variety of conditions on pipelines.
3) There must be a metallic path electrically connecting the anode and cathode.
4) The anode and cathode must be immersed in an electrically conductive electrolyte. The usual soil moisture or water surrounding pipelines normally fulfills this condition.

Once these conditions are met, a "corrosion cell" is created and an electric current will flow and metal will be consumed at the anode. Remove any one of the above four items and the corrosion is stopped.[2]

Any time we have two different metals that are physically or electrically connected and immersed in water, they become a "battery." A small amount of current will flow between the two metals. One of the metals gives up bits of itself in the form of metal ions to the current in the water. The more active metal is the anode, and the less active metal is the cathode, and is protected. This is called galvanic corrosion, and over time it destroys dissimilar metals in water. Galvanic corrosion is often misnamed "electrolysis."[3]

TROUBLESHOOTING, MAINTENANCE AND REPAIR

Water Quality

The quality of the water in the distribution system may have various negative effects on the backflow prevention assemblies. In each community the water chemistry may be slightly different. There will be variations in the pH and the chlorine residual. The disinfectant in the water system may be chlorine, chloramine, iodine, or some other product. Water with a high pH (greater than 7.3) may be depositing material on the interior of the pipes. Calcium carbonate is one example of material that is deposited on the interior of pipes. Backflow prevention assemblies exposed to depositing water may need to be cleaned more often. On the other hand, water with a low pH may tend to be more aggressive as a pH below 7.0 tends to be acidic.

Chloramine is the by-product of mixing ammonia with chlorine. Water systems that use chlorine as a disinfectant and also add ammonia in the treatment process will rely on the chloramine to control bacteria. One drawback to using chloramine as a disinfectant is that the black rubber parts (Buna-N), which are in contact with the chloramine will tend to disintegrate rapidly.

Remember that all water is considered corrosive. It is simply a matter of the degree of corrosivity.

Summary

In order to prevent backflow it is essential that backflow preventers be properly tested and maintained. If tests are performed correctly, they give an accurate indication of the working condition of the backflow preventer. A review of present and past test results can also indicate needed repairs and maintenance. Moreover, routine maintenance is essential to ensure that the backflow preventer will continue to work properly between testing periods.

Recommendations on routine maintenance that should be performed and a list of the tools needed to make basic repairs were discussed. In addition, safety issues and some hints for troubleshooting malfunctioning assemblies were provided.

CHAPTER SEVEN

Chapter Seven Review

7-1 How do you know whether the outlet valve leaks on an RP?

meter runs, hear it, RV won't open

7-2 What causes higher-than-normal readings on check valve #1 and check valve #2 of the RP?

water flowing through the assembly

7-3 You closed the outlet shut-off valve and the relief valve starts to drip. What's wrong?

Probably CV #1 leaks

7-4 If the sensing tube is clogged, how do you know? What does the relief valve do?

No water with zero on gauge

or heavy flow from vent

REFERENCES

1. Purzycki, J. Letter received April 5, 2004. BAVCO, Long Beach, CA

2. *ELECTROLYSIS*, retrieved April 2004 from http://www.multimedia-sa.gr/olympic-sun/electrolysisen.html, Olympic Sun, Souda, Greece.

3. "Why Do We Need Sacrificial Anodes in Our Pumps Today?", *Talking Points: Cathodic Protection*, retrieved April 2004 from http://www.waterousco.com/service/010723.htm, Waterous Company, South St. Paul, MN.

CHAPTER EIGHT
DEVELOPING A CROSS-CONNECTION CONTROL PROGRAM

In most states, the water purveyor has primary responsibility for the development of a backflow prevention or cross-connection control program. A backflow prevention program is essential to reduce and, hopefully, prevent the hazards associated with backflow. In some states, a greater responsibility for preventing backflow is delegated to other agencies, such as the health department, the plumbing official, and/or environmental protection agencies. No matter who is primarily responsible, everyone involved in backflow prevention, including the consumer, shoulders some of the responsibility for developing and implementing an effective program.

This chapter explains how to develop a backflow prevention program. The water purveyor is assumed to be the instigator of the program. However, the items discussed herein would be useful to any other interested agency or individual. In this chapter, the goals and objectives of an effective program will be presented, and existing programs will be assessed to determine program deficiencies. The administration of a backflow prevention program will be considered. Mechanisms (e.g., ordinances, user contracts, policies, and standard operating procedures) that can be employed to enact stated objectives will be discussed. Further, the essential elements of a good backflow prevention program will be reviewed to provide the body and substance of the program. Since each community has its own concerns and priorities, backflow prevention programs should be tailored to meet specific needs.

In most states, the water purveyor has primary responsibility for the development of a backflow prevention or cross-connection control program.

Getting Started

The first step in the development of a backflow prevention program is educating the managers and operators of the program on backflow prevention and cross-connection control, i.e., the hazards, the potential liabilities, and the methods of backflow prevention. For a program to be effective, the persons actually implementing the program need to understand the need for it. A successful program will be developed if the ideas and suggestions of these individuals are considered.

The next step is to develop goals and objectives for the program. A goal is defined as the ultimate objective of an endeavor, the end toward which the effort is directed. A goal is written as a statement. A stated goal of a backflow prevention program is to prevent backflow by preventing the installation of cross-connections and eliminating existing cross-connections or providing adequate backflow prevention. The goal(s) should establish a clear commitment to the program on the part of the administration. Objectives are more specific than goals. They clearly state specific requirements needed to achieve the ultimate goals. Objectives should describe a particular intended outcome. They should give a description of a specific behavior or performance. Objectives are written as a sentence beginning with a verb that defines observable behavior. For instance, "to understand" is vague and unobservable, while "to install" is more specific. Some possible objectives of a backflow prevention program are:

1) To prevent the installation of cross-connections in all new establishments.
2) To install reduced pressure principle assemblies at the point of service to all facilities that represent a high hazard.
3) To predict, when reviewing construction plans, where cross-connections might be created.

The specific objectives of the program are established in order to achieve the overall goals. Inclusion of seven essential elements is vital when developing these objectives. These essential elements are:

1) Establishing legal authority.
2) Plan review of new construction.
3) Utilization of standards and specifications to delineate "approved" assemblies.
4) Testing and maintenance responsibilities.
5) Record keeping.
6) Program for surveying and retrofitting existing facilities.
7) Training and educating utility personnel, water consumers, agencies, and individuals involved in backflow prevention.

These elements will be discussed in detail later in this chapter. At this point, the examples provided below should help illustrate how these seven elements can be incorporated into objectives. For instance, "legal authority" can be included through the following objective: construct an ordinance that establishes the water purveyor's authority to:

a) require backflow prevention assemblies,
b) inspect premises and review plans for potential cross-connections, and
c) discontinue water service to those facilities that have cross-connections.

"Testing and maintenance" could be included in objectives such as:

a) Conduct an annual test on reduced pressure principle assemblies, double check valve assemblies, reduced pressure detector assemblies, double check detector assemblies, pressure vacuum breakers, and spill-resistant vacuum breakers,
b) Troubleshoot problems with backflow prevention assemblies, and
c) Report backflow prevention assembly test results.

The *American Water Works Association Manual M14* provides minimum guidelines on what should be included in a cross-connection control program. In addition, it is important that the responsibilities outlined in

Chapter 3 be re-examined as the objectives are formulated. The limits of available funding should only be considered after the objectives have been prioritized with respect to public safety and potential liability.

Once the goals and objectives of the program have been determined, the current program should be reviewed to ascertain which objectives are being met. Once the program has been examined, it is important to determine how each of the deficiencies will be corrected. At this point, some new objectives may need to be developed in order to meet the overall goal. If no backflow prevention program currently exists, Appendices M and N provide some suggestions for developing objectives.

Administration of the Program

Once the objectives have been established, they need to be ranked or prioritized. This will help determine how the program will be implemented. Some utilities (especially smaller ones) may find it necessary to implement only portions of the program in stages. For instance, they may develop an ordinance that calls for plan review of new construction to begin immediately, but with phases in a program for surveying and retrofitting existing facilities starting with the highest hazards. This would give the utility time to hire more personnel to handle the increased workload or to rewrite current job descriptions. However, the seven essentials listed previously must be in every backflow prevention plan from the outset.

The next step is to determine how to realize the objectives. There are three different approaches that can be taken.

1) Develop a program without the aid of the other agencies involved in backflow prevention.
2) Develop a program jointly with other agencies.
3) Contract the development of a program (and possibly its implementation) either to an agency involved in backflow prevention (e.g., the health department) or to a private organization.

Each of these approaches has advantages and disadvantages. Some managers feel the development of separate programs results in unnecessary expense for water users. For example, a water purveyor may require a backflow preventer at the meter even though the health department or plumbing inspector has eliminated potential hazards by requiring backflow-prevention assemblies at each possible point of contamination. The development of a separate program might thus be considered a duplication of effort. This level of protection, however, could be very important if the protection at one level fails. On the other hand, joint program development will require compromises, as one agency may not rank a particular objective as highly as another. Some agencies are unable or unwilling to compromise the development of their program in any manner. Understandably, the water purveyors may not feel that they can compromise, because they bear the ultimate responsibility and thus liability for ensuring that the water meets all applicable standards. The third approach, contracting the development of the

The seven essential elements must be included in every backflow prevention plan.

program, may seem the simplest solution. However, it is important to remember that contracting the development (or implementation) of the program does not release the water purveyor from its responsibilities. The water purveyor must oversee the development of the program to ensure that all "reasonable and prudent" actions are taken to prevent backflows. The water purveyor must carefully spell out what should be included in the program, or the program may not adequately meet the demands of the water purveyor's responsibilities or special needs.

Deciding between the three alternatives depends on the particular situation. Cost analysis may help determine which approach should be utilized. In any case, it is important to be aware of what the other agencies involved in backflow prevention are doing. It is essential to communicate with them as much as possible to avoid duplication of services, but more importantly to ensure that backflow is prevented.

> It is important to be aware of what the other agencies involved in backflow prevention are doing.

Once the approach for meeting the objectives has been determined, the next step is to decide which of the following mechanisms will be employed to structure the program: an ordinance, service contract, policy, resolution, and/or standard operating procedure. All four of these mechanisms must be considered concurrently when developing the program.

An Ordinance

An **ordinance** is the most important component of any backflow prevention program. It provides the legal basis for enforcement of the program. Ordinances can be used simply to establish the existence of a program (with the specifics of the program provided in the form of separate policies), or they can be more detailed, with specific objectives and policies listed in the ordinance itself. For example, an ordinance could state that only backflow prevention assemblies approved by the administrators of the program may be installed. A separate policy would then be written to define "approved assemblies." Alternatively, the approved assemblies could be listed in the ordinance itself. Both methods are acceptable. Simple ordinances provide the administrator of the program with more flexibility to make changes or update the program, and allow the objectives of the program to be established by those who are the most informed on the topic of cross-connection control and backflow prevention. More detailed ordinances, however, provide a stronger legal basis. Objectives written in legal language help eliminate ambiguities. The advantages and disadvantages of each approach should be considered in light of the special needs of each community. The amount of detail to be contained in the ordinance must be determined prior to development of the ordinance itself.

> An ordinance is the most important component of any backflow prevention program.

Service Contracts

Service contracts are normally used in lieu of an ordinance when the water utility is privately, not publicly, owned. However, service contracts

> A service contract literally states what the water customer must do in order to receive service.

are also used by publicly-owned utilities for water customers that fall outside the jurisdiction of the local ordinance, for instance, when a city-owned utility serves residents outside the city limits. Service contracts serve much the same purpose as an ordinance, as they provide the legal basis on which the program must be built. The service contract literally states what the water customer must do in order to receive service. For instance, the service contract should prohibit unprotected cross-connections, and should provide the right to discontinue service until the cross-connection is eliminated or properly protected by the installation of the appropriate form of backflow prevention. This will also apply to required annual testing and any necessary maintenance.

Policies and Rules

An ordinance is the basis of a program, but it cannot stand alone. An administrative framework must also be in place to implement the program. This framework consists of the goals, objectives, and policies of the program. The development of goals and objectives has already been discussed. A policy is a selected course of action that helps to determine what decisions should be made under a certain set of conditions. **Policies** are used to guide the execution of the ordinance and the program's objectives. For example, policies could be established to determine:

1) Who will install backflow prevention assemblies.
2) Who will test backflow prevention assemblies (utility personnel, private contractors, or both).
3) Who will repair backflow prevention assemblies.
4) Whether testers must be certified.
5) What type of protection will be required for high *vs.* low hazards.
6) Which type of backflow prevention assembly will be used at each hazard location.
7) What records will be kept on each backflow preventer.
8) What specifications and standards will be used to delineate an "approved" assembly.
9) How training will be provided to employees and customers.
10) Which employees will receive training.
11) What timetable will be established for completing the surveying and retrofitting program.
12) How the surveying and retrofitting program will be conducted.
13) How the customers will be notified.

All of these items could be covered under written policies established by the administration. One of the advantages of policies is that city councils, county commissioners, or other elected officials need not be involved in modifications to policy decisions, whereas they would be involved if changes were proposed to an ordinance. Reliance on policies, rather than the ordinance, for the delineation of specific requirements allows more frequent modifications, which results in a more up-to-date program.

Policies are used to guide the execution of the ordinance and the program's objectives.

DEVELOPING A CROSS-CONNECTION CONTROL PROGRAM

Standard Operating Procedures

Standard operating procedures (SOPs) or written guidelines should also be used to augment ordinances. SOPs are written, step-by-step instructions for completing a specific task. SOPs should be written as short, direct sentences or phrases that start with an action verb. SOPs provide a consistent method for implementing policies established by the administration, as well as for completing other tasks necessary to the program. Because SOPs provide a standard method of operation, everyone involved is better informed on what is happening and what is expected. SOPs also serve as an excellent training tool. Many record-keeping functions could easily be managed through the use of SOPs. An SOP can be used to establish the "routing" of backflow test reports through the utility (from receiving to reviewing to filing). SOPs can also be used to deal with non-compliance of backflow ordinances (e.g., notifying the customer of non-compliance, re-notification after an established time period, and steps thereafter, to discontinue service, if necessary).

An SOP that deals with non-compliance would provide consistent treatment of all water customers. Consistent treatment of the customer is essential, especially if water service is disconnected, to avoid charges of

SOPs are written, step-by-step instructions for completing a specific task.

For SOPs to be successful, they need to be updated periodically.

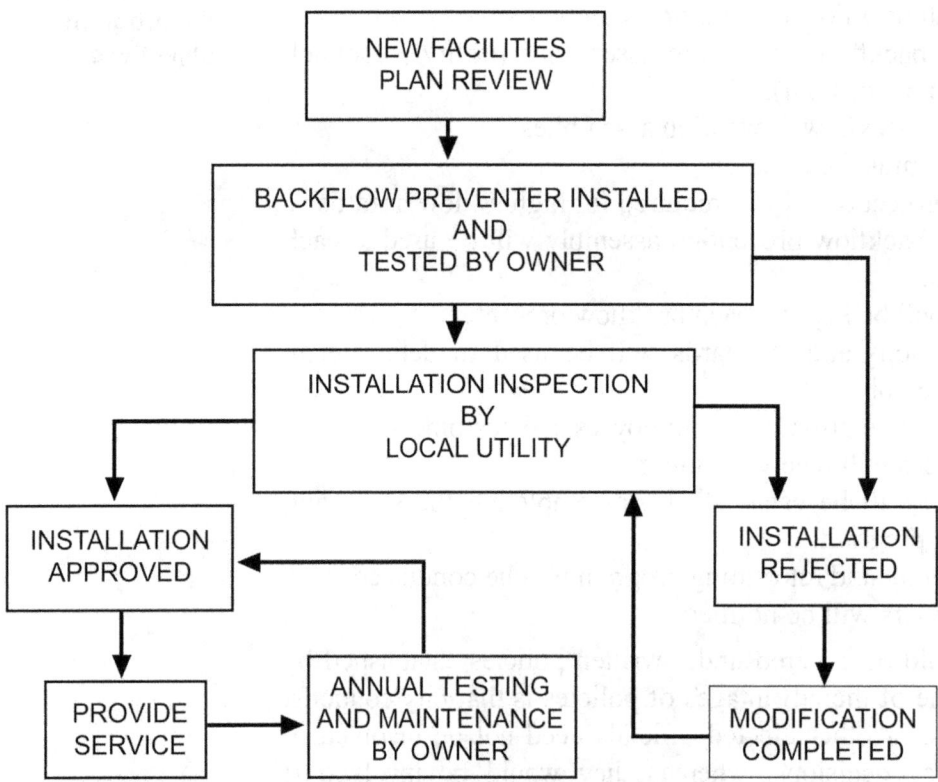

Figure 8-1 **Flowchart of Steps to Have Plumbing Plans Approved**
A flow chart can be employed to show the steps necessary to have new construction plumbing approved.

harassment and possible legal action. Emergency procedures for dealing with a backflow incident are also suitably addressed through the use of a standard operating procedure.

For SOPs to be successful, they need to be updated periodically so that they remain current with departmental policies. Once the SOP is outdated, it should no longer be followed. SOPs are most often followed when those who utilize the procedures in their everyday operations are involved in the SOP development.

Often, a particular objective could be adequately handled through one or more of the mechanisms discussed above. In addition, other methods besides those listed can be utilized. For instance, a flow chart like the one illustrated in Figure 8-1 can be employed to show the steps necessary to have new construction plumbing approved as it relates to backflow prevention.

Overall, written mechanisms yield a more consistent method of program implementation than reliance on verbal instruction, and they should, therefore, be utilized whenever possible.

Administrative Authority

Establishing a line of administrative authority is essential to every program. This will delineate who will make the final decisions on the program's policies and goals and who will settle disputes between the water purveyor and the customer. The top level of this administrative line might be the city or county manager, manager of the utility, or manager of the board of city or county commissioners. The top level could also be a separate board, such as the local code enforcement board or a special board consisting of health department officials, plumbing officials, utility representatives, and perhaps some citizens-at-large. The manager of the utility might seem the best choice, since he or she would probably be most supportive of the backflow prevention program. However, the water consumer may not feel satisfied with an appeal process that ends within the utility. An administrative line that terminates outside the utility may provide the water customer with greater satisfaction and thus reduce the number of cases that go beyond the appeal process and into the civil courts. One problem with utilizing an individual or board outside the utility is that they might not be well informed on the hazards of backflow.

Seven Elements of a Cross-Connection Control Program

As each of the seven essential elements is discussed, some of the mechanisms (ordinances, service contracts, policies, or standard operating procedures) that can be used to implement these elements will also be addressed. Often, any one of these mechanisms can be used. Which mechanism is used to implement each of the seven elements will be dependent on the particular circumstances of a given community.

Establishing Legal Authority

Florida Administrative Code 62-555.360 requires that the water purveyor "shall establish and implement a routine cross-connection control program to detect and control cross-connections and prevent backflow of contaminants into the water system." In order to effectively implement a program, legal authority must be available to enforce it. While written policies have been used to establish legal authority, generally either an ordinance or a service contract is used. Privately-owned utilities generally utilize service contracts, while publicly-owned utilities adopt an ordinance. Because of their legal nature, it is essential that both be reviewed by legal counsel prior to adoption or approval. Because both of these mechanisms serve the same basic purpose, the methods for their development and the items that should be contained within them will be very similar. The term "service contract" can be substituted for "ordinance" wherever it is found in the following discussion.

The next step is to determine what should be contained within the ordinance itself. This can be done by re-examining the goals and objectives, by obtaining the advice of legal counsel and by examining what others have included in their ordinances. The following items should be contained in the ordinance:

1) The general purpose or goal of the ordinance.
2) The State rules and regulations, the local plumbing code and any standards, specifications, or guidelines adopted by reference should be included.
3) Definitions of terminology contained within the ordinance.
4) Delineation of responsibilities.
5) Inspection authority for the purpose of discovering potential cross-connections (provisions should also be made relating to plan review of new facilities).
6) Requirements related to the proper applications for each type of backflow-prevention assembly.
7) Requirements related to the use of approved backflow prevention assemblies.
8) Provisions related to the maintenance and testing of the backflow preventers. At a minimum, testing should be done annually. Additionally, training and/or certification should be required of those who test backflow preventers.
9) Penalties for non-compliance, including termination of water service.

When establishing penalties for non-compliance, it is best to make the fine for non-compliance more costly than compliance. For instance, if the cost of installing the appropriate backflow preventer is $500.00, an appropriate fine for non-compliance would be $1,000.00. Public notification of non-compliance could be also an incentive for compliance. No matter what types of incentives are used to promote compliance, it is essential that

> **It is essential that service contracts and ordinances be reviewed by legal counsel prior to adoption or approval.**

> **It is essential that the water purveyor treat all customers the same.**

the water purveyor treat all customers the same. It is required by law that all customers receive equal protection.

While the previously mentioned list provides the essential items that should be included in every ordinance, many other items could be included, depending on the amount of detail desired. For instance, a list of those facilities that present a health hazard could be included. Some utilities also require, by ordinance, a permit for the installation of irrigation systems. This helps ensure that the appropriate form of backflow prevention is installed on the irrigation system. Installation standards might also be included as part of the ordinance, or they might be included under the policies of the program. Installation standards are utilized to make sure the backflow preventers are installed correctly. For instance, the standards might specify certain types of PVC pipe or certain sizes of thrust block. Recall that there are both advantages and disadvantages to including a great amount of detail in the ordinance, and that any details excluded in the ordinance must be included in the program elsewhere (i.e., in policies). However, it is important that the ordinance does not limit or restrict the implementation of the program. Periodically during the development of the ordinance, it is essential that it be reviewed by legal counsel. This will help ensure that it is both effective and enforceable.

The final step is to get the ordinance approved by the appropriate governing body. A well thought-out, carefully planned, and prepared presentation is essential. A series of workshops to inform the citizens and governing bodies about backflow prior to any formal presentation is usually a good idea. The presentation should include the hazards associated with backflow, the laws that require the development of a backflow prevention program, why the enactment of the ordinance is essential to the program, and the penalties and liabilities associated with the absence of an effective program.

A well thought-out, carefully planned and prepared presentation is essential.

The presentation should include the hazards associated with backflows, the laws that require the development of a program, why enactment is essential to the program, and the penalties and liabilities for non-compliance.

Plan Review of New Construction

Plan review is very important because it helps to eliminate cross-connections before they are created. Implementing this element as one of the initial steps in any backflow prevention program is essential. The plan reviewer identifies existing cross-connections and spots areas where cross-connections could conceivably be created through simple plumbing alterations, thereby improving the engineers' and contractors' awareness of these hazards. Thus, plan review not only helps eliminate cross-connections, but also serves as an educational tool.

While provisions requiring plan review of new construction should be contained within the ordinance itself, policies may be needed to spell out the details of this essential element. For instance, it must be clear who will do the plan review (e.g., the plumbing inspector, the fire department, the health department, the water purveyor or a combination of two or more). A number of tactics can be adopted, based on how the program as a whole is developed. Each of the agencies can do their own plan review, or the water

Plan review not only helps eliminate cross-connections, but also serves as an educational tool.

purveyor may be willing to rely on the plumbing inspector's review. Perhaps the water purveyor can rely on the plumbing inspector's review with certain exceptions (e.g., in high hazard cases), or the plumbing inspector may screen the plans and refer some of them to the water purveyor for further review. The health department could also play the role of chief reviewer or assistant reviewer. Relationships with the other agencies and their role in the implementation of the backflow prevention program will help determine who will review plans.

No matter who reviews the plans, it is essential that all plans for new construction be reviewed. Those submitting plans for new construction should be informed of the process for obtaining approval and the requirements that must be met to obtain approval. Policies should designate the proper application for each of the backflow prevention assemblies (if these are not detailed in the ordinance itself). In other words, it must be clear which type of assembly can be used under each type of backflow condition (e.g., high hazard or low hazard, backsiphonage or backpressure, isolation or containment).

The degree of hazard (high or low) should also be established in the ordinance or by policy for different types of businesses. Health officials may be of assistance in assessing the level of hazard. An additional benefit of this policy is that it promotes consistent treatment of all persons requesting plan approval, and thus helps to enhance community relations. Most people are willing to meet legal requirements *if* they have a clear idea of what the requirements are from the outset and if they understand the process for meeting these requirements.

After the initial plan review is completed, an inspection of the facility should be conducted at some point during construction. Conducting an inspection during construction often prevents the installation of cross-connections and, thereby, eliminates the need for costly backflow preventers. Once construction is complete, a final inspection should be made to detect if any cross-connections were created inadvertently and to ensure that the backflow preventer installed at the service connection was installed properly and is functioning. If everything is in compliance, the inspector should state that the customer is in compliance as of the date of the inspection. Thus, if cross-connections are made at a later date, the inspector is less likely to be charged with negligence for failing to spot cross-connections that did not exist at the time of the final inspection. Certificates of Occupancy should not be issued until all backflow prevention assemblies have been inspected and field tested.

Using Standards and Specifications to Define "Approved" Assemblies

In addition to establishing the correct application for each assembly, establishing some mechanism to regulate the quality of the assemblies themselves is essential. As suggested earlier, the most effective method might be to state in the ordinance that only "approved" assemblies be used, and then clarify through policy the definition of "approved." This allows the administrator of the program to more easily change the definition of "approved" as new assemblies replace old and as information on the reliability of the assemblies is gathered and analyzed.

Standards and specifications should be used to define an "approved assembly." Recall that several different agencies "approve" or set standards and specifications for backflow assemblies (Table 5-2). However, only FCCC & HR and A.S.S.E. actually publish a list of "approved" assemblies.

In addition, the water purveyor could also use standards and specifications to restrict the definition of "approved assemblies" so that those assemblies that require frequent maintenance, or that are not reliable under the specific conditions of the geographic region, are excluded. For instance, some assemblies may be more resistant to freezing, while others are more resistant to corrosive (or depositing) water conditions. Other items to be considered are the availability of replacement parts, the length of warranty and whether the assembly underwent field testing before approval. If replacement parts are not readily available, the whole assembly must be replaced to protect from the creation of a cross-connection. This results in unnecessary expense for the consumer. Field-tested assemblies give the purchaser some assurance that the model does work and does not have any mechanical failures. Recall also that only FCCC & HR tests backflow prevention assemblies in the field.

It is important that the water purveyor carefully define "approved assembly," because the water purveyor will shoulder some liability if an assembly fails to function properly and a backflow occurs. A plaintiff's attorney might argue that the water purveyor failed to take all "reasonable and prudent" steps to reduce the chances of backflow. The water purveyor's best defense against this charge is a program that periodically reviews and updates the standards and specifications set for the assemblies.

Testing and Maintenance

Once the backflow preventers are installed, it is important that they be maintained in good working condition. A backflow assembly that is not working provides the consumer and the water purveyor with a false sense of security. Therefore, the testing and maintenance portions of the backflow prevention program are very important.

The interval between required testing should be established either in the ordinance or through policy. All assemblies are required to be tested at least once a year. (Appendix I) Any model that has a higher-than-average failure rate should be tested more frequently. Also, air gaps should be checked to ensure that the free air space has not been reduced or the air gap eliminated.

Policies or standard operating procedures should be employed to assure that backflow preventers are tested by certain prescribed methods (e.g., according to FCCC & HR methodology), that gauge calibration is routinely

A backflow assembly that is not working provides the consumer and the water purveyor with a false sense of security.

performed and monitored, and that assemblies are repaired according to manufacturer's directions.

One of the first steps is to establish who will test and maintain the assembly—the consumer or the water purveyor. Or, a policy could be established that requires a licensed plumber to repair the assembly and a certified tester to test the assembly after the repair has been made. Still another option is to subcontract this duty. There are obviously a number of other options that could be developed. If the water purveyor decides to do the testing and repair, a standard price is recommended, whether the assembly requires no repairs or costly repairs to avoid accusations that repairs are being required unnecessarily. Obviously, the standard charge should be high enough to cover the costs of implementing this program.

No matter who tests, maintains, and repairs the assembly, the water purveyor cannot be alleviated from the responsibility of ensuring that the assemblies are adequately maintained. The water purveyor will determine who will test and maintain the assemblies based on liability concerns, available funding and manpower limitations. Liability concerns might induce the water purveyor to assume this responsibility. However, if the water purveyor decides to delegate this responsibility, verifying the competence of the tester could help to reduce the water purveyor's liability. For instance, the competence of the tester could be verified by requiring that only individuals who have had some specified amount of training be allowed to test and repair the assemblies. In fact, it is recommended that utilities require "certified testers." While certification may not be currently mandatory, this would clearly be looked upon as a "reasonable and prudent" action. Liability could also be limited by requiring that the tester endorse (via signature) a statement verifying that the test was performed according to the required procedures and that the assembly was not exercised prior to testing. It has been suggested that testers sign a "code of conduct" document so they understand what is expected of them. (See Chapter 3, Laws and Responsibility)

Small towns with few backflow preventers often elect to test and maintain the assemblies themselves and can accomplish this task without a significant increase in personnel. Larger cities may not have the manpower to test all of their assemblies, therefore, they may decide to test only the high hazard locations, or only those locations with reduced pressure principle assemblies and delegate the other testing to private testers. Funding and manpower limitations often support delegating the task of testing and maintenance to the consumer. If the responsibility for testing is delegated to the consumer, the water purveyor should develop a list of testers that can be distributed to the consumer when notices on retesting are mailed.

In short, even if water purveyors decide to delegate or subcontract testing and maintenance activities, they must ensure that these activities are being conducted in a satisfactory manner. This should be done through periodic spot checks and records review. One problem that sometimes occurs is that the tester exercises the assemblies prior to testing. For instance, an assembly is exercised by physically moving or causing the check valves or

One problem that sometimes occurs is that the tester exercises the assemblies prior to testing.

relief valves to operate in a manner that would prevent backflow. Frequently, pressure vacuum breakers are exercised prior to testing since the air-inlet disc is commonly stuck in the closed position. If this is being done, a realistic indication of whether the assembly will work under normal conditions cannot be obtained. If it appears that the assemblies are being exercised prior to testing, the testers should be reminded of the proper testing procedures prescribed by policy or SOP.

It is strongly recommended that the water purveyor test every assembly after installation whether installation was accomplished internally, contracted out, or delegated to the consumer. This allows the water purveyor to verify that the appropriate backflow preventer was installed, that it was installed properly, and that it is working properly. The ordinance should designate the individuals responsible for installing the backflow preventers. If certified testers are required, this should also be noted in the ordinance.

Record Keeping

In order to adequately monitor testing and maintenance of backflow prevention assemblies, a record keeping program must be established. The following information should be recorded for each backflow prevention assembly:

1) The type of assembly (RP, DCVA, or PVB), and its date of installation and the installation specifications.
2) The name of the manufacturer, model number, size, and serial number of the assembly.
3) The location of the assembly: the street address and the location on the grounds or within the building.
4) The customer's name, tap number, meter number, account number, or other identification numbers.
5) The type of actual or potential hazard (e.g. high or low hazard).
6) The test results before and after repair or maintenance.
7) The maintenance performed or the repairs that were made to the unit, including the replacement parts and part numbers and the date these repairs were made.
8) Information on backflows through the assembly, including any litigation resulting from failure of the assembly.

This information should be included on the test report form itself. A sample test report is provided in Appendix E.

The following information should be kept on an annual basis:

1) The total number of each type of backflow preventers that are installed.
2) The number of annual tests conducted.
3) The number of cross-connection control on-site surveys performed.

Other records that should be kept include the total number of field tests performed on all the backflow preventers and the total number of failures. This information will prove useful during the following year's budgeting process. For instance, assume that the utility performs the testing, and that last year two testers were needed to test 1,000 backflow preventers. If 500 more backflow preventers were installed this year, the need for additional personnel can more easily be justified.

The record keeping process can provide other valuable information that can be used to improve the program. Records review allows the administrators to spot problems with assemblies, monitor the testing and testers, and ensure that the testing frequency is adequate.

Keeping accurate information on the number and type of failures that occur on specific models allows the performance of different devices and assemblies to be monitored. Problems with certain parts or certain assembly model numbers can be identified by simply totaling the test results submitted by all testers and doing simple statistical analyses on these totals. This process can be made much simpler by computerizing the record keeping functions. For instance, a 60% failure rate of Brand "X" assembly indicates some type of problem with that brand of backflow preventer exists. Once this kind of information is obtained, a reasonable and prudent water purveyor would exclude the installation of that particular brand of assembly until the problem could be corrected. Further, steps should be taken to increase testing of the existing assemblies of this type to ensure that they function properly. If the assembly consistently fails when tested, that model may have to be removed from service.

Monitoring the test reports themselves could help to detect field testing errors. For instance, a tester who never fails a backflow preventer might not be properly testing the assembly and/or might be exercising the assembly prior to testing.

Another record keeping function is to ensure that testing is performed at the appropriate intervals as established by ordinance or policy. For instance, if policy requires that reduced pressure principle assemblies be tested annually, a system needs to be established to ensure that the testing is indeed being conducted annually. Some types of computer programs can generate a list of the facilities that are due for testing. Of course, the method used (computerized or non-computerized) will be shaped by financial limitations. In a simple non-computer system, the essential data on an assembly is recorded on a 3" × 5" card (other, more extensive information can be kept in a file). The cards are filed by the date that the current test expires. Thus, every assembly that needs retesting in any particular month can be located. If the consumer is responsible for testing, the card will be utilized to determine when to notify the customer that a retest on the backflow preventer is required. The utility will simply mail a standard form letter (Appendix O) stating that the consumer's backflow prevention assembly must be re-tested. If the utility is responsible for testing, the cross-connection control supervisor or tester could schedule the work chronologically. Larger utilities might find it more convenient to develop a system that allows for testing all assemblies in one area of town before they move on to the next section, in order to save travel time.

The potential for backflow through a non-functioning backflow preventer can be greatly reduced if accurate records are kept and reviewed. If records reveal that a backflow preventer has not been tested and is out of compliance, steps can be taken to correct the problem. Usually, these steps are outlined in policies or SOPs. The first step is to notify the customer of non-compliance (Appendix O). It is recommended that all non-compliance letters be sent certified mail with return receipt requested. A grace period of 30 days for compliance is usually given. If the consumer is responsible for testing, the next step is to determine if the consumer has complied by having the assembly tested. Normally, the tester would submit the test results to the water purveyor. Some water purveyors also require that the assembly be tagged as having passed or failed. If the consumer does not comply, the next step is to notify the customer of the penalties for non-compliance and provide the customer with another short grace period (10 to 30 days is common). Many water purveyors choose to notify the customer in the first letter that non-compliance will eventually result in termination of service. In any event, the customer should receive a final warning letter, before discontinuation of service, which commonly allows a very short grace period, (e.g., 10 days). Copies of these letters should also be sent to health department officials and the agency that enforces the SDWA. The health department officials might also try to persuade the water customer to comply with the ordinance to prevent potential health problems. Keep the local enforcement agency informed so they can be supportive of any necessary enforcement action. In addition, the clerk of the court should receive a copy of the final warning letter that states that non-compliance will result in the discontinuation of service. This is important if the case eventually reaches the civil court system.

Program for Surveying and Retrofitting Existing Facilities

Recall that over 100,000 cross-connections are currently created every day. Water purveyors must address the problem of these existing cross-connections, since each one can allow the backflow of contaminants into the potable water supply.

Before implementing a survey and retrofitting program, determine which method of backflow prevention will be utilized: isolation, containment or a combination of both. Recall that containment involves placing a backflow preventer at the point of service to the consumer, while isolation involves placing a backflow preventer at each potential location where a cross-connection could be created. In general, the water purveyor should at least utilize containment and should recommend the isolation of point sources of contamination. A recommendation is often all that can be given, because the water purveyor's authority inside private premises is extremely limited. If a joint program is established, either between the water purveyor and the health department or between the water purveyor and the plumbing authority, both containment and isolation may be utilized. Or, the water purveyor and health department may determine that all health hazards should be protected by both isolation and containment, while only containment (or isolation) will be utilized for low hazards. In rare situations, isolation methods might be relied upon as the sole form of protection. For instance, one state has elected to rely solely on isolation coupled with annual inspections of the premises, by a competent individual, to ensure that no new cross-connections are installed. Also where the surveying program is subcontracted to the health department or where the health department has primary enforcement responsibilities, isolation might be relied upon as the sole form of protection because health department officials have greater inspection privileges/authority inside a facility. Generally, however, isolation should not be relied on as the sole source of protection. No matter which method is utilized, a policy should outline which approach will be taken and who will be responsible for testing and inspecting the assemblies involved.

Many water purveyors elect to subcontract surveying and retrofitting duties because of the manpower requirement this program demands. Sometimes health departments are willing to take on this program, or at least are willing to work with the water purveyor to develop a comprehensive program. Other water purveyors elect to slowly phase in this element of the program over an extended time period. For instance, the highest hazard areas might be surveyed over the first 5 years, then low hazards over the next 6 years, thus, less manpower is needed to implement the program.

In order to speed up the retrofitting process, some water purveyors choose to require a backflow preventer based on the type of facility. For instance, because doctors' offices potentially present a high hazard, the ordinance or policy could require that doctors' offices have reduced pressure principle assemblies installed at the point of service. While this does help speed up the retrofitting process it should not be relied upon as a substitute for an actual inspection of the premises for cross-connections. For instance, a psychologist's office that only presents low hazards would have a justifiable complaint that they were being required to install a higher level of protection than necessary. An actual survey of the facility should be made whenever feasible to determine the current and potential hazards and the appropriate level of protection needed at the point of service. From this inspection, recommendations can also be made to the consumer, the health department, and/or plumbing official on what protection is needed at point sources within the facility.

Surveying and retrofitting programs should first target those facilities that present the highest hazard. Determining these facilities is the first step. A number of different lists identify health hazard facilities (Appendices P) and cross-connection locations (Appendix T). The next step is to determine if any of those facilities are in the service area. Probably the best aid is the local phone book. Other sources that could be of help include the local health department (especially if they have an occupational health program), business permitting agencies, or other local environmental agencies. The local fire department might also have a listing of establishments that

Surveying and retrofitting programs should first target those facilities that present the highest hazard.

contain flammable or explosive materials, and materials that most likely present a high hazard. If the fire department has not developed such a list, they should be encouraged to do so. A questionnaire (Appendix Q) could also be utilized to help determine which of the facilities presents a high hazard. The questionnaire could also request information on known cross-connections. To educate consumers and also inform them of the program, information on cross-connections, the hazards associated with backflow and consumer liability in the event of a backflow could also be sent along with the questionnaire.

Once a listing of the potential high hazard facilities in the area has been developed, it must be prioritized starting with the highest hazards. When doing this, special attention should be given to those facilities with auxiliary water supplies. For instance, hospitals are required to have two separate water sources. Top priority should also be given to those facilities that use reclaimed water, such as golf courses, school grounds, parks, tree farms, etc. More and more facilities are likely to take advantage of reclaimed water. Once they do, their systems should be re-checked to ensure that no new cross-connections have been created. Also, facilities that utilize other non-potable water sources for their fire sprinkler systems or for lawn irrigation systems should be closely scrutinized. Finally, any low-water-pressure areas of the city or town should be targeted as an area of concern.

Once a prioritized list of all the establishments that present a hazard has been developed, a schedule must be designed to ensure that the program progresses in a timely fashion. This compliance schedule can either be included in the ordinance or established by policy.

The next step is to actually survey the facilities (in order of priority) to determine what hazard level actually exists. The surveyor should schedule an appointment with the facility owner (Appendix S). Scheduling an appointment promotes good public relations with the consumer and allows the surveyor time to explain the program and the importance of preventing backflow. If at all possible, the inspection itself should be made by a two-member team. This allows one person to carefully inspect while the other person converses with the consumer about the program. When conducting the survey, the surveyor or inspector must look at all water outlets to determine whether a cross-connection currently exists, and also how easily one could develop. Appendix T lists common cross-connection locations for different types of establishments. Appendix R contains two survey forms that can be used when conducting an inspection. The format of one of the forms helps to remind the inspector/surveyor of cross-connections commonly found in that particular type of facility. It is also important for the surveyor to examine the installation of existing backflow preventers. For instance, air gaps should be checked to ensure that extensions to the water outlet have not circumvented the air gap, or that flood rims have not been elevated. The location of all backflow preventers should be recorded for future reference. If backflow preventers are installed vertically, manufacturer's specifications should be checked to ensure that this is permitted.

Backflow preventers installed on hot water lines should meet specifications for high-temperature operating conditions.

Once the survey is complete, the consumer should be informed as to what will be required and why. This should be followed with a letter to confirm the requirements (Appendix S).

Some municipalities may "grandfather" existing backflow preventers, even though they do not meet the specifications and standards established under the program. If "grandfathering" is allowed, policies related to these grandfathered assemblies should be developed. For instance, more frequent monitoring might be required, and a schedule for replacement should be established.

When alterations are made to plumbing lines, a new survey should be scheduled to ensure that no new cross-connections have been created. The process of surveying and retrofitting is a continuing one. While plumbing laws prohibit the installation of new cross-connections, reliance on this alone is not sufficient. All backflow prevention programs must include a plan for surveying and retrofitting in order to prevent backflow through existing cross-connections.

Training and Education

A training and education program is the final element that all backflow prevention programs must include. The main targets of the program are agency personnel and the water customers.

All of the employees involved in backflow prevention should receive some training. This includes the testers, repairers, plant operators, distribution personnel, utility inspectors, meter readers, customer service representatives, secretarial staff, and billing department personnel. Moreover, those who are responsible for managing the program need training in ordinance writing, emergency plan development and all the other aspects of managing a backflow prevention program. It is recommended that the administrator of the program be experienced in water works. Those individuals who actually test and repair the backflow preventers will also need extensive training. The water treatment plant personnel should receive training to ensure that there are no cross-connections within the plant itself. Surprisingly, most water treatment and wastewater treatment plants have a number of cross-connections. Undetected backflow of contaminants within the treatment plant has a great potential for creating very serious backflow problems.

All the distribution personnel, from those who lay and repair the pipes to meter readers and repairers, should be trained in backflow prevention. It is essential that the personnel who work in the water distribution department understand the dangers of cross-connections as well as the hydraulic conditions that create backflow, as they could create cross-connections. The meter readers, backflow assembly repairers, and inspectors should be able to spot cross-connections, because they are observing the distribution system most frequently. If meter readers or repairers are able to identify illegal connections such as irrigation systems, they not only reduce the amount of unmetered water, but also help eliminate hazardous cross-connections.

Do not overlook the importance of educating the administrative assistants. The administrative staff will have to deal with the comments, questions and concerns of the customers as implementation of the cross-connection program begins. They will be better able to address this responsibility if they have a thorough understanding of backflow and cross-connections and are knowledgeable about the backflow prevention program. In the same respect, the billing department personnel will be better able to answer questions and concerns if they are educated about the program. Being able to answer questions quickly and accurately promotes better public relations with the consumer and also saves the program manager time.

Educating the consumer is also one of the responsibilities of the water purveyor. Many backflow incidents could be avoided if the consumer is made aware of the hazards of backflow, their responsibilities for preventing backflows and the possible liability that they could incur if a backflow occurred. If a backflow does occur, ignorance of the law is not an adequate defense for the consumer. However, part of the liability might lie with the water purveyor if the water purveyor failed to educate consumers about their responsibilities. The water purveyor is assumed to be the authority on the subject of cross-connection control and backflow prevention.

Educating the consumer is also one of the responsibilities of the water purveyor.

Several different methods can be used to educate the consumer. A "mail stuffer" sent with the utility bill is a good method of educating a large number of consumers, on a very basic level, at a relatively low cost. Hose bib vacuum breakers are commonly addressed in mail stuffers, because estimates show that most cross-connections are created by garden hoses.[1] Frequently, flyers are hung on the doorknob of residential facilities after a backflow preventer is installed. The topic of thermal expansion is usually addressed in these flyers. Supplying hardware stores in the area with fliers or leaflets on the hazards of backflow is also a good idea. This information can be distributed with purchases of hoses, equipment, or supplies that might create a cross-connection or create the potential for the backflow of hazardous materials. At the same time, the information would serve to educate the hardware store personnel. The address and phone number of the program manager should be provided on these fliers in case the customer has any questions.

Another method for reaching a large number of people is a public service announcement on radio, television, or billboards. Information on the hazards of cross-connections can also be discussed when giving tours of the water treatment plant. Lectures or seminars can be utilized to educate groups of individuals, such as plumbers, lawn sprinkler contractors, fire sprinkler contractors, or others who might be involved in the installation or alteration of plumbing. On a smaller scale, some effort should be made to educate the consumer when surveying and retrofitting existing facilities. The important point to remember when educating consumers is to provide them with the information in a manner that does not unduly alarm or excite them.

Whether a joint program (with the health department or plumbing authority) or an independent program, training would be beneficial to provide to the other agencies involved in backflow prevention. In fact, educating these individuals might encourage taking a more active role in backflow prevention, or in some instances, motivating them to start in backflow prevention. Health department field inspectors and administrators, plumbing officials, fire system inspectors, firefighters, city or county business permitting agencies, and other city or county environmental agencies should be targeted. Education programs can enhance cooperation between the water purveyor and those who could make valuable contributions to the backflow prevention program. Good cooperation with firemen, for instance, would be beneficial because fire-fighting emergencies make large water demands and create hydraulic conditions that can potentially cause backflow. If the firefighters notify the water utility of emergencies, the water purveyor can monitor water pressure to determine if it is adequate to meet the demands of firefighting. At the very least, the firefighters can notify the water purveyor after an emergency, so that the area could be checked for signs of backflow. This would significantly prevent or reduce the damage from backflow. Educating fire marshals is also important, because they are usually the only ones who conduct inspections on fire sprinkler systems as fire sprinkler systems are exempt from most plumbing regulations.

Likewise, environmental agencies and health department officials might be able to provide information on potential high hazard facilities. Also, many health departments and the water management districts require a permit prior to the installation of a private well. This list could be cross-referenced against the water purveyor's customer list to determine if potable water is supplied to anyone with a private well. This would allow the water purveyor to identify those customers who do not have backflow protection at the meter to prevent the backflow of water from these private wells. Business permitting agencies could alert the water purveyor to new businesses that might present a high hazard if a backflow occurred.

The key to cooperation is sufficient education of all the parties involved. Once again, these education and training programs can either be developed in-house or at training centers such as the University of Florida Center for Training, Research and Education for Environmental Occupations (TREEO). Education is one of the best mechanisms for preventing and reducing backflow, as well as for aiding in the detection of potential cross-connections. Water purveyor personnel are likely to take their responsibilities for preventing backflow more seriously if they understand the reasons for implementing the program and if they are made aware of the important role they play in executing it. Likewise most consumers are willing to comply with backflow prevention laws once they are made aware of the hazards associated with backflow and the responsibility and liability they have for preventing backflow.

CHAPTER EIGHT

Developing a Program for Dealing with Emergencies

Because backflow sometimes occur even with a comprehensive backflow prevention program in place, the water purveyor must develop a plan to handle backflow emergencies. SOPs are one of the best tools available to implement this portion of the program. SOPs can outline what information should be gathered when receiving a complaint, how to investigate complaints, how to determine if a backflow has occurred and how to respond to the complaint or backflow emergency (including immediate, short-range, medium-range, and long-range responses). The following discussion provides information useful in developing a program for dealing with complaints resulting from an apparent backflow.

When a complaint is initially received, the person receiving the complaint should try to gather as much relevant information as possible. A form with a standard list of questions can be used to ensure that essential information is obtained.

While it is important to receive a good description of the problem, the person taking the complaint should try to refrain from suggesting problems if possible. People generally tend to agree with the suggestion instead of carefully assessing the real problem. The person taking the complaint should be as friendly and courteous as possible. Often, this is difficult because the person making the complaint is upset or irate. Listen to the problem intently before asking questions. This helps the person feel that someone is willing to listen and is concerned about their problem.

The next step is to investigate the complaint. SOPs can be used to establish what sort of response is appropriate under a certain set of conditions. For instance, if only one complaint occurs, the response would probably be somewhat different than if there were 10 complaints from the same area. However, no matter how many people complain, certain minimum steps should be taken. The water purveyor should at least send someone to the site of the complaint to examine the water. In addition, certain minimum tests should be performed, such as pH, chlorine, and bacteriological analysis. The pH and chlorine tests are good immediate indicators of potential problems and the bacteriological analysis gives an indication of the integrity of the system. A certified laboratory designated will be able to test samples quickly on short notice, including weekends or evenings in the event of an extreme emergency. SOPs can also help establish the circumstances for discontinuing water service.

If a backflow is suspected or reported, designating personnel who will manage the emergency is important. Inform these personnel in advance what will be expected of them. SOPs can help do this. Established lines of communication with the other agencies involved in backflow prevention (for instance, the health department, police department, and fire department). The health department can help with the epidemiology (determining the source of the problem) and may also be able to help with sample collection and distribution of a safe source of water until the distribution system can be used again. Some states require that their offices of environmental protection be notified of any violation of drinking water regulations within 48 hours. Also, establish contacts at local hospitals and clinics. This allows the water purveyor to quickly inform them of the type of the contaminant once it is determined. The police and fire departments can help notify customers of a backflow event. The local radio and television stations can provide invaluable assistance in notifying customers of a backflow incident. By informing the media about the hazards of backflow, they can provide the necessary information in the event of a backflow without unduly alarming the public. Journalists might be encouraged to report on the backflow prevention program before an emergency. Making contacts with these agencies *prior* to a backflow event allows time to establish who will be able to provide what services and to what extent.

Once a backflow has occurred, SOPs can be utilized to provide step-by-step guidance on what the person at the site should do to detect the source of the problem, eliminate the problem, and restore quality water. The following discussion gives some guidance on what should be done at each step in the process.

Immediate responses include the steps that should be taken at the site of the complaint, such as sample collecting, water testing, and potentially shutting off the water supply. The appropriate sample bottles and proper sampling techniques must be used. The sample results play an important role in establishing the cause of the backflow and minimizing its effects. Sample bottles should be located in the vehicle that is assigned to respond to a complaint, and extra sample bottles should be set aside for use in emergency situations. Generally, if the water is going to be shut off, the person at the site will probably need to report back to a supervisor.

The short-range responses include determining the cause of the problem (with or without the aid of other officials), eliminating the source of the problem, and minimizing the effects of the backflow through containment and public notification. Determining the source of the problem and isolating the source is much simpler if the distribution maps are up-to-date and provide all the necessary information. The distribution maps should include basic information such as the size of lines and the location of hydrants, valves, lines, and pumping stations. In addition, the type of backflow prevention assemblies at each location should be noted on the maps. This would help to identify health hazard locations. Areas of low pressure might be identified on the map, as well as any significant changes in elevation. Recall that the health department can also be called upon at this time to help establish the source of the problem. The media, as well as police and fire personnel, could assist with notifying water customers of the problem.

Medium-range responses include steps to restore the quality of the water. Meanwhile, however, those without water need to be supplied with a safe source of drinking water. Often, the health department can be called upon to help provide this service. Restoring potable water to the consumers may take several days to several months to accomplish, depending on the

CHAPTER EIGHT

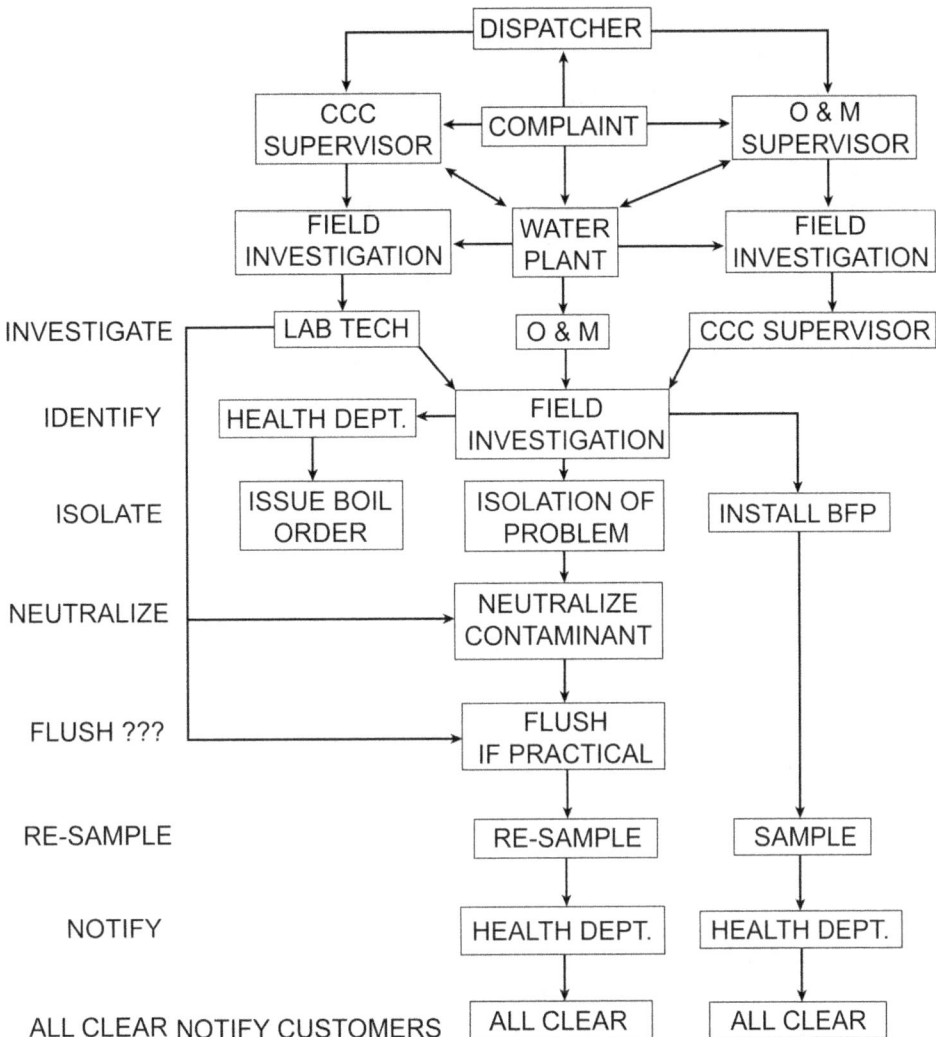

Figure 8-2 Suggested Emergency Response Flowchart
Whoever receives the initial complaint must be able to contact the proper persons to begin the investigation.

type of contaminant. Removing the contaminant might be as simple as flushing the lines or neutralizing the contaminant and retesting to ensure the contaminant has been adequately removed. On the other hand, in extreme cases, the water lines might need to be replaced. Recall the backflow incident that occurred at Allegheny County, PA, where the water purveyor had to replace the plumbing and water lines to 75 apartments (at a cost of $300,000) after chlordane, a chemical with a very low maximum contaminant level, contaminated them during a backflow event.[2]

Long-range responses include a review of how the emergency was handled in order to improve future responses to backflow or other emergencies. Assess what worked and what did not, and how SOPs should be modified to improve responses to future incidents. The performance of the other agencies and individuals who were involved in the backflow incident should be reviewed. For instance, did the media unnecessarily alarm the public? If so, re-education or improved lines of communication may be necessary.

A written procedure for handling an emergency is a valuable aid during the stressful crisis of the emergency itself. How a water purveyor responds to an emergency may be just as important as the steps taken to prevent backflow in terms of assessing liability. Pre-planning provides responses that are carefully thought out. Time is allowed for the development of contingency plans. Alternate backflow-event managers can be assigned. Thus, one person knowledgeable about backflow is always available even when leave time is taken. Emergency planning does not indicate a lack of confidence in the backflow prevention program, but rather contributes to the prevention of the hazardous effects of backflow.

Program Manual

A program manual can be used to assemble all the components of the program in an organized fashion. It should state the goal and objectives of the program and should also contain a copy of the ordinance or service contract. All the policies established under the ordinance should be contained therein, as well as all the SOPs, copies of form letters, and any other materials related to the program. For instance, the manual might include a listing of the high-hazard facilities in the area, along with their addresses, a list of certified testers, copies of the local plumbing codes, and telephone numbers of contacts at other agencies involved in backflow prevention, as well as those of media contacts and other emergency personnel. The manual should be updated periodically to stay current with any changes in the program. The development of the manual itself helps to organize the program, possibly pinpointing weaknesses, and provides a ready reference for information related to the cross-connection and backflow prevention program.

Summary

An effective cross-connection control program will contain at least seven essential elements: established legal authority through an ordinance or operating policy; a written plan for new construction plan review; a standard list of approved backflow assemblies; annual field testing and required maintenance procedures; a program for efficient record keeping; a written plan for inspecting and retrofitting existing customers; and training and education for employees and customers.

If all of these tasks are accomplished, there is still no guarantee that a cross-connection incident will not occur in your community. Accomplish as much as possible and document all actions. Be prepared for the day that the cross-connection control program has to be defended before a judge. If you have done an effective job, you should be proud that you have done everything in your power to protect the health and welfare of your customers.

REFERENCES

1. "50 Cross-Connection Questions, Answers & Illustrations: Relating to Backflow Prevention Products and Protection of Safe Drinking Water Supply," Watts Regulator, Watts Industries, Inc., Andover, MA.

2. "Stop Backflow: Typical Cases for Backflow Prevention," *Watts Regulator News*, Watts Regulator, Watts Industries, Inc., Andover, MA.

SUGGESTED SUPPLEMENTARY READINGS

ABPA News Articles

Hermsen, A. Fire Systems, A Threat to Health?, *ABPA News* (March/April 1991).

O'Brien, L. What Does It Take to Be an Effective Program Manager, *ABPA News* (May/June 1994).

Schwartz, P. TESTING – Exercise That Relief Valve?, *ABPA News* (May/June 1993).

Stinnett, L. Fire System Backflow Prevention: Yes, No, Who Cares?, *ABPA News* (Jan./Feb. 1991).

Stinnett, L. President's Message – Fire Systems Follow-up, *ABPA News* (March/April 1991).

Drinking Water & Backflow Prevention Articles

Editor. Backflow Preventer Test Kits – An Overview, *Drinking Water & Backflow Prevention* (March 1993).

Editor. Ballcock Hazards, *Drinking Water & Backflow Prevention* (Dec. 1990).

Editor. Black Water, *Drinking Water & Backflow Prevention* (April 1991).

Garza, G. Gauges: What's the Difference?, *Drinking Water & Backflow Prevention* (Oct. 1993).

Hassig, J. AWWA's M14 Challenge, *Drinking Water & Backflow Prevention* (Feb. 1991).

O'Brien, L. Utilities – Public Education, *Drinking Water & Backflow Prevention* (Sept. 1993).

O'Brien, L. Which Test Kit Should I Buy?, *Drinking Water & Backflow Prevention* (July 1992).

O'Brien, L. Why Bother with Public Education?, *Drinking Water & Backflow Prevention* (Sept. 1993).

Florida Water Resources Journal Articles

Ritland, R. & O'Brien, L. Reclaimed Water: How Does It Affect Your Backflow Prevention Program?, *Florida Water Resources Journal* (Aug. 1991).

Handbooks and Manuals

Accepted Procedures and Practice in Cross Connection Control Manual. Cross Connection Control Committee, Pacific Northwest Section AWWA and Cross Connection Control Committee, British Columbia Section AWWA. (4th ed.).

Angeles, G. J., Sr. *Cross Connections and Backflow Prevention.* AWWA, Denver, CO (2nd ed., 1974).

Cross-Connection Control Manual. United States Environmental Protection Agency, EPA 570/9 89-007. Washington, D.C. (1989).

Manual of Cross-Connection Control. Foundation for Cross-Connection Control and Hydraulic Research, University of Southern California, Los Angeles, CA (8th ed., 1988).

Recommended Practice for Backflow Prevention and Cross-Connection Control, Manual M14. AWWA, Denver, CO (2nd ed., 1990).

SUGGESTED SUPPLEMENTARY READINGS

Journal Articles

Committee Report - Prevention of Groundwater Backflow Into Distribution Systems. AWWA Distribution Division Committee on Prevention of Groundwater Backflow in Distribution Systems. *Jour. AWWA*, 71:2:76 (Feb. 1979).

Craun, G.F. Outbreaks of Waterborne Disease in the United States: 1971-1978. *Jour. AWWA*, 73:7:360 (July 1981).

Craun, G.F. et al. Waterborne Disease Outbreaks in the US - 1971-1978. *Jour. AWWA*, 68:8:420 (Aug. 1976).

Craun, G.F. & Gunn R.A., Outbreaks of Waterborne Disease in the United States: 1975 - 1976. *Jour. AWWA*, 71:8:422 (Aug. 1979).

David, J.H. & Murrell, L.R. Legal Aspects of Backflow and Cross Connection Control. *Jour. AWWA*, 68:8:397 (Aug.1976).

Gorden, S.F. et al. Panel Discussion - Cross Connection - Problems and Answers. *Jour. NEWWA*, 6:179 (June 1978).

Grady, R.P. Cross Connection Control - A Management Decision. *Jour. NEWWA*, 12:309 (Dec. 1980).

Grady, R.P. Portland Water District's Experience With a Cross Connection Survey, *Jour. NEWWA*, 9:234 (Sept. 1971).

Lee, R.D. Protecting Community Water Supplies, *Jour. AWWA*, 64:4:26 (April 1972).

McGuillan, R.G. & Spenst, P.G. The Addition of Chemicals to Apartment Water Supplies. *Jour. AWWA*, 68:8:415 (Aug. 1976).

Miller, K.J. Counterpoint - The Regulation and Prevention of Cross Connections in Water Distribution Systems. *Jour. AWWA*, 69:5:12 (May 1977).

Roller, J. A. Cross-Connection Control Practices in Washington State, *Jour. AWWA*, 68:8:407 (Aug. 1976).

Springer, E.K. Cross-Connection Control. *Jour. AWWA* 68:8:405 (Aug. 1976).

Springer, E.K. Viewpoint - Wanted: Comprehensive Cross Connection Control. *Jour. AWWA*, 72:8:17 (Aug. 1980).

TREEO-FS/AWWA Cross-Connection Control Committee. (1991, February 15) Backflow Prevention Program Survey Results (draft). (available from TREEO, 3900 SW 63rd Blvd., Gainesville, Fl 32608-3848) R. L. Ritland.

Woodhull, R.S. Viewpoint - The Regulation and Prevention of Cross Connections in Water Distribution Systems, *Jour. AWWA*, 69:5:12 (May 1971).

OpFlow Articles

Anderson, G.D. The Basics of Cross-Connection Control. *OpFlow*, 4:11:3 (Nov. 1978).

Back-siphonage: A Hazard to Public Health. *OpFlow*, 2:5:3 (May 1976).

Cross-Connection-Control Guide for Operators - I. Cross Connection Terminology. *OpFlow*, 9:1:4 (Jan. 1983).

Cross-Connection-Control Guide for Operators - II. Devices That Protect Against Cross Connections. *OpFlow*, 9:2:3 (Feb. 1983).

SUGGESTED SUPPLEMENTARY READINGS

Cross-Connection Control Guide for Operators - III. Cross Connection Control Programs. *OpFlow*, 9:3:5 (Mar. 1983).

Cross-Connection Control in Your Own Backyard. *OpFlow*, 6:6:6 (June 1980).

DDT in Water Supply. *OpFlow*, 5:3:1 (Mar. 1979).

Markwood, I.M. Cross-Connections-Legal or Lethal? *OpFlow*, 1:9:3 (Sept. 1975).

Miller, R. S. Cross Connection Control Down on the Farm. *OpFlow,* 5:11:1 (Nov. 1979).

O'Brien, L. Cross-Connection Control Supervisor's Checklist. *OpFlow* (July 1995).

Standards and List of Approved Devices

ASME Standard for Air Gaps in Plumbing Systems. Standard A112.1.2 (ANSI Approved). American Society of Mechanical Engineers, New York, NY (1973).

ASSE Standard for Performance Requirements for Pipe-Applied Atmospheric Type Vacuum Breakers. Standard 1001-2002 (ANSI Standard A112.1.). American Society of Sanitary Engineering, Bay Village, OH (2002).

ASSE Standard for Performance Requirements for Hose Connection Vacuum Breakers Standard 1011-2004 - ANSI approved American Society of Sanitary Engineering, Bay Village, OH (2004).

ASSE Standard for Performance Requirements for Backflow Preventers with Intermediate Atmosphere Vent. Standard 1012-2002. American Society of Sanitary Engineering, Bay Village, OH (2002).

ASSE Standard for Performance Requirements for Reduced Pressure Principle Backflow Preventers and Reduced Pressure Fire Protection Principle Backflow Preventers, Standard 1013-1999. American Society of Sanitary Engineering, Bay Village, OH (1999).*ANSI Approved – 1999.*

ASSE Standard for Double Check Backflow Prevention Assemblies and Double Check Fire Protection Backflow Prevention Assemblies, Standard 1015-1999. American Society of Sanitary Engineering, Bay Village, OH (1999). *ANSI Approved – 1999.*

ASSE Performance Standard for Vacuum Breaker Wall Hydrants, Freeze Resistant, Automatic Draining Type. Standard 1019-2004. American Society of Sanitary Engineering, Bay Village, OH (2004).

ASSE Performance Standard for Vacuum Breakers, Antisiphon, Pressure Type. Standard 1020-2004. American Society of Sanitary Engineering, Bay Village, OH (2004). *ANSI approved.*

ASSE Performance Standard for Dual Check Valve Type Backflow Preventers. Standard 1024-2004. American Society of Sanitary Engineering, Bay Village, OH (2004).

ASSE Performance Standard for Laboratory Faucet Backflow Preventers. Standard 1035-2002. American Society of Sanitary Engineering, Bay Village, OH (2002). *ANSI Approved – 2002.*

ASSE Performance Standard for Reduced Pressure Detector Fire Protection Backflow Prevention Assemblies. Standard 1047-1999. American Society of Sanitary Engineering, Bay Village, OH (1999). *ANSI Approved – 1999.*

ASSE Performance Standard for Double Check Detector Fire Protection Backflow Prevention Assemblies. Standard 1048-1999. American Society of Sanitary Engineering, Bay Village, OH (1999). *ANSI Approved – 1999.*

ASSE Performance Standard for Hose Connection Backflow Preventers. Standard 1052-2004. American Society of Sanitary Engineering, Bay Village, OH (2004).

SUGGESTED SUPPLEMENTARY READINGS

ASSE Professional Qualification Standards Backflow Prevention Assemblies, American Society of Sanitary Engineering, Bay Village, Ohio (1991).

ANSI/AWWA Standard for Backflow Prevention Devices - Double Check Valve Backflow Prevention Assemblies. Standard C510-97. AWWA, Denver, CO (1997).

ANSI/AWWA Standard for Backflow Prevention Devices - Reduced Pressure Principle Backflow Prevention Assemblies. Standard C511-97. AWWA, Denver, CO (1997).

List of Approved Backflow-Prevention Devices. Foundation for Cross Connection Control and Hydraulic Research, University of Southern California, Los Angeles, CA. (Available to members only-updated periodically).

Technical Papers

Anderson, D.C. Cross Connections - Their Importance and Control. Proc. AWWA Ann. Conf., Paper No 18-5 (1981).

Annual Summary of Water-Related Disease Outbreaks. Center for Disease Control, US Dept of Health & Human Services, PHS, Atlanta, GA (Published annually).

Klimko, R.G. Cross-Connection Control Program of the City of Cleveland. Proc. AWWA Ann. Conf., Paper No 22-b (May 1977).

Ongerth, H.J. Cross Connection Control- The California State Department of Health Viewpoint. Proc. AWWA Ann. Conf., Paper No 22-c (May 1977).

Springer, E.K. A Sip Could be Fatal. Proc. AWWA Distribution System Symposium. Paper No. 3-1 (Feb 1980).

Springer E. K. The Nuts and Bolts of Cross Connection. Proc. AWWA Distribution System Symposium, Paper No. 3-1 (1982).

Wubbena, R.L. Comprehensive State/Local Approach to Cross Connection Control. Proc. AWWA Ann Conf., Paper No 22-e (May 1977).

GLOSSARY

absolute pressure—The total pressure; gauge pressure plus atmospheric pressure. Absolute pressure is generally measured in pounds per square inch (psia).

air gap (air-gap separation)—The unobstructed vertical distance through the free atmosphere between the lowest opening from any pipe or outlet supplying water to a tank, plumbing fixture, or other device, and the flood-level rim of the non-pressurized receiving vessel or receptacle.

approved—Accepted by the Responsible Authority. Meets applicable specifications and standards.

atmospheric pressure—The pressure exerted by the weight of the atmosphere (14.7 psi at sea level). As the elevation above sea level increases, the atmospheric pressure decreases. 1.0 psi is equivalent to a 2.31 foot column of water at sea level.

atmospheric vacuum breaker (AVB)—A mechanical backflow prevention device consisting of a float check valve and an air inlet port; designed to prevent backsiphonage by allowing air to enter the downstream water line. This unit does not provide protection against backpressure or continuous pressure. A shut-off valve is not allowed downstream from the device. A shut-off valve downstream would allow the device to be subjected to continuous pressure.

auxiliary water supply—Any water supply on or available to the premises other than the potable water supply (e.g. community water system, non-community water system, or "other" water system). Auxiliary waters may include water from another potable water supply or any natural source(s) such as a well, lake, spring, river, stream, harbor, reclaimed water (reuse water), or industrial fluids.

backflow—The reversed flow of a non-potable source into a potable system, because a pressure differential exists where the pressure on the non-potable side is greater than the pressure on the potable side. There are two different types of backflow: backsiphonage and backpressure.

backflow prevention assembly—A mechanical backflow preventer (i.e., SVB, PVB, DCVA, RP), used to prevent the backward flow of contaminants or pollutants into a potable water distribution system. An assembly has a resilient seated, full-flow shut-off valve before and after the backflow preventer making it testable in-line. The assembly is shipped with the shut-off valves attached to the backflow preventer. An assembly is labeled with the manufacture's symbol, size, serial number, model number, the working pressure, and the direction of flow. The Foundation for Cross Connection Control and Hydraulic Research at the University of Southern California tests and approves backflow prevention assemblies.

backflow prevention device—A mechanical backflow preventer without the shut-off valves. An atmospheric vacuum breaker is a device. It does not have shut-off valves on the downstream side of the backflow prevention mechanism. Also, any backflow prevention assembly without the shut-off valves is called a device. The American Society of Sanitary Engineers (ASSE) approves backflow prevention devices.

backflow prevention method—A mechanism for preventing backflow that includes mechanical backflow prevention assemblies (SVB, PVB, DCVA, RP) and devices (AVB) as well as air gaps.

backpressure (superior pressure)—A condition in which the pressure in a non-potable system is greater than the pressure in the potable water distribution system. Superior pressure will cause non-potable liquids to flow into the potable water distribution system through cross-connections.

GLOSSARY

backsiphonage—Reverse flow of liquid caused by a partial vacuum in the potable water distribution system. A condition that occurs when the supply pressure drops below atmospheric pressure (sub-atmospheric).

bypass—Any arrangement of pipes, plumbing, or hoses designed to divert the flow around an installed device or assembly through which the flow normally passes.

certified backflow prevention assembly tester—A person who has attended and satisfactorily completed at least a 32-hour training course that is endorsed by the Florida Section of the American Water Works Association.

condition of service—An agreement between water supplier and the consumer that specifies the obligation and responsibilities of each in order for service to be provided.

consumer—Person or facility receiving service from a potable water system.

containment (policy)—To confine potential contamination caused by a cross-connection within the facility where it arises by installing a backflow prevention assembly at the point of service. Sometimes called premise isolation or service protection.

contamination—The introduction of any substance into water at levels that degrades the quality of the water, making it unfit for human consumption because it would adversely affect public health.

continuous pressure—A condition in which upstream pressure is applied continuously (more than 12 hours) to a device or assembly. Continuous pressure can cause mechanical parts within a backflow prevention device to become stuck or frozen in place thus causing the backflow preventer to malfunction.

critical level—The critical level is a reference line representing the level of the check valve seat within a spill-resistant vacuum breaker or atmospheric vacuum breaker. This line is used in measuring the elevation of the vacuum breaker above the highest point of water use, either the highest outlet or flood level rim.

cross-connection—Any arrangement of pipes, fittings, fixtures, or devices that directly or indirectly connects a non-potable system to a potable water system. Bypass arrangements, jumper connections, removable sections, swivel or change over assemblies, or any other temporary or permanent connecting arrangement through which backflow may occur are considered to be cross-connections.

cross-connection control—The use of assemblies, devices, methods, and procedures to prevent contamination or pollution of a potable water supply through cross-connections.

degree of hazard—The danger posed by a particular substance or set of circumstances. Generally, a low degree of hazard is one that does not affect health, but may be aesthetically objectionable thus it is termed a non-health hazard. A high degree of hazard is one that could cause serious illness or death thus it is termed a health hazard.

direct cross-connection—A link between the potable water supply and any other non-potable system, which is subject to both backsiphonage and backpressure.

distribution system—All pipes, fittings, and fixtures used to convey liquid or gas from one point to another.

double check valve assembly (DCVA)—A method of backflow prevention consisting of two independently operating check valves, 4 test cocks, and 2 shut-off valves. It is only used to protect against a non-health hazard. It can be subjected to backpressure.

dual check valve—A method of backflow prevention consisting of two independently operating check valves, without any test cocks or shut-off valves. It is considered appropriate for use against a non-health hazard. It can be subjected to backpressure. Sometimes called a residential dual check.

effective opening—The minimum cross-sectional area at the point of water supply discharge. The opening's dimensions are expressed in terms of the diameter of a circle or if the opening is not circular, the equivalent cross-sectional area. Used in determining the minimum vertical distance that should be provided to create an air gap.

flood level rim—That level from which liquid in plumbing fixtures, appliances, tanks, or vats will overflow to the floor, when all drain and overflow openings built into the equipment are obstructed.

gauge pressure—Pounds per square inch (psi) that are registered on a gauge. Gauge pressure measures only the amount of pressure above (or below) atmospheric pressure.

health hazard—A cross-connection or potential cross-connection involving a contaminant in sufficient concentration to spread disease or cause death. Another term for health hazard is high hazard.

indirect cross-connection—A temporary link between the potable water supply and any other non-potable system, which is subject to backsiphonage only.

isolation (policy)—To confine a potential source of contamina-tion to the non-potable system being served; to provide a backflow prevention mechanism at each actual or potential cross-connection. Sometimes called internal protection.

liability—Legally responsible for or being obligated by law for the protection of the potable water supply.

maximum contaminant level (MCL)—The maximum permissible level of a contaminant in water delivered to the free-flowing outlet of the ultimate user of a public water system. Contaminants added to the water under circumstances controlled by the user, except those resulting from corrosion of piping and plumbing caused by water quality, are excluded.

negative pressure—Pressure that is less than atmospheric pressure. Negative pressure in a pipe can induce a partial vacuum that can siphon non-potable liquids into the potable water distribution system.

negligence—Not meeting one's responsibilities and causing harm.

non-health hazard—A cross-connection or potential cross-connection involving any pollutant or contaminant (at low levels) that will not create a health hazard but will create a nuisance, or be aesthetically objectionable, if introduced into the potable water supply. Another term for non-health hazard is low hazard.

non-potable water or system—Any liquid, gas or solid that can be diluted, dissolved, suspended, or mixed with water that adversely affects the quality of the water.

non-pressurized receiving vessel—another term for open receptacle such as a sink or a bath tub.

GLOSSARY

non-toxic—Not poisonous; a substance what will not cause illness or discomfort if consumed.

pathogen—A disease causing agent or organism.

permanent cross-connection—A link between the potable water supply and any other non-potable system that is designed to remain in place.

physical disconnection (separation)—Removal of pipes, fittings, or fixtures that connect a potable water supply to non-potable system or one of questionable quality.

plumbing—Any arrangement of pipes, fittings, fixtures, and devices for the purpose of moving liquids or gases from one point to another, generally within a single structure.

poison—A substance that can injure, impair, or cause death to a living organism.

pollutant—A substance that deteriorates the aesthetic quality of water or other materials but is not harmful to health. Pollutants are considered non-health or low hazards.

potable water—Water that is safe for human consumption (meets the Safe Drinking Water Standards) and is aesthetically pleasing.

pressure—The force exerted on a surface, expressed in units of pounds per square inch (psi), feet of head, etc.

pressure vacuum breaker assembly (PVB)—An assembly consisting of one independently operating spring loaded check valve, an independently operating, loaded air-inlet valve, 2 test cocks, and 2 shut-off valves. This assembly is designed to prevent backsiphonage. It cannot be used where it may be subjected to backpressure. It can be operated under continuous pressure.

protected cross-connection—A cross-connection between a potable and non-potable system where adequate methods are provided to prevent backflow.

purveyor—See water supplier.

reduced pressure principle assembly (RP) or reduced pressure zone assembly (RPZ)—A mechanical assembly consisting of four test cocks, two shut-off valves, two independently operating, spring loaded check valves with a reduced pressure zone between the checks. The zone contains a relief port that will open to atmosphere if the pressure in the zone falls within 2 psi of the supply pressure. The assembly provides protection against both backpressure and backsiphonage.

refusal of service (shut off policy)—A formal policy adopted by a utility's governing board to enable the utility to refuse or discontinue service where a known or potential hazard exists and corrective measures are not undertaken to eliminate the hazard.

regulating agency—Any local, state, or federal authority given the power to issue rules or regulations having the force or law for the purpose of providing uniformity in details and procedures.

relief valve—A device designed to release air from a pipeline, or introduce air into a line if the internal pressure drops below atmospheric pressure.

GLOSSARY

service connection—A piping connection between the water purveyor's main and a user's system.

spill-resistant vacuum breaker (SVB)—An assembly designed to prevent backsiphonage that can be used under continuous pressure; the assembly includes an independently operating spring loaded check valve and an independently loaded air inlet valve located on the discharge side of the check with shut-off valves located on the inlet and outlet side of the assembly, a resilient seated test cock located upstream of the number one shut-off valve and downstream of the check with a properly located vent valve above the check valve and below the air inlet valve.

submerged inlet—An arrangement of pipes, fittings, or devices that introduces water into a non-potable system below the flood-level rim of a receptacle.

superior pressure—See backpressure.

temporary cross-connection—A link between the potable water supply and any other non-potable system created with removable sections, swivel or change-over devices, garden hoses, and other non-permanent methods.

test cock—An appurtenance on an assembly or valve used when testing the assembly.

thermal expansion—The increase in pressure in a closed water system due to heating and expanding water.

toxic—Poisonous; a substance capable of causing injury or death. A toxin may be ingested, inhaled, or absorbed through the skin.

unprotected cross-connection—A cross-connection between a potable and non-potable system where inadequate methods are provided to prevent backflow.

vacuum—Pressures below atmospheric pressure. A condition induced by negative (sub-atmospheric) pressure that causes backsiphonage to occur.

vacuum breaker—See atmospheric vacuum breaker and pressure vacuum breaker.

Venturi effect—As the velocity (speed) of water increases, the pressure decreases. The Venturi effect can create a vacuum (sub-atmospheric pressure condition) in a distribution system. Also known as Bernoulli's Principle.

water supplier (purveyor)—An organization that is engaged in producing and/or distributing potable water.

waterborne disease—Any disease that is primarily transmitted through water for example typhoid, cholera, giardiasis.

INDEX

A

absolute pressure 66
adhesion 108
air gap 106
air inlet valve 89
assembly 81
approval tag 160
approved assembly 82
atmospheric vacuum breaker 84

B

backflow 1
backflow prevention 27
backpressure 3
backsiphonage 3
barometric loop 69
biological contaminants 11
biological pollutants 11
bleed-off tee 144
breach of warranty 37
buffer 105
bypass 171

C

check valve 89
chemical contaminants 20
chemical pollutants ... 18 clogged sensing line 174
cohesion 108
commercial fire sprinkler systems 115
compensating tee 152
containment 18
contaminant 2
corrosion 185
critical level 90
cross-connection 1

D

detection time 16
direct cross-connection 5
device 81
differential pressure gauge 127
double check detector assembly 114
double check valve assembly 94
dual check valve 112
dump 99
dumping 105

E

electrolysis 185
enclosure 120
existence of a warranty 36
expressed warranty 37

F

federal regulation 31
flood rim level 107
full-flow characteristic 82

G

gate valve 82
gauge pressure 65

H

high hazard 2
hose bibb vacuum breaker 88

I

Ideal Gas Law 70
implied warranty 37
indirect cross-connection 53
isolation 18

J

INDEX

K

kinetic energy 60

L

lawn irrigation systems 90
low hazard 2

M

maximum contaminant level (MCL) 20
method 81

N

negligence 37
non-potable water 2

O

onset time .... 17 ordinance 192

P

parallel installation 133
pathogenicity 17
permanent cross-connection 2
personal injury 38
pit installation 95
policies 193
pollutant 2
potable water 1
potential energy 60
pressure 59
pressure vacuum breaker 88
primacy 6
prohibited cross-connection 34
purveyor 39

Q

R

reclaimed water 211
reduced pressure detector assembly
 115

reduced pressure principle backflow
prevention assembly 97
red tag 161
resilient seated 83

S

Safe Drinking Water Act (SDWA) 6
service contract 192
sight tube 131
single check valve 111
spill-resistant vacuum breaker 90
spit 101
spitting 104
standard of care 37
standard operating procedures 194
state regulations 33
sub-atmospheric 68
survival time 18
susceptibility 27

T

temperature and pressure valve 120
temporary cross-connection 5
test gauge 127
toxicity 27

U

V

vault installations 96
Venturi Effect 71
virulence 17

W

wedge valve 82

X

Y

Z

Table of Appendices

Appendix A	AWWA Policy Statement	229
Appendix B	Pertinent Sections of FDEP Code	231
Appendix C	Abbreviations	245
Appendix D	Field Test Procedures	247
Appendix E	Test and Maintenance Report Forms	253
Appendix F	Troubleshooting Guide	257
Appendix G	Test Kits: Suppliers and Repair Locations	261
Appendix H	Repair Parts Suppliers	265
Appendix I	Plumbing Code: Testing and Lawn Irrigation	269
Appendix J	FCCC & HR Approval Process	271
Appendix K	Testing the RP Chart	273
Appendix L	A.S.S.E. Numbers: Approved Assemblies and Devices	275
Appendix M	Building a Model Ordinance	277
Appendix N	Elements of Program Ordinance	279
Appendix O	Form Letters: Required Annual Testing, Follow-up, Final, and Repair	283
Appendix P	List of Health Hazard Facilities	289
Appendix Q	CCC Questionnaire	293
Appendix R	Survey Inspection Forms	295
Appendix S	Sample Form Letters: Survey Inspection	299
Appendix T	List of Common Cross-Connection Locations	303
Appendix U	Incident Report Form	303
Appendix V	Selecting Proper Backflow Prevention Discussion	311
Appendix W	Nomenclature Chart	315
Appendix X	Lawn Irrigation Chart	317
Appendix Y	Article on Reuse Water	319
Appendix Z	Freeze Protection	333

Appendix A

AWWA Policy Statement

AWWA Statement of Policy on Public Water Supply Matters
CROSS-CONNECTIONS
Adopted by the Board of Directors Jan. 26, 1970
Revised June 24, 1979
Reaffirmed June 10, 1984
Revised Jan. 28, 1990, and Jan. 21, 2001

The American Water Works Association (AWWA) recognizes water purveyors have the responsibility to supply potable water to their customers. In the exercise of this responsibility, water purveyors or other responsible authorities must implement, administer, and maintain ongoing backflow prevention and cross-connection control programs to protect public water systems from the hazards originating on the premises of their customers and from temporary connections that may impair or alter the water in the public water systems. The return of any water to the public water system after the water has been used for any purpose on the customer's premises or within the customer's piping system is unacceptable and opposed by AWWA.

The water purveyor shall assure that effective backflow prevention measures, commensurate with the degree of hazard, are implemented to ensure continual protection of the water in the public water distribution system. Customers, together with other authorities, are responsible for preventing contamination of the private plumbing system under their control and the associated protection of the public water system.

If appropriate backflow-prevention measures have not been taken, the water purveyor shall take or cause to be taken necessary measures to ensure that the public water distribution system is protected from any actual or potential backflow hazard. Such action would include the testing, installation, and continual assurance of proper operation and installation of backflow-prevention assemblies, devices, and methods commensurate with the degree of hazard at the service connection or at the point of cross connection or both. If these actions are not taken, water service shall ultimately be eliminated.

To reduce the risk private plumbing systems pose to the public water distribution system, the water purveyor's backflow prevention program should include public education regarding the hazards backflow presents to the safety of drinking water and should include coordination with the cross connection efforts of local authorities, particularly health and plumbing officials. In areas lacking a health or plumbing enforcement agency, the water purveyor should additionally promote the health and safety of private plumbing systems to protect its customers from the hazards of backflow.

Appendix B

Florida Department of Environmental Protection
Rule 62-555.360

**Pertinent Sections of
Florida Administrative Code 62-550 and 62-555**

CHAPTER 62-550 DRINKING WATER STANDARDS, MONITORING, AND REPORTING

PART II DEFINITIONS

62-550.200 Definitions for Public Water Systems:

(22) "CROSS-CONNECTION" means any physical arrangement whereby a public water supply is connected, directly or indirectly, with any other water supply system, sewer, drain, conduit, pool, storage reservoir, plumbing fixture, or other device which contains or may contain contaminated water, sewage or other waste, or liquid of unknown or unsafe quality which may be capable of imparting contamination to the public water supply as the result of backflow. By-pass arrangements, jumper connections, removable sections, swivel or changeable devices, and other temporary or permanent devices through which or because of which backflow could occur are considered to be cross-connections.

(56) "MAXIMUM CONTAMINANT LEVEL" (MCL) means the maximum permissible level of a contaminant in water which is delivered to any user of a public water system.

Specific Authority 403.861(9) FS. Law Implemented 403.853, 403.854, 403.8615, 403.862 FS. History–New 11-9-77, Amended 1-13-81, 11-19-87, Formerly 17-22.103, Amended 1-18-89, 5-7-90, 1-3-91, 1-1-93, Formerly 17-550.200, Amended 9-7-94, 12-9-96, 9-22-99, 8-1-00, 11-27-01, 4-3-03, 11-25-03, 10-14-04, 11-28-04, 1-17-05, 12-30-11.

62-550.720 Recordkeeping:

Suppliers of water shall retain on their premises, or at a convenient location near their premises, the following records:
(1) Copies of any written reports, summaries, or communications relating to cross-connection control program or sanitary surveys of the system conducted by the system itself, by a private consultant, or by any local, State or Federal agency, shall be kept for a period not less than 10 years after completion of the sanitary survey.
(2) Records concerning a variance or exemption granted to the system shall be kept for a period ending not less than 5 years following the expiration of the variance and exemption.
(3) Monthly operation reports shall be kept for a period of not less than 10 years.
(4) Any system subject to the requirements of Rule 62-550.800, F.A.C., shall retain, for no fewer than 12 years, original records of all sampling data and analyses, reports, surveys, letters, evaluations, schedules, Department determinations, and any other information required by Rule 62-550.800, F.A.C.

Specific Authority 403.861(9) FS. Law Implemented 403.861(16) FS. History–New 11-19-87, Formerly 17-22.820, Amended 1-18-89, 1-1-93, 7-4-93, Formerly 17-550.720, Amended 11-27-01.

CHAPTER 62-555 PERMITTING, CONSTRUCTION, OPERATION, AND MAINTENANCE OF PUBLIC WATER SYSTEMS

62-555.330 Engineering References for Public Water Systems:

In addition to the requirements of this chapter, the requirements and standards contained in the following technical publications are hereby incorporated by reference and shall be applied in determining whether permits to construct or alter public water system components, excluding wells (but including well pumping equipment and appurtenances), shall be issued or denied. Each of these publications is available from the publisher or source listed for the publication. The specific requirements contained in this chapter supersede the requirements and standards contained in these publications.

(1) *Recommended Practice for Backflow Prevention and Cross-Connection Control*, AWWA Manual M14, Third Edition, 2004, American Water Works Association (AWWA). Published by the AWWA, 6666 W. Quincy Avenue, Denver, CO 80235.

(2) *Ultraviolet Disinfection Guidelines for Drinking Water and Water Reuse*, December 2000, National Water Research Institute (NWRI) and American Water Works Association Research Foundation. Published by the NWRI, P.O. Box 20865, Fountain Valley, CA 92728-0865.

(3) *Water Distribution Systems Handbook*, 1999, Larry W. Mays, Editor in Chief. Published by McGraw-Hill, Post Office Box 182604, Columbus, OH 43218-2605.

Specific Authority 403.861(9) FS. Law Implemented 403.861(7) FS. History–New 11-19-87, Formerly 17-22.630, Amended 1-18-89, 1-3-91, 1-1-93, Formerly 17-555.330, Amended 9-22-99, 8-28-03, 5-5-2014

62-555.360 Cross-Connection Control for Public Water Systems:

(1) Cross-connections, as defined in Rule 62-550.200, F.A.C., are prohibited unless appropriate backflow protection is provided to prevent backflow through the cross-connection into the public water system. This does not prohibit a public water system from being interconnected to another public water system of the same type without backflow protection (i.e., a community water system [CWS] may be interconnected to another CWS without backflow protection, a non-transient non-community water system [NTNCWS] may be interconnected to another NTNCWS without backflow protection, and a transient non-community water system [TWS] may be interconnected to another TWS without backflow protection).

(a) Appropriate backflow protection for various applications is described in *Recommended Practice for Backflow Prevention and Cross-Connection Control: AWWA Manual M14,* Third Edition, as clarified and modified in paragraphs (b) and (c) below and in Table 62-555.360-2, which appears at the end of this section. The third edition of *AWWA Manual M14* is incorporated herein by reference; is available from the American Water Works Association, 6666 West Quincy Avenue, Denver, CO 80235, www.awwa.org; and is available for review at the Department of Environmental Protection, Source and Drinking Water Program, MS 3520, 2600 Blair Stone Road, Tallahassee, Florida 32399-2400, at the Department of Environmental Protection district offices, and at the Approved County Health Departments.

(b) Except for the temporary cross-connections described in paragraph (c) below, cross-connections between a public water system and a wastewater system or reclaimed water system are prohibited (i.e., an air gap shall be maintained between any public water system and any wastewater system or reclaimed water system). The Department shall allow an exception to this requirement if the supplier of water provides justification for the exception and provides alternative backflow protection that achieves a level of reliability and public health protection similar to that achieved by an air gap (e.g., two biannually-tested reduced-pressure principle assemblies installed in series); however, in no case shall the Department allow a single, annually-tested mechanical backflow preventer to be used as the only protection against backflow of wastewater or reclaimed water into a public water system.

(c) Temporary cross-connections may be made between a public water system and a wastewater system or reclaimed water system for either of the following purposes:

1. To supply water for flushing or testing a new wastewater force main or new reclaimed water main, in which case a double check valve assembly or reduced-pressure principle assembly shall be provided at the cross-connection.

2. To supply water for temporarily operating a new reclaimed water main that has not yet been connected to a reclaimed water supply, in which case a reduced-pressure principle assembly shall be provided at the cross-connection.

(2) Each community water system (CWS) shall establish and implement a cross-connection control program utilizing backflow protection at or for service connections from the CWS in order to protect the CWS from contamination caused by cross-connections on customers' premises. This program shall include a written plan that is developed using recommended practices of the American Water Works Association set forth in *Recommended Practice for Backflow Prevention and Cross-Connection Control: AWWA Manual M14,* Third Edition, as clarified and modified in paragraph (a) below. The third edition of *AWWA Manual M14* is incorporated herein by reference and is available as indicated in paragraph 62-555.360(1)(a), F.A.C.

(a) The minimum components that each CWS shall include in its written cross-connection control plan are listed and described in Table 62-555.360-1, which appears at the end of this section. The categories of customers for which each CWS shall ensure backflow protection is provided at or for the service connection from the CWS to the customer are listed in Table 62-555.360-2, which appears at the end of this section.

(b) Each CWS serving more than 10,000 persons shall prepare and submit cross-connection control program annual reports. The first annual report shall cover calendar year 2016, and subsequent annual reports shall cover each calendar year thereafter. These reports shall be prepared using Form 62-555.900(13), Cross-Connection Control Program Annual Report, effective 5-5-14, which is incorporated herein by reference and which is available as described in Rule 62-555.900, F.A.C., and at http://www.flrules.org/Gateway/reference.asp?No=Ref-04104. These reports shall be submitted to the appropriate Department of Environmental Protection district office or Approved County Health Department within three months after the end of the calendar year covered by the report.

(3) Upon discovery of a prohibited or inappropriately protected cross-connection, public water systems either shall ensure that the cross-connection is eliminated, shall ensure that appropriate backflow protection is installed to prevent backflow into the public water system, or shall discontinue water service. If the discovered cross-connection is on the premises of a customer of a community water system (CWS) and if the customer's premises is in a category described in

Table 62-555.360-2, which appears at the end of this section, the CWS shall ensure that appropriate backflow protection is provided at or for the water service connection to the customer regardless of whether the cross-connection is eliminated or whether internal backflow protection is installed at the cross-connection to the customer's plumbing system.

Table 62-555.360-1: Minimum Components that Each Community Water System (CWS)
Shall Include in Its Written Cross-Connection Control (CCC) Plan (Effective 5-5-14)

Component Number and Description
I. Legal authority for the CWS's CCC program – i.e., an ordinance, a bylaw or resolution, or water service rules and regulations. The legal authority shall include or reference Components 2 and 3 below.
II. The CWS's policy establishing where backflow protection at or for service connections from the CWS is mandatory. A. This policy shall identify categories of customers for which the CWS is requiring backflow protection at or for the service connection to the customer and shall specify the minimum backflow protection that the CWS is requiring for each such category of customers. B. This policy shall be no less stringent than Table 62-555.360-2, which appears at the end of Rule 62-555.360, F.A.C.
III. The CWS's policy regarding ownership, installation, inspection/testing, and maintenance of backflow protection that the CWS is requiring at or for service connections from the CWS. A. This policy shall specify whether the CWS or customer is responsible for installation, inspection/testing, and maintenance of backflow protection being required at or for service connections. B. This policy shall specify design and performance standards, and shall specify installation criteria, for new backflow protection being required at or for service connections. Installation criteria shall be consistent with installation criteria in *AWWA Manual M14* as incorporated into subsection 62-555.360(2), F.A.C., and shall assure the backflow protection is installed as close as practical to the CWS's meter or customer's property line but, in all cases, before the first distribution line off of the customer's water service line. C. This policy shall specify the frequency for inspecting air gaps (AGs) being required at or for service connections and shall specify qualifications for persons inspecting such AGs. All AGs being required at or for service connections pursuant to Table 62-555.360-2, which appears at the end of Rule 62-555.360, F.A.C., shall be inspected at least annually. D. This policy shall specify the frequency for testing backflow preventer assemblies[1] being required at or for service connections, shall specify qualifications for persons testing such assemblies, and shall specify test procedures for such assemblies. Assemblies being required at or for non-residential service connections[2] pursuant to Table 62-555.360-2, which appears at the end of Rule 62-555.360, F.A.C., shall be tested after installation or repair and at least annually thereafter and shall be repaired if they fail to meet performance standards. Assemblies being required at or for residential service connections[2] pursuant to Table 62-555.360-2 shall be tested after installation or repair and at least biennially thereafter and shall be repaired if they fail to meet performance standards. E. This policy shall specify the frequency for refurbishing or replacing dual check devices (DuCs) being required at or for service connections. DuCs being required at or for service connections pursuant to Table 62-555.360-2, which appears at the end of Rule 62-555.360, F.A.C., shall be refurbished or replaced at least once every 5 to 10 years or at a lesser frequency determined by the CWS if the CWS documents that the lesser frequency is appropriate based on data from spot-testing DuCs in its system or based on data from backflow sensing meters in its system.

IV. The CWS's procedures for evaluating customers' premises to establish the category of customer and the backflow protection being required at or for the service connection(s) from the CWS to the customer.[3] A. The CWS shall evaluate the customer's premises at a newly constructed service connection before the CWS begins supplying water to the service connection. B. The CWS shall evaluate the customer's premises at an existing – i.e., previously constructed – service connection whenever the customer connects to a reclaimed water distribution system, whenever an auxiliary water system is discovered on the customer's premises, whenever a prohibited or inappropriately protected cross-connection is discovered on the customer's premises, and whenever the customer's premises is altered under a building permit in a manner that could change the backflow protection required at or for a service connection to the customer.	
V. The CWS's procedures for maintaining CCC program records.[4] A. The CWS shall maintain a current inventory of backflow protection being required at or for service connections from the CWS. B. The CWS shall maintain records of the installation, inspection/testing, and repair of backflow protection being required at or for service connections from the CWS.	

[1] Backflow preventer assemblies include the following: double check valve assemblies (DCs) and double check detector assemblies (DCDAs); pressure vacuum breaker assemblies (PVBs); and reduced-pressure principle assemblies (RPs) and reduced-pressure principle detector assemblies (RPDAs).

[2] For the purpose of this table, "residential service connection" means any service connection, including any dedicated irrigation or fire service connection, that is two inches or less in diameter and that supplies water to a building, or premises, containing only dwelling units; and "non-residential service connection" means any other service connection.

[3] CWSs may evaluate customers' premises using questionnaires, reviews of construction plans or pertinent records, on-site inspections, or any combination thereof.

[4] CWSs may maintain all records in either electronic or paper format.

Table 62-555.360-2: Categories of Customers for Which Each Community Water System (CWS) Shall Ensure Minimum Backflow Protection Is Provided at or for the Service Connection from the CWS to the Customer (Effective 5-5-14)

Category of Customer	Minimum Backflow Protection[1] to Be Provided at or for the Service Connection from the CWS to the Customer
Beverage processing plant, including any brewery	DC if the plant presents a low hazard[2]; or RP if the plant presents a high hazard[2]
Cannery, packing house, rendering plant, or any facility where fruit, vegetable, or animal matter is processed, excluding any premises where there is only restaurant or food service facility	RP
Car wash	RP
Chemical plant or facility using water in the manufacturing, processing, compounding, or treatment of chemicals, including any facility where a chemical that does not meet the requirements in paragraph 62-555.320(3)(a), F.A.C., is used as an additive to the water	RP
Dairy, creamery, ice cream plant, cold-storage plant, or ice manufacturing plant	RP[3]
Dye plant	RP

Film laboratory or processing facility or film manufacturing plant, excluding any small, noncommercial darkroom facility	RP
Hospital; medical research center; sanitarium; autopsy facility; medical, dental, or veterinary clinic where surgery is performed; or plasma center	RP
Laboratory, excluding any laboratory at an elementary, middle, or high school	RP
Laundry (commercial), excluding any self-service laundry or Laundromat	RP
Marine repair facility, marine cargo handling facility, or boat moorage	RP
Metal manufacturing, cleaning, processing, or fabricating facility using water in any of its operations or processes, including any aircraft or automotive manufacturing plant	DC if the facility presents a low hazard[2]; or RP if the facility presents a high hazar[2]
Mortuary	RP
Premises where oil or gas is produced, developed, processed, blended, stored, refined, or transmitted in a pipeline or where oil or gas tanks are repaired or tested, excluding any premises where there is only a fuel dispensing facility	RP
Premises where there is an auxiliary or reclaimed water system[4,5]	A. At or for a residential service connection[6]: DuC[7] B. At or for a non-residential service connection[6]: DC if the auxiliary or reclaimed water is a low hazard[8,9]; or RP if the auxiliary or reclaimed water is a high hazard[8,9]
Premises where there is a cooling tower	RP
Premises where there is an irrigation system that is using potable water and that... I. Is connected directly to the CWS's distribution system via a dedicated irrigation service connection II. Is connected internally to the customer's plumbing system	I. At or for a residential or non-residential dedicated irrigation service connection[6]: PVB if backpressure cannot develop in the downstream piping[10]; or RP if backpressure could develop in the downstream piping[10] II. None[11]
Premises where there is a wet-pipe sprinkler, or wet standpipe, fire protection system that is using potable water and that... I. Is connected directly to the CWS's distribution system via a dedicated fire service connection[12] II. Is connected internally to the customer's plumbing system	I. A. At or for a residential didicated fire service connection[6]: DuC if the fire protection system contains no chemical additives and is not connected to an auxiliary water system[4]; or RP or RPDA the fire protection system contains chemical additives or is connected to an auxiliary water system[4,13] I. B. At or for a non-residential dedicated fire service connection[6]: DC or DCDA if the fire protection system contains no chemical additives and is not connected to an auxiliary water system[4]; or RP or RPDA if the fire protection system contains chemical additives or connected to an auxiliary water system[4,13] II. None[11]

Radioactive material processing or handling facility or nuclear reactor	RP
Paper products plant using a wet process	RP
Plating facility, including any aircraft or automotive manufacturing plant	RP
Restricted-access facility	RP
Steam boiler plant	RP
Tall building – i.e., a building with five or more floors at or above ground level	DC if the customer has no potable water distribution lines connected to the suction side of a booster pump; or RP if the customer has one or more potable water distribution lines connected to the suction side of a booster pump
Wastewater treatment plant or wastewater pumping station	RP
Customer supplied with potable water via a temporary a permanent service connection from a CWS fire hydrant	Varies[14]

[1] Means of backflow protection, listed in an increasing level of protection, include the following: a dual check device (DuC); a double check valve assembly (DC) or double check detector assembly (DCDA); a pressure vacuum breaker assembly (PVB); a reduced-pressure principle assembly (RP) or reduced-pressure principle detector assembly (RPDA); and an air gap. A PVB may not be used if backpressure could develop in the downstream piping.

[2] The CWS shall determine the degree of hazard. "Low hazard" or "non-health hazard" and "high hazard" or "health hazard" are defined in *AWWA Manual M14* as incorporated in paragraph 62-555.360(l)(a), F.A.C., and subsection 62-555.360(2), F.A.C.

[3] A DC may be provided if it was installed before 5-5-14; and if such a DC is replaced on or after 5-5-14, it may be replaced with another DC.

[4] For the purpose of this table, "auxiliary water system" means a pressurized system of piping and appurtenances using auxiliary water, which is water other than the potable water being supplied by the CWS and which includes water from any natural source such as a well, pond, lake, spring, stream, river, etc., includes reclaimed water, and includes other used water or industrial fluids described in *AWWA Manual M14* as incorporated in paragraph 62-555.360(l)(a), F.A.C., and subsection 62-555.360(2), F.A.C.; however, "auxiliary water system" specifically excludes any water recirculation or treatment system for a swimming pool, hot tub, or spa. (Note that reclaimed water is a specific type of auxiliary water and a reclaimed water system is a specific type of auxiliary water system.)

[5] The Department shall allow an exception to the requirement for backflow protection at or for a residential or non-residential service connection from a CWS to premises where there is an auxiliary or reclaimed water system if all of the following conditions are met:

•The CWS is distributing water only to land owned by the owner of the CWS.
•The owner of the CWS is also the owner of the entire auxiliary or reclaimed water system up to the points of auxiliary or reclaimed water use.
•The CWS conducts at least biennial inspections of the CWS and the entire auxiliary or reclaimed water system to detect and eliminate any cross-connections between the two systems.

[6] For the purpose of this table, "residential service connection" means any service connection, including any dedicated irrigation or fire service connection, that is two inches or less in diameter and that supplies water to a building, or premises, containing only dwelling units; and "non-residential service connection" means any other service connection.

[7] A DuC may be provided only if there is no known cross-connection between the plumbing system and the auxiliary or reclaimed water system on the customer's premises. Upon discovery of any cross-connection between the plumbing system and any reclaimed water system on the customer's premises, the CWS shall ensure that the cross-connection is eliminated. Upon discovery of any cross-connection between the plumbing system and any auxiliary water system other than a reclaimed water system on the customer's premises, the CWS shall ensure that the cross-connection is eliminated or shall ensure that

the backflow protection provided at or for the service connection is equal to that required at or for a non-residential service connection.

[8] Reclaimed water regulated under Part III of Chapter 62-610, F.A.C., is a low hazard unless it is stored with surface water in a pond that is part of a stormwater management system, in which case it is a high hazard; well water is a low hazard unless determined otherwise by the CWS; industrial fluids and used water other than reclaimed water are high hazards unless determined otherwise by the CWS; reclaimed water not regulated under Part III of Chapter 62-610, F.A.C., and surface water are high hazards.

[9] Upon discovery of any cross-connection between the plumbing system and any reclaimed water system on the customer's premises, the CWS shall ensure that the cross-connection is eliminated.

[10] A DC may be provided if both of the following conditions are met:
• The dedicated irrigation service connection initially was constructed before 5-5-14.
• No chemicals are fed into the irrigation system.

[11] The CWS may rely on the internal backflow protection required under the *Florida Building Code* or the predecessor State plumbing code. The CWS may, but is not required to, ensure that such internal backflow protection is inspected/tested and maintained the same as backflow protection provided at or for service connections from the CWS.

[12] The Department shall allow an exception to the requirement for backflow protection at or for a residential or non-residential dedicated fire service connection from a CWS to a wet-pipe sprinkler, or wet standpipe, fire protection system if both of the following conditions are met:
• The fire protection system was installed and last altered before 5-5-14.
• The fire protection system contains no chemical additives and is not connected to an auxiliary water system as defined in Footnote 4.

[13] Upon discovery of any cross-connection between the fire protection system and any reclaimed water system on the customer's premises, the CWS shall ensure that the cross-connection is eliminated.

[14] The CWS shall ensure that backflow protection commensurate with the degree of hazard is provided at or for the service connection from its fire hydrant.

Rulemaking Authority 403.086(8), 403.853(3), 403.861(9) FS. Law Implemented 403.086(8), 403.852(12), 403.853(1), 403.855(3), 403.861(17) FS. History–New 11-19-87, Formerly 17-22.660, Amended 1-18-89, 1-3-91, 1-1-93, Formerly 17-555.360, Amended 8-28-03, 5-5-14.

62-610.419 Application/Distribution Systems and Cross-Connection Control:

(1) There are no above ground hose bibbs (spigots or other hand-operated connections).
(2) Subsurface application systems may be used if the reclaimed water is made available to the plant root zone and the hydraulic loading rates and cycles comply with Rule 62-610.423, F.A.C.
(3) No cross-connections to potable water systems shall be allowed. For systems permitted under subsection 62-610.418(2), F.A.C., the permittee shall develop and obtain Department acceptance for a cross-connection control and inspection program as discussed in Rules 62-610.469 and 62-555.360, F.A.C.
(4) For all systems, there shall be readily identifiable "non-potable" notices, marking, or coding on application/distribution facilities and appurtenance.

Rulemaking Authority 403.051, 403.061, 403.064, 403.087 FS. Law Implemented 403.021, 403.051, 403.061, 403.062, 403.064, 403.085, 403.086, 403.087, 403.088 FS. History–New 4-4-89, Formerly 17-610.419, Amended 1-9-96, 11-19-07.

62-610.469 Application/Distribution Systems and Cross-Connection Control:

(3) Except as specifically allowed in this paragraph, above ground hose bibbs (spigots or other hand operated connections) shall not be present. Hose bibbs shall be located in locked vaults, service boxes, or compartments which shall be clearly labeled as being of nonpotable quality (bearing the words in English and Spanish: "Do not drink" together with the equivalent standard inter-

national symbol). Hose bibbs which can only be operated by a special tool may be placed in nonlockable vaults, service boxes, or compartments clearly labeled as nonpotable water (bearing the words in English and Spanish: "Do not drink" together with the equivalent standard international symbol). Vaults, service boxes, and compartments meeting the requirements of this rule may be located above or below grade. For restricted access sites, the Department shall approve the use of hose bibbs that are not in vaults, service boxes, or compartments, if the applicant provides an affirmative demonstration in the engineering report that alternate means of securing the hose bibb will preclude unauthorized use of the hose bibb. If the Department approves alternate measures for securing hose bibbs for restricted access sites, the alternate control measures and the hose bibb shall be color coded and clearly labeled as being of nonpotable quality (bearing the words in English and Spanish: "Do not drink" together with the equivalent standard international symbol).

7. Assessment of the physical condition and integrity of facilities to be converted.

8. Reasonable assurance that cross-connections will not result, public health will be protected, and the integrity of water, wastewater, and reclaimed water systems will be maintained when the conversion is made.

(7) Cross-connection control.

(a) No cross-connections to potable water systems shall be allowed. The permittee shall submit documentation of Department acceptance for a cross-connection control and inspection program, pursuant to Rule 62-555.360, F.A.C., for all public water supply systems located within the area to be served by reclaimed water.

(b) Reclaimed water shall not enter a dwelling unit or a building containing a dwelling unit except as allowed by Rules 62-610.476 and 62-610.479, F.A.C.

(c) Maximum obtainable separation of reclaimed water lines and domestic water lines shall be practiced. A minimum horizontal separation of three feet (outside to outside) shall be maintained between reclaimed water lines and either potable water mains or sewage collection lines. The Department shall approve smaller horizontal separation distances if one of the following conditions is met:

1. The top of the reclaimed water line is installed at least 18 inches below the bottom of the potable water line.

2. The reclaimed water line is encased in concrete.

3. The applicant provides an affirmative demonstration in the engineering report that another alternative will result in an equivalent level of protection.

(d) The provisions of Chapter 62-604, F.A.C., are applicable to in-ground crossings. No vertical or horizontal separation distances are required for above-ground crossings.

(e) Separation distance requirements in paragraphs 62-610.469(7)(c) and (d), F.A.C., apply to transmission and distribution systems located in rights-of-ways. Similar separation distances are recommended, but are not required on properties where reclaimed water is being used.

(f) All reclaimed water valves and outlets shall be appropriately tagged or labeled (bearing the words in English and Spanish: "Do not drink" together with the equivalent standard international symbol) to warn the public and employees that the water is not intended for drinking. All piping, pipelines, valves, and outlets shall be color coded, or otherwise marked, to differentiate reclaimed water from domestic or other water. Effective January 1, 1996, underground piping which is not manufactured of metal or concrete, shall be color coded for reclaimed water distribution systems using Pantone Purple 522C using light stable colorants. Underground metal and concrete pipe shall be color coded or marked using purple as a predominant color. If tape is used to mark the pipe, the tape shall be permanently affixed to the top and each side of the pipe (three locations parallel to the axis of the pipe). For pipes less than 24 inches in diameter, a single tape may be used along the top of the pipe. Visible, above-ground portions of the reclaimed water distribution system shall be clearly color coded or marked. New systems and expansions of existing systems for which permit applications are submitted to the Department on or after January 1, 1996, shall comply with this color-coding standard. It is recommended, but shall not be required, that distribution and application facilities located on private properties, including residential properties, be color coded using Pantone Purple 522C.

(g) The return of reclaimed water to the reclaimed water distribution system after the reclaimed water has been delivered to a user is prohibited.

(h) The permittee is responsible for conducting inspections within the reclaimed water service area to verify proper connections, monitor proper use of reclaimed water, and minimize the potential for cross-connections. Inspections are required when customers first connect to the reclaimed water distribution system. Periodic inspections are required as specified in the cross-connection control and inspection program.

Rulemaking Authority 403.051, 403.061, 403.087 FS. Law Implemented 403.021, 403.051, 403.061, 403.062, 403.085, 403.086, 403.087, 403.088 FS. History–New 4-4-89, Amended 4-2-90, Formerly 17-610.469, Amended 1-9-96, 8-8-99.

62-610.660 Cross-Connection Control and Protection of the Reclaimed Water Supply:

(1) No cross-connections to potable water systems shall be allowed

(2) For all systems, there shall be readily identifiable "non-potable" or "do not drink" notices, marking, or coding on application/distribution facilities and appurtenances.

(3) Protection of Reclaimed Water Supply.

(a) The return of reclaimed water to the reclaimed water distribution system after the reclaimed water has been delivered to an industrial facility is prohibited. This prohibition shall not apply to industrial sites which were using reclaimed water before January 1, 1996, or which were identified as future users of reclaimed water in a complete permit application received by the Department before January 1, 1996.

(b) The permittee shall conduct an evaluation of the potential for cross-connections and backflow to the reclaimed water distribution system. This analysis shall include an evaluation of the types of substances present at the industrial site which could potentially backflow into the reclaimed water system and the risk associated with possible backflow. The applicant shall evaluate the need for backflow prevention devices on the reclaimed water connection to the industrial facility. This analysis shall be included in the engineering report. A backflow prevention device shall be provided on the reclaimed water service connection to the industrial site, unless the evaluation in the engineering report provides reasonable assurances that there is minimal risk of cross-connection or backflow with contamination of the reclaimed water supply. This requirement for backflow prevention devices shall not apply to industrial sites which were using reclaimed water before January 1, 1996 or which were identified as future users of reclaimed water in a complete permit application received by the Department before January 1, 1996.

Rulemaking Authority 403.051, 403.061, 403.087 FS. Law Implemented 403.021, 403.051, 403.061, 403.062, 403.085, 403.086, 403.087, 403.088 FS. History–New 4-4-89, Amended 4-2-90, Formerly 17-610.660, Amended 1-9-96.

CHAPTER 62-610 REUSE OF RECLAIMED WATER AND LAND APPLICATION

62-610.419 Application/Distribution Systems and Cross-Connection Control:

(1) New reclaimed water application/distribution systems (and replacements of existing systems) shall be designed such that:
(a) Drawdown of holding ponds shall be accomplished as soon as is appropriate. For this purpose, a minimum hydraulic capacity of 1.5 times the maximum daily flow (at which adequate treatment can be provided) of the treatment plant is required; the actual hydraulic criterion selected shall be justified in the engineering report on the basis of holding pond storage capacity, assimilative capacity of the soil-plant system, and similar considerations;
(b) The system design facilitates maintenance and harvesting of the irrigated areas and precludes damage from the use of maintenance equipment or harvesting machinery;
(c) The system is designed to prevent clogging with algae;
(d) Exposed pipes are labeled;
(e) Spray equipment is designed and located to minimize aerosol carry-over from the application area (e.g., low pressure sprays) to areas beyond the setback distances described in Rule 62-610.421(2), F.A.C.; and
(f) There are no above ground hose bibbs (spigots or other hand-operated connections).
(2) Subsurface application systems may be used if the reclaimed water is made available to the plant root zone and the hydraulic loading rates and cycles comply with Rule 62-610.423, F.A.C.
(3) No cross-connections to potable water systems shall be allowed.
(4) For all systems, there shall be readily identifiable "non-potable" notices, marking, or coding on application/distribution facilities and appurtenance.

Specific Authority 403.051, 403.061, 403.087 FS. Law Implemented 403.021, 403.051, 403.061, 403.062, 403.085, 403.086, 403.087, 403.088 FS. History–New 4-4-89, Formerly 17-610.419, Amended 1-9-96.

62-610.469 Application/Distribution Systems and Cross-Connection Control:

(1) New slow-rate land application systems, expansions of existing distribution systems, and replacement of existing systems shall be designed to provide, at a minimum, hydraulic capacity of 1.5 times maximum daily flow (at which adequate treatment can be provided) of the treatment facility. The actual hydraulic criterion selected shall be justified in the engineering report on the reclaimed water.
(2) Application of reclaimed water on public access facilities shall be controlled by agreement with the wastewater management entity or by local ordinance.
(3) Except as specifically allowed in this paragraph, above ground hose bibbs (spigots or other hand operated connections) shall not be present. Hose bibbs shall be located in locked vaults, service boxes, or compartments which shall be clearly labeled as being of nonpotable quality (bearing the words in English and Spanish: "Do not drink" together with the equivalent standard international symbol). Hose bibbs which can only be operated by a special tool may be placed in nonlockable vaults, service boxes, or compartments clearly labeled as nonpotable water (bearing the words in English and Spanish: "Do not drink" together with the equivalent standard international symbol). Vaults, service boxes, and compartments meeting the requirements of this rule may be located above or below grade. For restricted access sites, the Department shall

approve the use of hose bibbs that are not in vaults, service boxes, or compartments, if the applicant provides an affirmative demonstration in the engineering report that alternate means of securing the hose bibb will preclude unauthorized use of the hose bibb. If the Department approves alternate measures for securing hose bibbs for restricted access sites, the alternate control measures and the hose bibb shall be color coded and clearly labeled as being of nonpotable quality (bearing the words in English and Spanish: "Do not drink" together with the equivalent standard international symbol).

(4) Reclaimed water shall not be used to fill swimming pools, hot tubs, or wading pools.

(5) Cross-connection control.

(a) No cross-connections to potable water systems shall be allowed. The permittee shall submit documentation of Department acceptance for a cross-connection control and inspection program, pursuant to Rule 62-555.360, F.A.C., for all public water supply systems located within the area to be served by reclaimed water.

(b) Reclaimed water shall not enter a dwelling unit or a building containing a dwelling unit except as allowed by Rules 62-610.476 and 62-610.479, F.A.C.

(c) Maximum obtainable separation of reclaimed water lines and domestic water lines shall be practiced. A minimum horizontal separation of three feet (outside to outside) shall be maintained between reclaimed water lines and either potable water mains or sewage collection lines. The Department shall approve smaller horizontal separation distances if one of the following conditions is met:

1. The top of the reclaimed water line is installed at least 18 inches below the bottom of the potable water line.

2. The reclaimed water line is encased in concrete.

3. The applicant provides an affirmative demonstration in the engineering report that another alternative will result in an equivalent level of protection.

(d) The provisions of Chapter 62-604, F.A.C., are applicable to in-ground crossings. No vertical or horizontal separation distances are required for above-ground crossings.

(e) Separation distance requirements in Rules 62-610.469(7)(c) and (d), F.A.C., apply to transmission and distribution systems located in rights-of-ways. Similar separation distances are recommended, but are not required on properties where reclaimed water is being used.

(f) All reclaimed water valves and outlets shall be appropriately tagged or labeled (bearing the words in English and Spanish: "Do not drink" together with the equivalent standard international symbol) to warn the public and employees that the water is not intended for drinking. All piping, pipelines, valves, and outlets shall be color coded, or otherwise marked, to differentiate reclaimed water from domestic or other water. Effective January 1, 1996, underground piping which is not manufactured of metal or concrete, shall be color coded for reclaimed water distribution systems using Pantone Purple 522C using light stable colorants. Underground metal and concrete pipe shall be color coded or marked using purple as a predominant color. If tape is used to mark the pipe, the tape shall be permanently affixed to the top and each side of the pipe (three locations parallel to the axis of the pipe). For pipes less than 24 inches in diameter, a single tape may be used along the top of the pipe. Visible, above-ground portions of the reclaimed water distribution system shall be clearly color coded or marked. New systems and expansions of existing systems for which permit applications are submitted to the Department on or after January 1, 1996, shall comply with this color coding standard. It is recommended, but shall not be required, that distribution and application facilities located on private properties, including residential properties, be color coded using Pantone Purple 522C.

(g) The return of reclaimed water to the reclaimed water distribution system after the reclaimed water has been delivered to a user is prohibited.

(h) The permittee is responsible for conducting inspections within the reclaimed water service area to verify proper connections, monitor proper use of reclaimed water, and minimize the potential for cross-connections. Inspections are required when customers first connect to the reclaimed water distribution system. Periodic inspections are required as specified in the cross-connection control and inspection program.

Specific Authority 403.051, 403.061, 403.087 FS. Law Implemented 403.021, 403.051, 403.061, 403.062, 403.085, 403.086, 403.087, 403.088 FS. History–New 4-4-89, Amended 4-2-90, Formerly 17-610.469, Amended 1-9-96, 8-8-99.

62-610.660 Cross-Connection Control and Protection of the Reclaimed Water Supply:

(1) No cross-connections to potable water systems shall be allowed

(2) For all systems, there shall be readily identifiable "non-potable" or "do not drink" notices, marking, or coding on application/distribution facilities and appurtenances.

(3) Protection of Reclaimed Water Supply.

(a) The return of reclaimed water to the reclaimed water distribution system after the reclaimed water has been delivered to an industrial facility is prohibited. This prohibition shall not apply to industrial sites which were using reclaimed water before January 1, 1996, or which were identified as future users of reclaimed water in a complete permit application received by the Department before January 1, 1996.

(b) The permittee shall conduct an evaluation of the potential for cross-connections and backflow to the reclaimed water distribution system. This analysis shall include an evaluation of the types of substances present at the industrial site which could potentially backflow into the reclaimed water system and the risk associated with possible backflow. The applicant shall evaluate the need for backflow prevention devices on the reclaimed water connection to the industrial facility. This analysis shall be included in the engineering report. A backflow prevention device shall be provided on the reclaimed water service connection to the industrial site, unless the evaluation in the engineering report provides reasonable assurances that there is minimal risk of cross-connection or backflow with contamination of the reclaimed water supply. This requirement for backflow prevention devices shall not apply to industrial sites which were using reclaimed water before January 1, 1996 or which were identified as future users of reclaimed water in a complete permit application received by the Department before January 1, 1996.

Specific Authority 403.051, 403.061, 403.087 FS. Law Implemented 403.021, 403.051, 403.061, 403.062, 403.085, 403.086, 403.087, 403.088 FS. History–New 4-4-89, Amended 4-2-90, Formerly 17-610.660, Amended 1-9-96.

Source: 2004 Florida Department of Environmental Protection. All rights reserved.

Appendix C

Abbreviations

Abbreviations

AG — Air Gap

RP — Reduced Pressure Principle Assembly
RPZ - RPPA - RPBFP

RPDA — Reduced Pressure Detector Assembly

DCVA or **DC** — Double Check Valve Assembly

DCDA — Double Check Detector Assembly

PVB — Continuous Pressure Vacuum Breaker

SVB — Spill-resistant Vacuum Breaker

AVB — Atmospheric Vacuum Breaker

Appendix D

Field Test Procedures
RP, DCVA, PVB, SVB

RP FIELD TEST

Prep	NOTIFY CUSTOMER INSPECT AREA FLUSH TESTCOCKS *(open 4, 3, 2, 1, then close 1, 2, 3, 4)* INSTALL FITTINGS INSPECT TEST KIT - CLOSE ALL NEEDLE VALVES
Observe CV 1 Leaks or Closed Tight	ATTACH HIGH HOSE TO TESTCOCK 2 ATTACH LOW HOSE TO TESTCOCK 3 OPEN TESTCOCK 3 SLOWLY then OPEN LOW BLEED OPEN TESTCOCK 2 **SLOWLY** then OPEN HIGH BLEED CLOSE OUTLET SHUT-OFF VALVE CLOSE BLEEDS - HIGH FIRST, LOW LAST OBSERVE CV 1 - (RECORD as CLOSED TIGHT or LEAKING)
Record Relief Valve	OPEN HIGH CONTROL VALVE 1 FULL TURN OPEN LOW CONTROL **SLIGHTLY** - NO MORE THAN 1/4 TURN RECORD RV OPENING > or = 2.0 psi CLOSE LOW CONTROL
Observe CV 2 Leaks or Closed Tight	BLEED VENT (BY-PASS) HOSE ATTACH VENT HOSE TO TESTCOCK 4 CLOSE VENT (BY-PASS) CONTROL OPEN TESTCOCK 4 RESET GAUGE - (LOW BLEED) OPEN VENT CONTROL ONE FULL TURN OBSERVE WHETHER RELIEF VALVE VENT DRIPS *(IF THE RELIEF VENT DRIPS, RESET GAUGE [low bleed], IF RELIEF VALVE DRIPS A **SECOND TIME**, THEN CHECK VALVE 2 HAS FAILED AND MUST BE RE-PAIRED)* CLOSE VENT (BY-PASS) CONTROL **ONLY** IF CV 2 LEAKS (RECORD as CLOSED TIGHT or LEAKING)
Record CV 1	RESET GAUGE - (LOW BLEED) RECORD CV 1 DIFFERENTIAL (at least 5.0 PSI and > RV opening)
Record Outlet Shutoff Valve	CLOSE TESTCOCK 2 – **WAIT** & CHECK GAUGE FOR LEAKS IN OUTLET SHUT-OFF VALVE (RECORD as CLOSED TIGHT or LEAKING)
Record CV 2	CLOSE VENT CONTROL CLOSE TESTCOCKS 3 & 4 REMOVE VENT HOSE FROM TESTCOCK 4 MOVE LOW HOSE TO TESTCOCK 4 MOVE HIGH HOSE TO TESTCOCK 3 OPEN TESTCOCK 4 SLOWLY then OPEN LOW BLEED OPEN TESTCOCK 3 SLOWLY then OPEN HIGH BLEED CLOSE HIGH BLEED FIRST, CLOSE LOW BLEED SLOWLY RECORD CV 2 DIFFERENTIAL > or = 1.0 PSI
Final	CLOSE TESTCOCKS - REMOVE ALL EQUIPMENT OPEN ALL NEEDLE VALVES ON TEST KIT OPEN OUTLET SHUT-OFF VALVE SLOWLY

12/10/2001

RP Field Test Procedures with Numbers

Prep	NOTIFY CUSTOMER INSPECT AREA FLUSH TESTCOCKS *(open 4, 3, 2, 1, then close 1, 2, 3, 4)* INSTALL FITTINGS INSPECT TEST KIT - CLOSE ALL NEEDLE VALVES
Observe CV 1 Leaks or Closed Tight **1**	ATTACH HIGH HOSE TO TESTCOCK 2 ATTACH LOW HOSE TO TESTCOCK 3 OPEN TESTCOCK 3 SLOWLY then OPEN LOW BLEED OPEN TESTCOCK 2 **SLOWLY** then OPEN HIGH BLEED CLOSE OUTLET SHUT-OFF VALVE CLOSE BLEEDS - HIGH FIRST, LOW LAST OBSERVE CV 1 - (<u>RECORD</u> as CLOSED TIGHT or LEAKING)
Record Relief Valve **2**	OPEN HIGH CONTROL VALVE 1 FULL TURN OPEN LOW CONTROL **SLIGHTLY** - NO MORE THAN 1/4 TURN <u>RECORD</u> RV OPENING > or = 2.0 psi CLOSE LOW CONTROL
Observe CV 2 Leaks or Closed Tight **3**	BLEED VENT (BY-PASS) HOSE ATTACH VENT HOSE TO TESTCOCK 4 CLOSE VENT (BY-PASS) CONTROL OPEN TESTCOCK 4 RESET GAUGE - (LOW BLEED) OPEN VENT CONTROL ONE FULL TURN OBSERVE WHETHER RELIEF VALVE VENT DRIPS *(IF THE RELIEF VENT DRIPS, RESET GAUGE [low bleed], IF RELIEF VALVE DRIPS A **SECOND TIME**, THEN CHECK VALVE 2 HAS FAILED AND MUST BE REPAIRED)* CLOSE VENT (BY-PASS) CONTROL **ONLY** IF CV 2 LEAKS (<u>RECORD</u> as CLOSED TIGHT or LEAKING)
Record CV 1 **4**	RESET GAUGE - (LOW BLEED) <u>RECORD</u> CV 1 DIFFERENTIAL (at least 5.0 psi and > RV opening)

Record Outlet
Shut-Off Valve
 	
5
	CLOSE TESTCOCK 2 –
	    WAIT & CHECK GAUGE FOR LEAKS IN OUTLET SHUTOFF VALVE
	(<u>RECORD</u> as CLOSED TIGHT or LEAKING)
	    If the shut-off valve is leaking, **stop testing** the RP (do not proceed to test #6)

Record CV 2 only if NO FLOW thru Shut-Off Valve **6**	CLOSE VENT CONTROL CLOSE TESTCOCKS 3 & 4 REMOVE VENT HOSE FROM TESTCOCK 4 MOVE LOW HOSE TO TESTCOCK 4 MOVE HIGH HOSE TO TESTCOCK 3 OPEN TESTCOCK 4 SLOWLY then OPEN LOW BLEED OPEN TESTCOCK 3 SLOWLY then OPEN HIGH BLEED CLOSE HIGH BLEED FIRST, CLOSE LOW BLEED <u>SLOWLY</u> <u>RECORD</u> CV 2 DIFFERENTIAL > or = 1.0 PSI
Final	CLOSE TESTCOCKS - REMOVE ALL EQUIPMENT OPEN ALL NEEDLE VALVES ON TEST KIT OPEN OUTLET SHUT-OFF VALVE SLOWLY

12/10/2001

DCVA FIELD TEST with DIFFERENTIAL GAUGE - SINGLE HOSE

Prep
: NOTIFY CUSTOMER
INSPECT AREA
FLUSH TESTCOCKS
INSTALL FITTINGS
INSPECT TEST KIT - CLOSE ALL NEEDLE VALVES

CV 1
: INSTALL COMPENSATING TEE (BLEED VALVE) ON TESTCOCK 2
INSTALL SHORT TUBE ON TESTCOCK 3
INSTALL TEST GAUGE AND END OF LOW HOSE AT SAME HEIGHT WATER
  DISCHARGES FROM SHORT TUBE
ATTACH HIGH HOSE TO TESTCOCK 2
OPEN TESTCOCK 3 TO FILL TUBE
CLOSE TESTCOCK 3
OPEN TESTCOCK 2 SLOWLY
OPEN HIGH BLEED - BLEED AIR FROM GAUGE
CLOSE HIGH BLEED
CLOSE OUTLET SHUTOFF VALVE
CLOSE INLET SHUTOFF VALVE
OPEN TESTCOCK 3 (TESTCOCK 2 MUST BE OPEN)
Note: GAUGE MUST READ 1.0 psi OR GREATER TO PASS
<u>RECORD</u> VALUE OF CV 1

CV 2
: CLOSE TESTCOCK 2 AND TESTCOCK 3
MOVE SHORT TUBE FROM TESTCOCK 3 TO TESTCOCK 4
REMOVE HIGH HOSE FROM TESTCOCK 2
OPEN INLET SHUTOFF VALVE
ATTACH HIGH HOSE TO TESTCOCK 3
OPEN TESTCOCK 4 TO FILL TUBE
CLOSE TESTCOCK 4
OPEN TESTCOCK 3 SLOWLY
OPEN HIGH BLEED - BLEED AIR FROM GAUGE
CLOSE HIGH BLEED
CLOSE INLET SHUTOFF VALVE
OPEN TESTCOCK 4 (TESTCOCK 3 MUST BE OPEN)
Note: GAUGE MUST READ 1.0 psi OR GREATER TO PASS
<u>RECORD</u> VALUE OF CV 2

Final
: CLOSE TESTCOCKS - REMOVE ALL EQUIPMENT
OPEN INLET SHUTOFF VALVE; OPEN OUTLET SHUT-OFF VALVE
OPEN ALL NEEDLE VALVES ON TEST KIT

12/10/2001

PVB FIELD TEST with DIFFERENTIAL GAUGE

Prep
- NOTIFY CUSTOMER
- INSPECT AREA
- REMOVE CANOPY
- FLUSH TESTCOCKS
- INSTALL FITTINGS
- INSPECT TEST EQUIPMENT - CLOSE ALL NEEDLE VALVES

Air Inlet
- INSTALL TEST GAUGE AND END OF LOW HOSE LEVEL WITH TESTCOCK 2
- ATTACH HIGH HOSE TO TESTCOCK 2
- OPEN TESTCOCK 2 SLOWLY
- OPEN HIGH BLEED - BLEED AIR FROM GAUGE
- CLOSE HIGH BLEED
- CLOSE OUTLET SHUTOFF VALVE
- CLOSE INLET SHUTOFF VALVE
- FINGER IN TOP - OPEN HIGH BLEED
- <u>RECORD</u> WHEN AIR INLET OPENS > or = 1.0 PSI

Check Valve
- CLOSE TESTCOCK 2
- REMOVE HIGH HOSE FROM TESTCOCK 2
- OPEN INLET SHUTOFF VALVE
- ATTACH COMPENSATING (BLEED-OFF) TEE TO TESTCOCK 1
- ATTACH HIGH HOSE TO TESTCOCK 1
- OPEN TESTCOCK 1 SLOWLY
- OPEN HIGH BLEED - BLEED AIR FROM GAUGE
- CLOSE HIGH BLEED
- CLOSE INLET SHUTOFF VALVE
- OPEN TESTCOCK 2
- WHEN WATER STOPS RUNNING FROM TESTCOCK 2
  - <u>RECORD</u> CHECK VALVE > or = 1.0 PSI

Final
- CLOSE TESTCOCKS - REMOVE ALL EQUIPMENT
- REPLACE CANOPY
- OPEN INLET SHUTOFF VALVE
- OPEN ALL NEEDLE VALVES ON TEST KIT
- OPEN OUTLET SHUTOFF VALVE SLOWLY

5/7/2002

SVB FIELD TEST with DIFFERENTIAL PRESSURE GAUGE

Prep	NOTIFY CUSTOMER
	INSPECT AREA
	REMOVE CANOPY
	FLUSH TESTCOCK AND VENT VALVE
	INSTALL FITTING
	INSPECT TEST EQUIPMENT - CLOSE ALL NEEDLE VALVES
Check Valve	CLOSE INLET SHUT-OFF VALVE
	OPEN VENT VALVE TO LOWER PRESSURE IN BODY
	WHEN WATER STOPS RUNNING FROM VENT VALVE
	<u>RECORD</u> VALUE OF CHECK VALVE (must be > or = 1.0 psi)
Air Inlet	INSTALL TEST GAUGE AND LOW HOSE LEVEL WITH BODY
	ATTACH HIGH HOSE TO TESTCOCK
	OPEN TESTCOCK SLOWLY
	OPEN HIGH BLEED - BLEED AIR FROM GAUGE
	CLOSE HIGH BLEED
	CLOSE OUTLET SHUT-OFF VALVE
	CLOSE INLET SHUT-OFF VALVE
	OPEN VENT VALVE TO LOWER OUTLET PRESSURE TO ATMOSPHERIC
	(AIR INLET VALVE MAY OPEN - BE PREPARED TO RECORD VALUE)<$>
	FINGER IN TOP - OPEN HIGH BLEED NO MORE THAN 1/4 TURN
	<u>RECORD</u> WHEN AIR INLET OPENS (must be > or = 1.0 psi)
	CLOSE VENT VALVE
	CLOSE HIGH BLEED
	OPEN INLET SHUT-OFF VALVE **SLOWLY**
Final	CLOSE TESTCOCK AND VENT VALVE
	REMOVE ALL EQUIPMENT
	REPLACE CANOPY
	OPEN INLET SHUT-OFF VALVE **SLOWLY**
	OPEN OUTLET SHUT-OFF VALVE SLOWLY

8/23/2004

Appendix E

Test and Maintenance Report Form

TEST AND MAINTENANCE REPORT

CUSTOMER: _____

STREET ADDRESS: _____

MAILING ADDRESS: _____

LOCATION OF ASSEMBLY: _____

TYPE OF ASSEMBLY:　　RP ☐　　　DC ☐　　　PVB ☐　　　SVB ☐　　　SIZE: _____

MANUFACTURER: _____　MODEL: _____　SERIAL NO: _____

GAUGE MANUF _____　SERIAL # _____　DATE CALIBRATED: _____

Check Valve #1	Relief Valve	Check Valve #2	Pressure Vacuum Breaker
☐ leaked or ☐ closed tight	opened at: _____ psi or did not open ☐	☐ leaked or ☐ closed tight	**Air Inlet:** did not open ☐ or opened at _____ psi
gauge pressure across check valve _____ psi	**Outlet shut-off valve:** ☐ leaked　☐ closed tight	gauge pressure across check valve _____ psi	**Check Valve:** leaked ☐ or held at _____ psi
☐ cleaned only Replaced: 　rubber kit　☐ 　CV assembly　☐ 　or 　disc　☐ 　O-rings　☐ 　Seat　☐ 　spring　☐ 　stem/guide　☐ 　retainer　☐ 　lock nuts　☐ 　Other　☐	☐ RV cleaned only Replaced: 　RV rubber kit　☐ 　RV assembly　☐ 　or 　disc　☐ 　diaphragm(s)　☐ 　seat　☐ 　spring　☐ 　guide　☐ 　O-rings　☐ 　Other　☐	☐ cleaned only Replaced: 　rubber kit　☐ 　CV assembly　☐ 　or 　disc　☐ 　O-rings　☐ 　seat　☐ 　spring　☐ 　stem/guide　☐ 　retainer　☐ 　lock nuts　☐ 　Other　☐	☐ cleaned only Replaced: 　rubber kit　☐ 　CV assembly　☐ 　disc, air inlet　☐ 　disk, CV　☐ 　seat, CV　☐ 　spring, air inlet　☐ 　spring, CV　☐ 　retainer　☐ 　guide　☐ 　O-rings　☐ 　Other　☐
Gauge pressure across check valve _____ psi	Relief valve opened at _____ psi	Gauge pressure across check valve _____ psi	air inlet _____ psi check valve _____ psi

NOTE: All repairs shall be completed within five (5) working days.

REMARKS: _____
_____

I hereby certify that this data is accurate and reflects the proper operation and maintenance of the assembly.

TESTER: _____　CERT. No: _____　DATE: _____

　　　　　　　　　　　　　　　　　　　　　　　　　　　　　　　　TIME: _____

This Assembly:　　☐ PASSED　　　☐ FAILED

TEST AND MAINTENANCE REPORT

CUSTOMER: _____

STREET ADDRESS: _____

MAILING ADDRESS: _____

LOCATION OF ASSEMBLY: _____

TYPE OF ASSEMBLY: RP ☐   DC ☐   PVB ☐   SVB ☐   SIZE: _____

MANUFACTURER: _____ MODEL: _____ SERIAL NO: _____

GAUGE MANUF _____ SERIAL # _____ DATE CALIBRATED: _____

Check Valve #1	Relief Valve	Check Valve #2	Pressure Vacuum Breaker
☐ leaked or ☐ closed tight	opened at: _____ psi or did not open ☐	☐ leaked or ☐ closed tight	**Air Inlet:** did not open ☐ or opened at _____ psi
gauge pressure across check valve _____ psi	*Outlet shut-off valve:* ☐ leaked ☐ closed tight	gauge pressure across check valve _____ psi	**Check Valve:** leaked ☐ or held at _____ psi

☐ cleaned only
Replaced:
  rubber kit ☐
  CV assembly ☐
  or
  disc ☐
  O-rings ☐
  Seat ☐
  spring ☐
  stem/guide ☐
  retainer ☐
  lock nuts ☐
  Other ☐
	☐ RV cleaned only
	Replaced:
	  RV rubber kit ☐
	  RV assembly ☐
	  or
	  disc ☐
	  diaphragm(s) ☐
	  seat ☐
	  spring ☐
	  guide ☐
	  O-rings ☐
	  Other ☐
		☐ cleaned only
		Replaced:
		  rubber kit ☐
		  CV assembly ☐
		  or
		  disc ☐
		  O-rings ☐
		  seat ☐
		  spring ☐
		  stem/guide ☐
		  retainer ☐
		  lock nuts ☐
		  Other ☐
			☐ cleaned only
			Replaced:
			  rubber kit ☐
			  CV assembly ☐
			  disc, air inlet ☐
			  disk, CV ☐
			  seat, CV ☐
			  spring, air inlet ☐
			  spring, CV ☐
			  retainer ☐
			  guide ☐
			  O-rings ☐
			  Other ☐

Gauge pressure across check valve _____ psi	Relief valve opened at _____ psi	Gauge pressure across check valve _____ psi	air inlet _____ psi check valve _____ psi

NOTE: All repairs shall be completed within five (5) working days.

REMARKS: _____
_____

I hereby certify that this data is accurate and reflects the proper operation and maintenance of the assembly.

TESTER: _____ CERT. No: _____ DATE: _____

TIME: _____

This Assembly: ☐ PASSED   ☐ FAILED

Appendix F

Troubleshooting Guide

Troubleshooting the RP

When you arrive	Action	Observation	Diagnostics	Action
Relief valve is not dripping.	Test #1-close outlet shut-off valve.	RV does not drip.	CV #1 is tight.	Continue testing RP*
Relief valve is not dripping.	Test #1-close outlet shut-off valve.	RV begins to drip.	CV #1 is leaking.	Repair CV #1.
Relief valve is dripping.	Test #1-close outlet shut-off valve.	RV stops dripping.	CV #2 leaks with backpressure.	Repair CV #2.
Relief valve is dripping.	Test #1-close outlet shut-off valve.	RV continues to drip.	CV #1 or RV leaks.	Open test cock #4. *

* If leak stops, then CV #1 leaks
* If leak continues, then Relief Valve leaks

TROUBLESHOOTING FLOWCHART

CUSTOMER IS CONNECTED TO WATER SUPPLY

Test #1 – last step - Close outlet shut-off valve

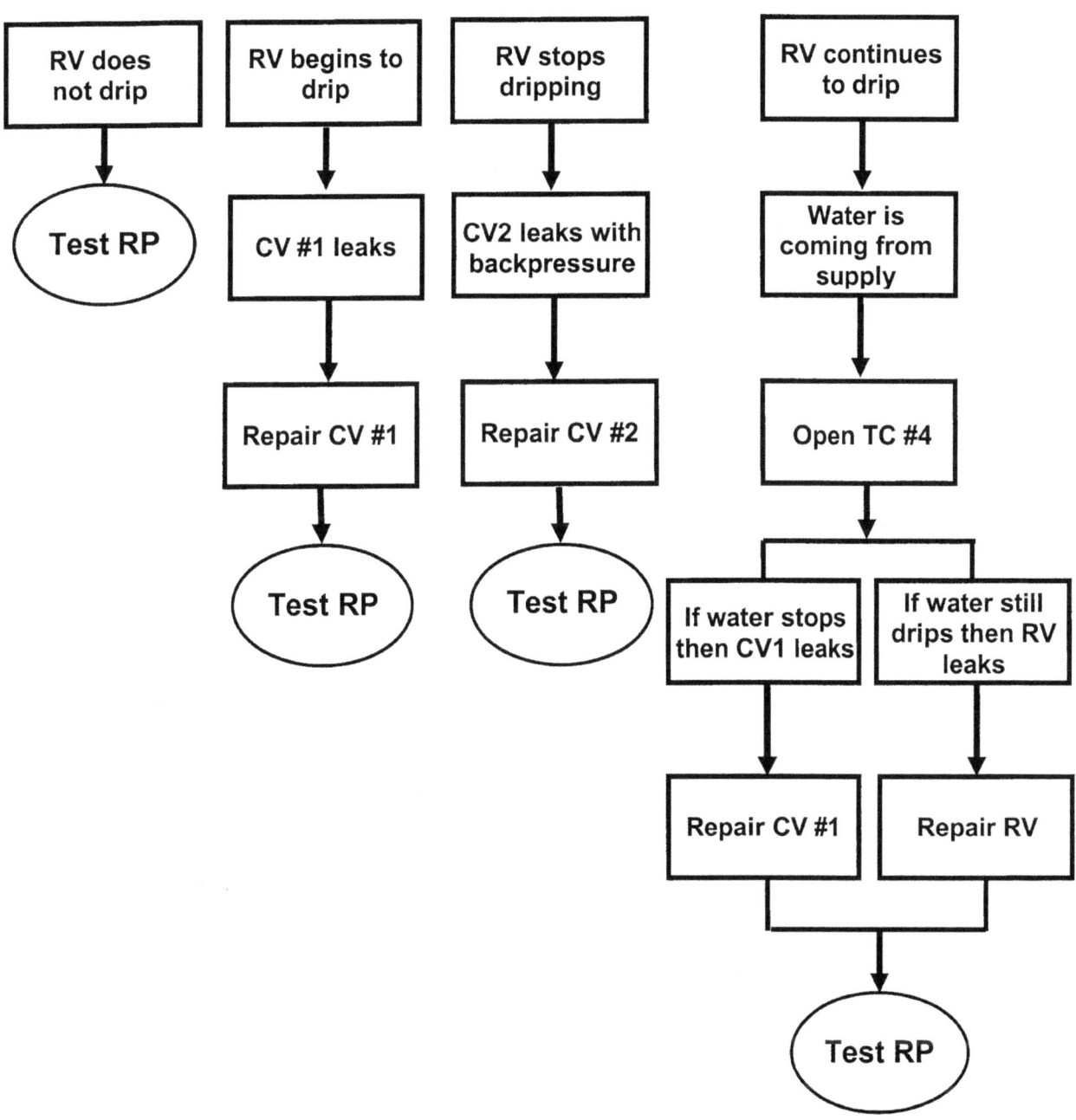

Appendix G

Test Kits
Suppliers and Repair Locations

Gauges and Manufacturers

Manufacturer	Gauge Model
Barton 1304 John Reed Court City of Industry, CA 91745 Ph: (818) 336-4502 Fax: (818) 968-8907 Web site: http://www.barton-instruments.com/	Model 226 (head only), 246 (3 inch face), and 247 (6 inch face)
Duke Products PO Box 16007 Irvine, CA 92713 Ph: (714) 581-7200 Fax: (714) 552-9368	Model E-Z 900, 246 BFT, 247 BFT, 226A-BFT, and 227A-BFT
Meriam Process Technologies 10920 Madison Avenue Cleveland, OH 44102 Ph: (216) 281-1100 Fax: (216) 281-0228 Web site: http://www.meriam.com/catalog.pdf	Model 1124
Mid-West Instrument 6500 Dobry Drive Sterling Heights, MI 48078-7298 Ph: (810) 254-6500 Fax: (810) 254-6509 Web site: http://www.midwestinstrument.com/	Model 830, 844P, and 845
Astra Industrial Services PO Box 888 3525 Old Conejo Road, Suite 104 Newbury Park, CA 91320 Ph: (800) 776-1464 Fax: (805) 794-1848	Model ASRP-4, ASRP-4PM
Watts Regulator Company 815 Chestnut Street North Andover, MA 01845 Ph: (978) 688-1811 Fax: (978) 794-1848 Web site: http://www.wattsreg.com/	Model TKDP, TKDL, TK99D, and TK99E

Note: There may be more test gauge manufacturers which are not listed on this page.

Test Gauge Repair and Calibration

Mid-West Instrument
    Florida

        Instrument Specialties, Inc
        51 Coastline Road
        Sanford, FL 32771
        Ph: 407-324-7800
        Fax: 407-324-1104
        attn: Shawn Kane

    North Carolina

        Specialty Valve & Controls Co.
        3001 Griffith Street
        Charlotte, NC 28203
        Ph: 704-522-9873
        Fax: 704-522-9875
        attn: Larry Bryan

    Mississippi

        Backflow Control
        262 Boatner Road
        Potts Camp, MS 38659
        Ph: 662-333-9007
        Fax: 662-333-7544
        attn: Roger Crane

    Michigan

        Mid-West Instrument
        6500 Dobry Drive
        Sterling Heights, MI 48314
        Ph: 810-254-6500
        Fax: 810-254-6509

Barton
    Florida

        AMJ Equipment Corp.
        PO Box 1648
        Lakeland, FL 33802
        Ph: 863-682-4500 or 1-800-881-1487
        Fax: 863-687-0077

All gauges:

        BAVCO
        20435 South Susana Road
        Long Beach, CA 90810-1136
        Ph: 310-639-5231
        Fax: 310-639-0721

        American Backfl ow Products, Inc
        7580-A Tennessee Street
        Tallahassee, FL 32304
        Ph: (850) 576-1814
        Fax: (850) 575-6508

Note: *There are many more authorized calibration centers not listed on this page. Check with gauge manufacturer for a calibration center in your area.*

Appendix H

Repair Parts Suppliers

Repair Parts Suppliers

Note: There are many more parts suppliers not listed on this page. Check with your manufacturer for a supplier in your area.

**AMERICAN BACKFLOW
SPECIALTIES, INC**
3940 HOME AVENUE
SAN DIEGO, CA 92105-5909
(619) 527-2525   FAX: (619) 527-2527

AMERICAN BACKFLOW PRODUCTS, INC
7580-A TENNESSEE STREET
TALLAHASSEE, FL 32304
(850) 576-1814   FAX: (850) 575-6508

ASTRA INDUSTRIAL SERVICES, INC.
PO BOX 888
NEWBURY PARK, CA 91320-0888
(805) 499-8729   (800) 776-1464
FAX (805) 499-9084

**BACKFLOW APPARATUS & VALVE CO.
(BAVCO)**
20435 S SUSANNA ROAD
LONG BEACH, CA 90248
(310) 639-5231   (800) 458-3492
FAX (310) 639-0721

BACKFLOW PREVENTION SUPPLY
962 EAST 900 SOUTH
SALT LAKE CITY, UT 84105
(801) 355-6736   (800) 733-6730
FAX (801) 355-9233

IMSCO
PO BOX 68
TAYLORS, SC 29687
(864) 268-2891   (800) 476-2212
FAX (864) 268-8912 OR (888) 587-0575

QUALITY SUPPLY CORPORATION
97 TOTTON ROAD
GRAY, MO 04039
(207) 657-4322   (800) 433-3136
FAX (207) 657-4313

PALM BEACH PLUMBING PARTS
2511 WESTGATE AVENUE
WEST PALM BEACH, FL 33409
(561) 687-3034   FAX (561) 687-3077

THE PART WORKS, INC
3545 INTERLAKE AVENUE N.
SEATTLE, WA 98103
(206) 632-8900   (800) 336-8900
FAX (206) 632-5180

WATER SPECIALTIES COMPANY, INC.
8 INDUSTRIAL PARK DRIVE, UNIT 13
HOOKSETT, NH 03106
(603) 668-0088   (800) 336-6530
FAX (603) 668-0080

Appendix I

Florida Building Code – Plumbing

Florida Building Code – Plumbing

Effective March 2002

Section 312 – TESTS AND INSPECTIONS

312.9 Inspection and testing of backflow prevention assemblies. Inspections shall be made of all backflow prevention assemblies to determine whether they are operable. Reduced pressure principle backflow preventer assemblies, double check-valve assemblies, double-detector check-valve assemblies and pressure vacuum breaker assemblies shall be tested. The frequency of testing shall be determined in accordance with the manufacturer's installation instructions. Where the manufacturer of the assembly does not specify the frequency of testing, the assembly shall be tested at least annually. The testing procedure shall be performed in accordance with one of the following standards:

ASSE 5010 1013-1. Sections 1 and 2	RP (USC – 9th Edition)
ASSE 5010 1015-1. Sections 1 and 2	DCVA – Duplex Gauge
ASSE 5010 1015-2	DCVA – two hose method (NEWWA)
ASSE 5010 1015-3. Sections 1 and 2	DCVA – single hose (USC)
ASSE 5010 1015-4. Sections 1 and 2	DCVA – sight tube method
ASSE 5010 1020-1. Sections 1 and 2	PVB (USC – 9th Edition)
ASSE 5010 1047-1. Sections 1, 2, 3 and 4	RPDA (USC – 9th Edition)
ASSE 5010 1048-1. Sections 1, 2, 3 and 4	DCDA – Duplex gauge
ASSE 5010 1048-2	DCDA – two hose method (NEWWA)
ASSE 5010 1048-3. Sections 1, 2, 3 and 4	DCDA – single hose (USC)
ASSE 5010 1048-4. Sections 1, 2, 3 and 4	DCDA – sight tube method
CSA B64.10	

CHAPTER P6
WATER SUPPLY AND DISTRIBUTION

§P608.16.5 Connections to lawn irrigation systems. The potable water supply to lawn irrigation systems shall be protected against backflow by an atmospheric-type vacuum breaker, a pressure-type vacuum breaker or a reduced pressure principle backflow preventer. A valve shall not be installed downstream from an atmospheric vacuum breaker. Where chemicals are introduced into the system, the potable water supply shall be protected against backflow by a reduced pressure principle backflow preventer.

Appendix J

**Approval Process
Foundation for Cross-Connection Control and Hydraulic Research
at the University of Southern California**

Foundation for Cross-Connection Control and Hydraulic Research
University of Southern California

Summary of Evaluation Process: (3 year approval)

1. Manufacturer submits set of working drawings for review.

   - Conformance with general design, material, and operational requirements.
   - If USC engineering staff has any corrections, the manufacturer usually makes changes at this point.
   - Recommended modifications may be incorporated into the design.

2. Manufacturer submits prototype assembly(s) for lab evaluation.
   Less than 2-1/2" - 3 units required.
   Greater than 2" - 1 unit required (see Chapter 10 of 9th edition).

   - Staff check to confirm that the assembly is in conformance with the general, design, material, and operational requirements.
   - Additionally, conformance to the working drawings and materials specifications is confirmed:
     - spring constants
     - hardness of elastomer materials
     - types of materials used

3. Laboratory Evaluation Tests - General: (each type of assembly may have additional tests)

   - Hydrostatic Tests - twice maximum working pressure for ten minutes.
   - Pressure Loss vs. Flow Rate: Pressure loss curve generated.
   - Pressure Drop Across Check Valve(s): direction of flow
   - Thermal Loop: Assemblies must be tested at their rated temperature and not less than 140°F, for 100 hours at the rated pressure of the assembly.
   - Life Cycle Test: The assembly must cycle through 5000 cycles at various flow rates.

4. The Field Evaluation: Manufacturer must install at least three (3) assemblies of each size and model in active field sites in different applications.

   - A minimum of three assemblies must provide twelve months of trouble-free service simultaneously. The assemblies are tested on a nominal thirty day schedule.
   - If there are no problems during the year the assemblies are disassembled at the end of the twelve month field evaluation to determine if there has been any deterioration of parts, undue wear, or any problems which may affect the assembly's ability to prevent backflow or meet all of the evaluation requirements.

5. Upon successful completion of Laboratory and Field Evaluation approval granted for a period of three (3) years.

8/14/95

Appendix K

Testing the RP Chart

Testing the RP

RP	NORMAL	CV1 LEAKS	CV2 LEAKS	OUTLET SHUT-OFF LEAKS	OUTLET SHUT-OFF AND CV2 LEAKS
Test 1 CV1	Observe CV1 - leaks or tight	STOP & repair then re-test	Observe CV1 leaks or tight	Observe CV1 leaks or tight	Observe CV1 leaks or tight
Test 2 RV	Record RV opening > 2.0 psi		Record RV opening, Reset	RV will not open - postpone RV test	RV will not open - STOP RV test
Test 3 CV2	Backpressure test on CV2 - leaks or tight RV does not drip		Backpressure test Record CV2 leaks, RV drips	Backpressure test on CV2	Backpressure test on CV2 - If CV2 leaks - **record RV opening**
Test 4 CV1	Reset & record CV1 5.0 psi or greater		Close vent control, reset & Record CV1	Reset & Record CV1	Close vent control, reset & record CV1
Test 5 Outlet Valve	Close testcock 2			Record that Outlet Shut-off Valve leaks	
Test 6 CV2	Move hoses & Record CV2 > 1.0 psi		No action!	You cannot perform differential test on CV2	You cannot perform differential test on CV2
Retest RV				Retest & record RV. Use by-pass hose between TC 1 and TC 4 if necessary	

Appendix L

A.S.S.E Numbers
Approved Backflow Prevention Assemblies and Devices

Approved Backflow Prevention <u>Assemblies</u>

BFP	Description	Approved by: FCCC & HR	ASSE	L.A. Lab
RP, RPPA, RPZ	Reduced Pressure Principle Assembly	✔	1013	
RPDA	Reduced Pressure Detector Assembly	✔	1047	
DCVA	Double Check Valve Assembly	✔	1015	
DCDA	Double Check Detector Assembly	✔	1048	
PVB	Pressure Vacuum Breaker Assembly	✔	1020	
SVB	Spill Resistant Vacuum Breaker	✔	1056	

Approved Backflow Prevention <u>Devices</u>

BFP	Description	Approved by: FCCC & HR	ASSE	L.A. Lab
AVB	Atmospheric Vacuum Breaker	✔	1001	✔
HCVB or HBVB	Hose Connected Vacuum Breaker		1011	
DuC w/ atmo vent	BFP w/ Atmospheric Vent		1012	
TSPV	Trap Seal Primer Valves		1018	
DuC or RDC	Dual Check Valve, (Residential)		1024	
DuC-CO_2	Dual Check Valve - Post mix for CO_2 dispensers		1032	
LFVB	Lab Faucet Vacuum Breakers		1035	
HCBFP or HCBP	Hose Connected Backflow Preventer		1052	
Meter w/DuC	Meter w/ Dual Check Valve		—	

Appendix M

Building a Model Backflow Prevention Program

Building a Model Backflow Prevention Program

STEP 1:   DEVELOP AN ORDINANCE OUTLINE

Use the following resource materials:

- *Backflow Prevention: Theory and Practice*
- Ordinances and policies adopted by other utilities.
- Course presentations.
- Determine what will be contained within the ordinance and what will be covered under the program policies.
- Recall that the ordinance you write and the program you carry out must meet the prime objective of safeguarding the public. On the other hand, it must not be so economically damaging to your community that it is unacceptable to local authorities.

STEP 2:   PRESENT YOUR ORDINANCE FOR ADOPTION

Each group will present its ordinance to the class for adoption. Class members assume the roles of local governing body and customers.

Your group should address the following questions during the presentation:

1. Why should this locality have a backflow prevention program and what will it involve?
2. Why is it necessary to enact an ordinance to implement the backflow prevention program?
3. Who will administer or have authority over the program and why?
4. How much will the program cost?
5. What is the source of funding for this program, e.g., user fees, rate adjustment, etc.?
6. When addressing the local governing body be specific about what will be contained in the ordinance and what will be covered under your program policies and be able to justify what is contained within each.

Your group should also be prepared to answer any questions that might be raised by the governing body or the customers. Questions you may want to prepare for:

1. What type of facilities will need backflow preventers?
2. What will be the implementation time for the program?
3. What types of backflow preventers are needed and what are the costs of the different types of backflow preventers?
4. Who will determine what type of backflow preventer is needed by a facility?
5. What recourse does a facility manager have if a customer chooses to challenge a decision made by utility personnel?
6. Who will install, test and repair the backflow preventers?
7. Will those individuals who install, test and repair the backflow preventers have to meet any special qualifications?
8. How often will the backflow preventers need to be tested and who will maintain the test records?
9. How will the ordinance be enforced?
10. Who will be liable if a backflow preventer is not tested and a backflow occurs?

Appendix N

Elements of CCC Program Ordinance

Cross-Connection Control
Elements of Program Ordinance/Policy/Service Contract

1. Written Legal Authority containing:

    a. the purpose of ordinance or contract

    b. a description of the public water system's and the premise owners' responsibilities

    c. appropriate definitions

    d. a policy requiring premise owners' water systems to be open for inspection to the public water system

    e. a policy specifying which service connections shall be equipped with premises-isolating backflow preventers and specifying exactly where in the service pipe these backflow preventers shall be located

    f. a policy specifying the types of premises-isolating backflow preventers that shall be used in different cases

    g. a policy defining approved backflow preventers and requiring that all premises-isolating backflow preventers installed after a certain date be approved backflow preventers

    h. a policy specifying that all required premises-isolating backflow preventers shall be tested at least annually by competent backflow preventer testers approved by the public water system and shall be repaired when necessary, and

    i. penalties for non-compliance with the ordinance or contract

2. A written schedule and written procedures for surveying and retrofitting existing facilities

3. Written procedures for plan review and inspection of all new construction

4. Written schedule and procedures for at least annual testing of backflow-prevention assemblies and for repair when necessary

5. Written procedures for approving competent backflow preventer testers and insuring that required premises-isolating backflow preventers are tested only by approved, competent backflow preventer testers

6. Written procedures for keeping installation, testing, and repair records for each required premises-isolating backflow preventer (these records shall be kept for not less than 10 years)

7. Written procedures for educating premise owners about:
(a) the need to have registered professional engineers or certified fire-protection system contractors check the hydraulics of existing fire-protection systems when premises-isolating backflow preventers are added at existing service connections to which existing fire-protection systems are in turn connected and
(b) the need to install thermal expansion devices and/or pressure relief valves within closed-loop plumbing systems created by the installation of premises-isolating backflow preventers, and

8. Written procedures for handling backflow complaints and emergencies

Cross-Connection Control
Program Ordinance/Service Contract

ELEMENTS OF ORDINANCE/SERVICE CONTRACT	COMMENTS	PAGE
Written Legal Authority containing:		
a. the purpose of ordinance or contract		
b. a description of the public water system's and the premise owners' responsibilities		
c. appropriate definitions		
d. a policy requiring premise owners' water systems to be open for inspection to the public water system		
e. a policy specifying which service connections shall be equipped with premises-isolating backflow preventers and specifying exactly where in the service pipe these backflow preventers shall be located		
f. a policy specifying the types of premises-isolating backflow preventers that shall be used in different cases		
g. a policy defining approved backflow preventers and requiring that all premises-isolating backflow preventers installed after a certain date be approved backflow preventers		
h. a policy specifying that all required premises-isolating backflow preventers shall be tested at least annually by competent backflow preventer testers approved by the public water system and shall be repaired when necessary		
i. penalties for non-compliance with the ordinance or contract		
A written schedule and written procedures for **surveying and retrofitting existing facilities**		
Written procedures for **plan review** and inspection **of all new construction**		
Written schedule and procedures for at least **annual testing** of backflow-prevention assemblies and for **repair when necessary**		
Written procedures for **approving competent backflow preventer testers** and insuring that required premises-isolating backflow preventers are tested only by approved, competent backflow preventer testers		
Written procedures for **keeping installation, testing, and repair records** for each required premises-isolating backflow preventer (these records shall be kept for not less than 10 years)		
Written procedures for **educating premise owners** about (a) the need to have registered professional engineers or certified fire-protection system contractors **check the hydraulics of existing fire-protection systems** when premises-isolating backflow preventers are added at existing service connections to which existing fire-protection systems are in turn connected and (b) the **need to install thermal expansion devices** and/or pressure relief valves within closed-loop plumbing systems created by the installation of premises-isolating backflow preventers, and		
Written procedures for handling backflow **complaints and emergencies**		
(A **program manual** containing all of the above mentioned written material)		

Recommendation of the Technical Advisory Committee for Rule 62-555.360 F.A.C.

Appendix O

Form Letters:

Required Testing of Backflow Preventers Letter
Second Notice
Final Notice
Program Non-Compliance Letter
Repair Letter

Required Testing of Backflow Preventers Letter

[YOUR LETTERHEAD]

Date

Customer's Name
Customer's Address
City, State, Zip

SUBJECT: **Required Annual Testing of Backflow Preventers**

Dear Customer:

    In order to continue to maintain the quality of ***[City or Utility Name's]*** water supply at the highest level possible, backflow preventers are required on all services where the water may come in contact with a contaminant.

    Connected to your water system is a backflow preventer which, by ***[local ordinance or state regulation]*** is required to be tested and maintained annually by a certified tester. It should be noted that failure to comply with this ordinance will eventually result in discontinuance of water service.

    I am enclosing the "Test and Maintenance Report" forms. The white copy must be returned to me by ***[date]***. I am also enclosing a list of approved certified backflow prevention device testers located in the area.

    I encourage you to have the tests performed as soon as possible. If any of your devices should need repair, the parts may take a few weeks to arrive.

    If you have any questions, please contact me at ***[your phone number]***.

Sincerely,

[Your Name]
Cross-Connection Control Supervisor

enclosures

Annual Test Notification – Second Notice

[YOUR LETTERHEAD]

Date

Customer's Name
Customer's Address
City, State, Zip

SECOND NOTICE

Subject: Required Annual Field Testing of Backflow Prevention Assemblies

Dear Customer:

On ***[date of mailing]*** a letter was mailed to you as notification of the required annual testing of existing backflow prevention assembly(s) installed at the above location.

Our records show that as of this date the necessary "Test and Maintenance Report" form has not been received. For assemblies to continue to operate efficiently, they must be tested and serviced when required.

You are hereby given notice to comply with the testing requirements set forth in our letter. Failure to comply will eventually result in your water service being terminated.

Enclosed are a list of certified backflow prevention testers and a "Test and Maintenance Report" form. Please help us insure the accuracy of our records by noting any corrections directly on the form, then have it completed by the tester of your choice. A copy of the completed form must be received at our office via fax or mail no later than ***[due date]***.

If you have any questions, please contact me at ***[your phone number]***.

Sincerely,

[Your Name]
Cross-Connection Control Supervisor

enclosures

Annual Test Notification – Final Notice

[YOUR LETTERHEAD]

Date

Customer's Name
Customer's Address
City, State, Zip

FINAL NOTICE

Subject: Required Annual Field Testing of Backflow Prevention Assemblies

Dear Customer:

On *[date of mailing]* a second letter was mailed to you as notification of the testing requirements of existing backflow prevention assembly(s) installed at the referenced address. City *[or county]* Ordinance requires annual field testing and maintenance of your backflow prevention assembly(s). In order to insure that testing and maintenance is performed properly, we have enclosed the "Test and Maintenance Report" form, which must be completed by a certified backflow prevention tester and returned to this utility. Our records indicate that we have not received the "Test and Maintenance Report."

You are hereby given notice to comply with the testing requirements set forth in City *[or county]* Ordinance Sec. *[section number]*.

Within 30 days of the date of this letter, please provide us with the following information:
1) A copy of the completed test form with completed field test results; or
2) A written explanation on why the field test has not been performed; or
3) Written confirmation that you will attend a Show Cause hearing at the Utility Administrative Building, located at *[address]* on *[date]*.

If you fail to respond to this letter, the Utility will arrange for the water service to your facility to be terminated on this *[date]*.

Please contact me at *[your phone number]*.

Sincerely,

[Your Name]
Cross-Connection Control Supervisor

enclosures

Program Non-Compliance Letter

[YOUR LETTERHEAD]

Date

Customer's Name
Customer's Address
City, State, Zip

SUBJECT: **Discontinuance of Water Supply**

Dear Customer:

You are hereby notified that in accordance with ***[local ordinance or state regulation]***, ***[number]***, the water supply to your premises, located at ***[address]***, will be disconnected at the end of 60 days, upon receipt of this letter, if you have not complied with the notifications of ***[City or Utility Name]***. The discontinuance of your water service will remain in effect until you have complied as required by state law.

If you have any questions, please contact me at ***[your phone number]***.

Sincerely,

[Your Name]
Cross-Connection Control Supervisor

enclosures

Repair Letter

[YOUR LETTERHEAD]

Date

Customer's Name
Customer's Address
City, State, Zip

Subject: Required maintenance of backflow prevention assembly

Dear Customer:

    The State Administrative Code and the local Plumbing Code require backflow prevention assemblies be field tested at least annually. Your backflow prevention assembly, which has serial number _____, has been reported as failing to perform according to the manufacturer's minimum specifications.

    A copy of the completed form must be received at our office via fax or mail within ten (10) days of the date of this letter. Failure to comply with this requirement may result in your water service being disconnected. Please contact me to as soon as this repair has been completed.

    I have enclosed a list of certified backflow prevention testers and a "Test and Maintenance Report" form.

    If you have any questions, please contact me at *[your phone number]*.

Sincerely,

[Your Name]
Cross-Connection Control Supervisor

enclosures

Appendix P

List of High Hazard Facilities

List of High Hazard Facilities

Aircraft
Amusement parks
Automotive plants
Autopsy facilities
Auxiliary water systems
Beverage bottling plants
Boilers (large) or hot water systems
Breweries
Buildings with sewer ejectors
Buildings with water storage tanks or non-potable water sources
Canneries
Car wash facilities
Centralized heating and air conditioning plants
Chemical plants using a water process
Chemical plants-manufacturing, processing, compounding, or treatment
Civil works
Clinics
Cold storage plants
Colleges
Commercial laundries
Convalescent homes
Creameries
Dairies
Dental buildings
Dye works
Fabricating plants
Film laboratories
Food processing plants
Gas production properties, storage, or transmission facilities
Gravel plants
Hospitals
Laboratories
Laundries
Manufacturing plants
Medical buildings
Metal plating industries
Metal processing
Metal manufacturing, or cleaning facilities
Metal fabricating plants
Missile plants
Morgues
Mortuaries
Motion picture studios
Nursing homes
Oil storage facilities, properties, production facilities, or transmission facilities
Packing houses
Paper and paper products plants
Petroleum storage plants, or processing facilities

Piers and docks
Plating plants
Power plants
Processing plants
Radioactive materials or substances-plants or facilities handling
Reclaimed wastewater areas
Recreational facilities using water (swimming pools, water slides, etc.)
Reduction plants
Restricted, classified, or other closed facilities
Rubber plants - natural or synthetic
Sand plants
Sanitariums
Schools
Tanneries
Wastewater pumping stations
Wastewater treatment plants
Water treatment plants
Waterfront facilities and industries

This list was compiled from the following sources:
1. *The Manual of Cross Connection Control,* 1982. City of Gainesville Regional Utilities, Gainesville, FL.
2. *Manual of Cross-Connection Control, (10th Edition),* 2009. Foundation for Cross-Connection Control and Hydraulic Research, Univ. of Southern CA, Univ. Park,
3. *Cross-Connection Control Training Package,* 1985. American Water Works Assn., New York, NY.
4. *Manual of Cross-Connection Control Procedures and Practices,* 1981. State of California Health and Welfare Agency, Department of Health Services, CA.
5. *Manual of Cross-Connection Control,* 1987. Florida Keys Aqueduct Authority, Key West, FL.

Appendix Q

Cross-Connection Control Questionnaire

Cross-Connection Control Questionnaire

		YES	NO
1.	Is there another source of water to your property other than the service connection to the Public Water System of (_____*Utility Name*_____), i.e., a private well?	( )	( )
2.	Is any equipment (such as a booster pump, elevated tank, etc.) used to increase water pressure above the supply pressure presently provided by the Public Water System of (_____*Utility Name*_____)?	( )	( )
3.	Are toxic chemicals used in your operation?	( )	( )
4.	Are ejectors used in your operation?	( )	( )
5.	Is water recycled during the operation of your air conditioner or other equipment in your plant?	( )	( )
6.	Are water supply lines submerged in tanks, vats, etc.?	( )	( )
7.	Are backflow prevention devices installed in your plumbing?	( )	( )

Data furnished by:

_____

Customer   Date

_____

Address   Account#

_____

                                                          Phone

Reported by: _____

                                                          Title

Remarks: _____

_____

Appendix R

Survey Inspection Forms

Cross-Connection Survey Form

Place: _____ Date: _____

Location: _____ Investigator(s): _____

Building Representative(s) and Title(s): _____

_____

_____

Water Source(s): _____

_____

Piping System(s): _____

_____

Points of Interconnection: _____

_____

Special Equipment Supplied with Water and Source: _____

_____

_____

Remarks or Recommendations: _____

_____

_____

Note: If necessary for clarity of description, attach sketches of cross-connections and their locations.

Inspected by: _____ Owner: _____

Cross-Connection Inspection Check List

Name of Firm: _____
Address: _____
Contact: _____ Phone: _____
Title: _____ Time: _____ Date: _____

MEDICAL / LABORATORY		KITCHEN / RESTAURANT	
Aspirator, medical		Coffee Urn	
Aspirator, hydro		Cooking Kettle	
Autoclave		Dishwasher	
Autopsy Table		Garbage Can Washer	
Auxiliary Water System		Garbage Disposal	
Bedpan Washer		Grease Trap	
Bottle Washer		Ice Maker	
Colonic Irrigator		Mop Sink	
Condenser		Overhead Spray Hose	
Cup Sink		Potato Peeler	
De-ionized Water System		Pressure Cooker	
Dental Cuspidor		Steam Table	
Digester			
Distillation Equipment		**DOMESTIC**	
Electron Spectroscope		Bidet	
Fermentation Tank		Dishwasher	
Flushing Floor Drains		Fertilizer Injector	
Gas Chromatograph		Fish Pond	
Heat Exchanger		Hose Bibbs	
Hydro-Therapy Bath		Lawn Irrigation	
KJELDAHL		Photo Lab	
KJECTEC Analyzer		Solar System: Passive	
Laundry Equipment		Active	
Mass Spectrograph		Toilet Ballcock, proper	
Pipette Washer		Water Softener	
Retort		Well	
Sitz bath			
Spectrophotometer		**SPECIALTY**	
Thermal Energy Analyzer		Aspirator. weedicide	
Ultrasonic Bath		Auto Shampoo & Wax	
X-Ray Equipment		Baptismal Font	
		Blueprint Machine	
HVAC		Compressors, Water Cooled	
Boiler Feed Line		Dye Vats	
Cooling Tower		Etching Tanks	
Expansion Tank, Chilled Water		Overhead Fill Tube/Hose	
Expansion Tank, Boiler Water		Radiator Flushing Equipment	
Solar System: Passive		Starch Tanks	
Active		Steam Cleaner	
		Welder, Water Cooled	
		MISCELLANEOUS	
PHOTO LAB		Fire Sprinkler System	
Automatic Developer		Fountain, Ornamental	
Film Washer		Hose Bibbs	
KIS Photo Processor		Lawn Irrigation	
Print Washer		Water Softener	
Photostat Equipment			
Rinse Sinks			

Appendix S

Sample Inspection Letters:

Initial Letter
Follow-up Letter

Initial Inspection Letter

[YOUR LETTERHEAD]

Date

Customer's Name
Customer's Address
City, State, Zip

Subject: On-site inspection of your facility and your water-using equipment

Dear Customer:

      The State Administrative Code and the local Plumbing Code require backflow prevention assemblies or devices be installed on all plumbing fixtures. These codes also require backflow prevention assemblies be installed on the service line immediately after the water meter on your type of facility.

      We will be glad to provide an on-site inspection to assist you in complying with the codes. We can make recommendations on how to maintain safe drinking water for your employees and visitors to your facility. We will have an inspector in your area on ***[due date]***. Please contact me to schedule a time that is convenient for you.

      I have enclosed some literature that will explain the importance of code compliance and cross-connection control.

      I have also enclosed a list of certified backflow prevention testers and a "Test and Maintenance Report" form.

      A copy of the completed form must be received at our office via fax or mail within ten (10) days of the date of this letter.

      If you have any questions, please contact me at ***[your phone number]***.

Sincerely,

[Your Name]
Cross-Connection Control Supervisor

Follow-Up Inspection Letter

[YOUR LETTERHEAD]

Date

Customer's Name
Customer's Address
City, State, Zip

Subject: Previous on-site inspection of your facility

Dear Customer:

We visited your facility on *[date]* and performed an on-site inspection of your water using equipment. A list of the items that require attention is enclosed. *[enclose list]*

Please understand that there may be other cross-connections that were not found during this inspection. It is your responsibility to comply with the local building and plumbing codes. We recommend that the following assemblies and devices be installed:

[list assemblies and devices]

In addition, we require that a backflow prevention assembly (type) be installed on the main water service to your facility. We will have thirty (30) days to comply with this requirement.

We have attached for your convenience a list of certified backflow prevention testers and several "Test and Maintenance Report" forms.

A copy of the completed form must be received at our office via fax or mail within ten (10) days of the date of this letter.

If you require additional information or time to complete the required work, please contact me at *[your phone number]*.

Sincerely,

[Your Name]
Cross-Connection Control Supervisor

enclosures

cc: Local Health Department
        Local Plumbing Inspector

Appendix T

List of Common Cross-Connection Locations

List of Common Cross-Connection Locations

Air conditioning systems
Air compressors
Air conditioning chill water
Air liners
Air conditioning cooling tower
Air washers
Air conditioning condenser water
Aquariums
Aspirators
Autoclaves
Auxiliary systems
Baptismal fonts
Baptisteries
Bathing tanks
Bathtubs
Bedpan washers
Below-the-rim or inverted supply water inlets
Bidets
Bird baths
Boiler industrial feeder lines
Boilers
Bottle washer
Brine tank
Carbonators
Cheese tanks
Chemical feeders
Chiller tanks
Chlorinators
City water and sewer pumps direct connections
Coffee urns
Commercial pressure cookers
Commercial dishwashing machines
Compressors
Condensate tanks
Cooking kettles
Cooling systems
Cooling towers
Culture vats
Cuspidor (gym)
Dairy and stable watering troughs
Degreasing equipment
Demineralizer system
Dental cuspidors
Dental saliva ejectors
Developing tanks
Dishwashers
Drain lines
Drinking fountains

Dye jiggs, washers, vats & tanks
Etching tanks
Extractors
Fire standpipes
Fire drain lines
Fire protection sprinkler systems
Fish ponds
Floor drains
Flush tanks
Flushing rim
Food mixing tanks
Foot tubs
Fountains
Garbage can washers
Garbage disposals
Holding tanks
Hose faucets
Hospital laundry machines
Hospital digesters
Hot tubs
Hot water heater & tanks
Humidifier tanks & boxes
Hydraulic equipment
Hydro-therapy baths
Ice makers
Industrial plants protection meter
Industrial in-plant plumbing systems
Industrial condensers
Irrigation systems
Janitor closets
Kitchen equipment
Laboratory equipment
Laundry and other tubs
Lavatories
Lawn sprinkler systems
Liquid handling systems
Make up tank
Medical condensers
Medical aspirators
Medical equipment
Photostat equipment
Pipette washer
Plumber's friend (an attachment for non-threaded pipes and fixtures that allows a hose to be attached)
Pneumatic ejectors
Ponds
Potato peeler
Prime lines
Private wells
Processing tanks

Pumps
Re-circulated water
Rubber hoses equipped with hand controls or self closing faucets
Serrated faucets
Sewer, sanitary and storm (bypasses, sump pumps, blow offs)
Shampoo basins
Showers
Sinks
Siphon flush tanks
Sitz bath
Sizing vats and cones
Slop sinks
Solar heating systems
Solution tanks
Spring loaded glass washers
Starch tanks
Steam tables
Steam lines
Steam cleaner
Steam tables
Sterilizers
Stills
Suction side pump chlorinators
Swimming pools, commercial
Swimming pools, home
Tanks
Therapeutic baths
Threaded hose bibbs
Toilets (flush tank, ball cocks, flush valve, siphon jet)
Ultrasonic baths
Urinals (trough or siphon jet blowout)
Vacuum systems (water-operated)
Vats
Vegetable peelers
Wash tanks
Water treatment tanks
Water closets, tanks
Water troughs with vaccine or other substances added for poultry or other live stock
Water closets, flush
Water operated ejectors
Water street mains draining to sewers or storm drains
Water softening systems
Water operated ejectors
Water-jacketed tanks, vats, and pots
X-ray developing tanks

This list was compiled from the following sources:
1. *The Manual of Cross Connection Control,* 1982. City of Gainesville Regional Utilities, Gainesville, FL.
2. *Cross-Connection Control Training Package,* 1985. American Water Works Assn., New York, NY.

Appendix U

Incident Report

UNIVERSITY OF FLORIDA

TREEO Center

BACKFLOW INCIDENT REPORT

Reporting Agency _____ Report Date _____

Reported by _____ Title _____

Mailing Address _____

City _____ State _____ Zip _____ Phone _____

Date of Incident _____ Time of Occurrence _____

General Location (street, block) _____

Backflow Originated from:

    Name of Premise _____

    Street Address _____ City _____

    Contact Person _____ Phone _____

    Type of Business _____

Description of Contaminants: (attach chemical analysis or MSDS if available)

_____

_____

Distribution of Contaminant:

    Contained within customer's premise? Yes ☐   No ☐

    Number of persons affected _____

Effect of Contamination:

    Illness reported _____

    Physica lirritation reported _____

Cross-connection Source of Contaminant: (boiler, chemical pump, irrigation systems, etc)

_____

_____

(over)

Cause of Backflow: (main break, fire flow, etc.)

_____

_____

Corrective Action Taken to Restore Water Quality? (disinfection, main flushing, etc.)

_____

_____

Corrective Action Ordered to Eliminate or Protect Cross-connection:

_____

_____

Previous Cross-connection Survey of Premise:

Date _____ By _____

Type of Backflow Preventer at Meter Service:

    RP _____ RPDA _____ DCVA _____ DCDA _____ None _____

Type of Backflow Preventer Isolating Source of Contaminant:

    RP _____ RPDA _____ DCVA _____ DCDA _____ PVB _____

    AVB _____ AIR GAP _____ None _____

Date of last test of assembly _____

Notification of State Health Department:

    Date _____ Time _____ Person Notified _____

    Notified by _____

Attach sheets with additional remarks, sketches and/or media information.

Mail to: UF/TREEO Center
          3900 SW 63 Blvd
          Gainesville, FL 32608
          352/392-9570
          FAX: 352/392-6910

Appendix V

Selecting the Proper Backflow Prevention Assembly

Selecting the Proper Backflow Prevention Assembly
Jim Purzycki, BAVCO

An important task in an effective cross-connection control program is to assure the proper selection, installation, testing and maintenance of equipment to assure that a backflow condition will not effect the water quality being delivered to the water users. To properly select a correct level of backflow protection, a survey of the water user's facility and water use must first be conducted. This survey should evaluate the customer's use of water using equipment and establish the level of protection that must be achieved. To select the correct type of backflow preventer, several criteria that have been discovered in the cross connection control survey must be evaluated.

What type of backflow prevention assembly should be used?

Backflow preventers can be installed at the water user's point of service to assure the water user and their facilities do not have any water that is delivered from the customer back into the distribution system. This is referred to as service protection or containment. The assemblies typically used for service protection are the air gap (AG), the reduced pressure principle assembly (RP), the double check valve assembly (DC) and with some exceptions, the pressure vacuum breaker (PVB). The second type of installation is installed within the water user's plumbing system after the point of service. This location may be at the water using equipment or the point of hazard where backflow could happen. This installation is referred to as internal protection. The internal protection assemblies usually come under the jurisdiction of the local Plumbing Code and Building regulations. The types of assemblies utilized for internal protection are many depending on the recognized Plumbing Code in your area. Some of the backflow preventers and devices that are used are the AG, RP, DC, PVB, the spill-resistant vacuum breaker (SVB), and the atmospheric vacuum breaker (AVB). Water purveyors are usually involved with service protection applications to protect the quality in their distribution system. A purveyor may accept internal protection instead of service protection. A common exception for the installation of internal protection in lieu of service backflow preventers occurs when the purveyor can be assured that the internal backflow preventers are properly installed to protect all actual or potential cross connections within the water user's plumbing system. The purveyor must be further assured that these internal backflow preventers are periodically tested and maintained. The purveyor must be assured that no changes to the internal plumbing system will take place without a thorough review by the purveyor to assure the proper backflow protection is maintained.

What is the degree of hazard?

The next criteria to evaluate in selecting backflow preventers is the degree of hazard the actual or potential cross connection represents. The degree of hazard can be either high or low. A high hazard is one that if the substance is introduced in the distribution system it would seriously impair the water quality and if ingested could represent a health hazard that would lead to sickness or death to the water user. The usual backflow preventers for a high hazard application are AG, RP, PVB, and AVB. A low hazard is one that occurs when the substance is introduced into the distribution system and would not present a health hazard but could change the water quality by introducing an aesthetically objectionable substance. The usual backflow preventer for low hazard applications is DC.

What type of backflow is present?

The next variable that must be determined is the type of backflow condition present at this cross connection. Backflow can happen by backpressure and or backsiphonage. Backpressure is a hydraulic condition where a greater pressure is generated on the outlet side of a piece of equipment than the incoming pressure. To protect against backpressure the RP, DC can be used. Backsiphonage is a hydraulic condition where sub-atmospheric pressure is present at the inlet of a piece of equipment. AG, RP, DC, PVB and SVB can be used to protect against backsiphonage.

The specific piping conditions present at the installation location must also be evaluated as some types of backflow preventers cannot be under continual pressure or will have very precise height requirements in relationship to the piping in the system and/or surrounding grade.

What special hydraulic conditions exist?

Hydraulic conditions of the piping system must also be determined. Backflow preventers will reduce the water pressure into the water user's system. This affect of pressure reduction from an installed backflow preventer on the plumbing system must be evaluated. Another hydraulic criterion to evaluate is to assure the maximum working water pressure of a backflow preventer is not exceeded. Pressure present in a piping system is not always static. When fluctuations are present or water hammer conditions are present, these changes to pressure need to be evaluated. When a backflow preventer is installed, this creates a closed system within the water user's plumbing system and the hydraulic effects of a closed system must be evaluated. The effects of a closed system can subject the water user's plumbing system to a hydraulic condition called thermal expansion. When water is heated, the volume of water will expand, which will create a rise in pressure in a closed system. A closed system could contain pressure surges from many conditions such as thermal expansion or water hammer. If the plumbing system does not have a method to relieve this increase in pressure, the rise in pressure could cause damage to the piping, water using equipment or the backflow preventer. Proper plumbing code compliance to address the negative effects of a closed system must be followed by the water user.

What is the temperature range of the water being used?

The maximum and minimum working water temperature of the backflow preventer must also be considered for proper protection. The temperature of water can be heated by appliances such as water heaters. The effect of heat from other conditions such as radiant heat can also cause the temperature of the water to increase. Whatever the cause of the temperature increase, the temperature of the water must be within the operating temperature ranges of the backflow preventer. The lower limits of the backflow preventer's temperature limitations must not be exceeded.

Backflow preventers are considered mechanical equipment. Like any mechanical equipment they can fail if not properly maintained. Once the correct type of backflow preventer is chosen and installed, the backflow preventer must be periodically examined to assure it is continuing to protect against the identified actual or potential hazard. A backflow preventer does require periodic testing to assure the assembly is continuing to protect against backflow. It is important that the test procedures utilized are able to show when the backflow preventer is beginning to provide less than optimal performance but can still protect against backflow. Once the backflow preventer has degraded to a pre-determined point it must be repaired and returned to its original ability to protect against backflow.

Appendix W

Nomenclature Chart
Valves, Shut-Off, Check, Needle and Test Cocks

Valves: Shut-Off, Check, Needle and Text Cocks

item	also named	location	description
No 1 Shut-off Valve or Shut-off Valve #1	inlet valve, supply valve, upstream valve	on the supply side of the assembly.	Resilient seated, full flow characteristic. (ball or resilient wedge valves) Valves before 1986 are gate valves.
No 2 Shut-off Valve or Shut-off Valve #2	outlet valve, customer's valve, downstream valve	on the customer's side of the assembly.	Resilient seated, full flow characteristic. (ball or resilient wedge valves) Valves before 1986 are gate valves.
Check Valve No. 1	Check valve #1 or CV#1 or CV1	first check valve in assembly	Spring loaded, soft-seated, drip tight in direction of flow.
Check Valve No. 2	Check valve #2 or CD#2 or CV2	second check valve in assembly	Spring loaded, soft-seated, drip tight in direction of flow.
Test Cock No. 1	Test Cock #1 or TC#1 or TC1	before No. 1 shut-off valve.	Blow-out proof valves - sometimes ball valves.
Test Cock No. 2	Test Cock #2 or TC#2 or TC2	after No. 1 shut-off valve but before check valve #1.	Blow-out proof valves - sometimes ball valves.
Test Cock No. 3	Test Cock #3 or TC#3 or TC3	between check valve #1 and check valve #2.	Blow-out proof valves - sometimes ball valves.
Test Cock No. 4	Test Cock #4 or TC#4 or TC4	after check valve #2 but before No 2 shut-off valve.	Blow-out proof valves - sometimes ball valves.
High Bleed Needle Valve	High bleed	on gauge or test kit; usually on top and marked; might be left or right side.	High quality needle valves. Usually stainless steel needles in brass seats.
Low Bleed Needle Valve	Low bleed	on gauge or test kit; usually on top and marked; might be left or right side.	High quality needle valves. Usually stainless steel needles in brass seats.
High Pressure Needle Valve	High control High by-pass	on gauge or test kit; usually on bottom and marked; might be left or right side.	High quality needle valves. Usually stainless steel needles in brass seats.
Vent or By-pass Control Valve	Vent control Low control	on gauge or test kit; usually on bottom and marked; might be in middle or right side or left side.	High quality needle valves. Usually stainless steel needles in brass seats.
Low-Pressure Control Valve	Low by-pass	on gauge or test kit; usually on bottom and marked; might be left or right side.	High quality needle valves. Usually stainless steel needles in brass seats.

Appendix X

Lawn Irrigation Systems

Lawn Irrigation Systems

Use the proper Backflow Preventer for your particular system.

Description of system	AVB	PVB	SVB	DCVA	RP
Single zone system No valves downstream	✔				
Multi-zone systems		✔	✔		✔
Protects against backsiphonage only	✔	✔			
Must be installed at least 12" above highest outlet		✔	✔		
Must be installed at least 6" above highest outlet	✔				
Shrub Spray Heads *(low hazard)*	✔	✔	✔		✔
Pop-up sprinklers – flush mounted *(high hazard)*		✔	✔		✔
Chemical Injection - Aspirator or Venturi		✔	✔		✔
Chemical Injector Pump					✔
Able to test annually		✔	✔		✔

Note: Backflow preventers must be A.S.S.E. approved.

Appendix Y

Article on Reclaimed Water (Re-use)

Submitted to:

Florida Water Resources Journal

RECLAIMED WATER:

HOW DOES IT AFFECT YOUR BACKFLOW PREVENTION PROGRAM?

Submitted by:

Robin L. Ritland and Les O'Brien

May 30, 1991

About the Authors:

Robin L. Ritland is a training specialist at the Center for Training, Research and Education for Environmental Occupations (TREEO) and the author of *Backflow Prevention: Theory & Practice.*

Les O'Brien a senior training specialist at Center for Training, Research and Education for Environmental Occupations (TREEO), has been instructing courses in backflow prevention at for since 1983. He is the past president of the American Backflow Prevention Association (ABPA).

RECLAIMED WATER:
HOW DOES IT AFFECT YOUR BACKFLOW PREVENTION PROGRAM?

Water is one of those commodities we just can't do without. Unfortunately, most of us tend to take our water for granted—at least until recently. Water restrictions created by the recent drought and saltwater intrusion, common to coastal areas of the state, have forced us to look at how we are using our drinking water. That is why, in principle, reusing treated wastewater is a great idea. Nonetheless, water reuse will have a significant impact on your backflow prevention program when public health concerns are considered.

Potential for Backflows

The greatest threat posed by reuse of wastewater is the potential for cross-connections between reuse and potable water systems. Of course, there are rules that prohibit cross-connections: rules 17-610 and 17-555 Florida Administrative Codes (F.A.C.), and all the national plumbing codes prohibit cross-connections in plumbing systems. However, history shows that creating an environmental law or code does not necessarily ensure safety or compliance. This spring, there was a cross-connection between the reclaimed water system and a family's potable water plumbing in St. Petersburg; fortunately, no one became ill.[1] In Arizona, a cross-connection between the potable water system and a reclaimed water system at a private campground resulted in diarrheal illness in approximately 1,850 persons.[2] The United States Environmental Protection Agency reports that backflows were the cause of over 60,000 reported cases of illness from 1920-1980.[3] The potential for hazardous cross-connections can be expected to increase where reuse and potable systems are close together.

Backflow Prevention Programs

The best defense against liability from hazardous backflows is a good backflow prevention program. In fact, an acceptable program, prosecuted diligently and effectively, is the only legal defense according to a court precedent set in California.[4] Since 1976, the Florida Department of Environmental Regulation (FDER) has required community water systems to have a cross-connection control program although the agency has admitted enforcement is not always consistently diligent. Rule 17-610, F.A.C., requires applicants for reuse systems have their cross-connection control and inspection program approved by FDER. However, FDER's Drinking Water Section has no formal approval procedure. FDER indicated that the word 'approved' will probably be changed to 'acceptable' in future rule revisions. Meanwhile, FDER's Wastewater Section has provided assurances that permits are not issued for reclaimed systems until they are confident that the area served has an acceptable cross-connection control and inspection program. Program evaluation guidance is being developed and should be available to the districts soon.

Potential conflict or confusion may also occur between the permittee, usually the wastewater facility, and the distribution section which routinely manages the backflow prevention program. These conflicts can be magnified when these two systems have different owners.

TREEO and the Florida Section of the American Water Works Association (FS/AWWA) conducted a backflow prevention survey in 1989 of utilities that serve 5,000 connections or more. Forty-three percent of the utilities responded. The majority of respondents (82%) described their utility as having an adequate backflow prevention program. However, 10% reported that no one is testing backflow preventers in their area, 32% indicated they did not have any type of surveying and retrofitting program, and 39% indicated they do not conduct on-site inspections prior to providing water service.[5]

Preventing Cross-connections

What can a utility do to help prevent backflows from reclaimed water systems? Prevent cross-connections. According to rule 17-555.360(5)(b), F.A.C., cross-connection control programs with reuse systems "shall *consider* the following: (a) Enhanced public education efforts toward prevention of cross-connections. (b) Enhanced inspection programs, for portions of the distribution system in areas of reuse, for detection and elimination of cross connections." These should not be considerations; they *are* essential components of any "adequate" backflow prevention program. Surveys should not be restricted to areas of reuse; routine surveys are needed to protect the potable water distribution system and the customers served by it. In areas served by reuse water, make sure that the re-use water piping, valves and outlets are color coded, tagged, labeled or otherwise marked to differentiate them from other piping systems. This is required under rule 17-610.470, F.A.C. A variety of labeling/coding methods are being used: potable water PVC pipe marked "RE-USE WATER"; potable water PVC pipe marked with lavender marking tape, labeled "CAUTION-RECLAIM"; lavender PVC pipe, buried with yellow magnetic tape; brown PVC pipe. This variety demonstrates the potential for confusion between one system and the next. This is especially a problem in areas served by more than one utility.

Public education is critical because the public creates most cross-connections. Utilize bill stuffers. Speak at womens' clubs, business luncheons, and schools, or hold a seminar for the public. When the general public is being "sold" the merits of reclaimed water, make sure they also learn about the potential hazards of cross-connections. Currently, public education efforts vary significantly between utilities. Some reuse systems simply require customers to sign an agreement outlining their requirements and responsibilities. Another utility provides speakers to community service groups and schools; they also handout brochures and information packets to new reclaimed-water customers. One utility *requires* reclaimed water customers to attend a seminar that includes a tour of their virus lab. However, it is unclear how much information related to backflow prevention is provided as part of the public education process.

Backflow Preventers

A few cross-connections will occur no matter how diligently we try to prevent them and, backflow preventers remain the best defense against backflow. FDER regulations currently allow air gaps; reduced pressure principle assemblies (RPs); double check valve assemblies (DCVA); pressure vacuum breakers (PVB); atmospheric vacuum breakers (AVB); and residential dual check valves. Dual checks, however, are *only* acceptable if the service has no other hazards requiring greater protection.

AWWA's M14 manual is the best source of guidance for matching the backflow preventer to the application or the hazard conditions at the site; rule 17-555.360(2) F.A.C., states programs should be developed using practices outlined in M14. The second edition of M14 recommends reduced pressure backflow preventer for premises served by reclaimed water systems. AWWA does not recognize the dual check as a backflow preventer.[2] This recommendation appears to be in conflict with Florida regulations that allow a residential dual check. However, state regulations do allow local codes, ordinances, or regulations to require a higher level of protection.

Hazard Level of Reclaimed Water

Just what is the hazard level of reclaimed water? Frequently we hear that reclaimed water is "just about as pure as drinking water". There are different treatment levels for reclaimed water depending upon the intended use. Table 1 outlines the general treatment requirements based on the intended use for reclaimed-water most liable to be cross-connected to potable water systems. Reclaimed water systems supplying residential lawn irrigation systems, edible crops, toilets (in commercial or industrial facilities), fire protection systems, etc., are probably most susceptible to cross-connections. For this reason, this type of reclaimed water receives a fairly high degree of treatment: secondary treatment, high-level disinfection, and filtration. High level disinfection means that 75% of the daily monitoring samples are required to be free of fecal coliforms; the remaining 25% are required to contain no more than 25 fecal coliforms per 100 ml sample.

By comparison, community drinking water must have no more than one positive total coliform sample per month if the system serves ≤33,000 clients, and less than 5% positive total coliform samples per month if the system serves >33,000 clients; detection of coliforms beyond these limits poses a *"non-acute health risk"* and repeat samples must be taken. Any fecal coliform-positive, *E. coli*-positive, or total coliform-positive repeat sample is considered a violation that poses an *"acute risk to health"*.

Clearly, even reclaimed water that receives high level disinfection could often pose an acute health risk if drinking water standards were applied. This doesn't even take into consideration the long list of chemical contaminants tested for in potable water systems. It should be noted that several cities are conducting additional tests on their reclaimed water. Items being monitored include: heavy metals, nitrogen, pathogens, viruses, priority pollutants, phosphous, and volatile organics. An analysis of these test results would help in clarifying the risks associated with reclaimed-water. This analysis could also provide valuable information about the level of substances commonly found in reuse water. The potable water distribution system could then be routinely monitered for substances, e.g., phosphorus, that `indicate' a cross-connection between the reuse and potable water systems.

In a potable water distribution system, monitoring is done at representative points. However, for reuse water, the regulations require monitoring samples be taken prior to discharge to holding ponds or the reuse system. Since the regulations do not require maintenance of a chlorine residual within the reclaimed water distribution system, regrowth of microorganisms seems certain. In addition, because reclaimed water is often stored in golf course lakes, lagoons, ponds, or uncovered tanks, the potential for post-treatment contamination exists. The amount of disinfectant maintained in potable water distribution systems is provided mainly to prevent regrowth and is frequently not sufficient to effectively deal with contaminants added through cross-connections. Those systems utilizing chloramines face an even bigger problem, since at least one study has shown that chloramines are relatively inefficient at protecting against contaminates introduced after the treatment process and chloramines fail to act as an indicator of contamination.[6] A "total chlorine residual is present even after the addition of sizeable amounts of contaminatory materials", thus "the detection of a combined chlorine residual does not assure water potability".[6]

Is reclaimed water a health hazard or a non-health hazard? AWWA feels it is a health hazard and recommends the use of a reduced pressure principle assembly.[2] California has also taken the position that it presents a health hazard; the wastewater treatment level for irrigation of public access areas in California is similar to Florida (Table 1).[7 & 8]

Dual Check vs. Double Check

What is the difference between a dual check valve and a double check valve assembly? Double check valve assemblies that are "approved" by the Foundation for Cross-Connection Control and Hydraulic Research at the University of Southern California (the Foundation) must be testable and repairable in-line. Approved double check assemblies also must have resilient-seated, full-flow characteristic shut-off valves on each end of the assembly, as well as four test cocks. Dual checks, on the other hand, are not normally supplied with valves or test cocks. Most are repairable, but many must be removed from service in order to be repaired or tested. For approval, double check valve assemblies must meet rigorous specifications and pass extensive tests including a one-year field evaluation. By contrast, dual check valves are not field tested and are not subjected to the same rigorous testing required of approved backflow preventers.

Proponents of reuse systems frequently point out that there is a big difference in the initial cost of dual checks and double checks. A basic three-quarter inch dual check usually costs approximately $15.00, while an equivalent approved double check valve assembly costs around $50.00. Some manufacturers market dual checks that can be tested and serviced in-line. However, the costs increase significantly when test cocks and shut-off valves are added to these devices. Some resetters are designed to make servicing dual checks more convenient, but again, this increases the cost of the installation. A resetter and a dual check valve can cost as much as $40.00.

Maintenance and repair costs must also be taken into consideration. Approved backflow prevention assemblies should be tested at least annually, as outlined by AWWA M14, the Foundation, and all manufacturers' literature. It seems logical that dual checks should receive more frequent testing, since they do not have to meet the same rigorous approval requirements. In a TREEO telephone survey, several utilities were asked how often they test or repair dual checks. A number of them indicated that there were no plans to test or replace these valves. Only one utility indicated they plan to test them annually. Most utilities responded that they plan to replace the valve when water meters are changed. Their regular water meter change-out programs will result in dual check valve change frequencies ranging from 4 to 15 years. The annual failure rates of approved assemblies varies from 10% to 40% and manufacturers of these valves recommend testing at least every two years. Based on the failure rates of approved assemblies, it should be assumed that both check valves in most dual check valves will fail sometime before 5 years.

Of course, the big problem utilities face is how to possibly test and service all these valves annually? How many extra employees will they have to hire? Or how much will they have to pay a private contractor? Another viable option is to make the customer responsible for hiring a properly trained individual to service the valve.

Just installing these devices and never testing or replacing them gives the public and the purveyor a false sense of security. Under these circumstances, it is questionable whether dual checks really are "better than nothing", as is commonly stated.

Discussion

The potential for cross-connections and backflows will increase as reclaimed water lines are installed near potable water lines. The best defense against backflows is a well developed backflow prevention program. Preventing cross-connections, via plan and site review of new construction, and surveying and retrofitting of existing facilities, should be a major focus of that program. Selection of the appropriate backflow preventer should also be made with care; the decision made should be defendable in court. This is also true for the maintenance and repair program. While reclaimed water does receive a fairly high degree of treatment, it is important to bear in mind, post-treatment contamination and regrowth can affect the quality of the water once it leaves the wastewater treatment plant.

Recommendations

* Cross-connection control supervisors need to work closely with managers of reuse systems.

* It is essential that both the reuse system manager and the public be properly educated not only about the benefits of reclaimed water systems, but also the hazards of cross-connections.

* A consistent method should be adopted for color-coding and labeling reclaimed water piping and appurtenances.

* The potable water distribution system should be routinely monitered for substances that can `indicate' a cross-connection between the reuse and potable water systems.

* The costs of dual checks and double checks are really not that different, when maintenance and repair considerations are taken into account. The basic decision is whether the method of backflow prevention is "window dressing" or real.

Reclaiming wastewater is a great idea; however, there are risks associated with its use. These risks increase significantly if we fail to recognize and acknowledge them. Wastewater reclamation programs should "assure health protection without unnecessarily discouraging wastewater reclamation".[8] Definitely this valuable resource should be utilized, however, the emphasis should be placed on doing so in a safe manner.

References

1. Mix-up had family drink wastewater. (1991, March 30) *Gainesville Sun.*

2. Starko, K. M., E. C. Lippy, L. B. Dominguez, C. E. Haley, and H. J. Fisher. (1986) Campers' diarrhea outbreak traced to water-sewage link. *Public Health Reports* 101 (5) p. 527.

3. American Water Works Association. (1990) *Recommended Practice for Backflow Prevention and Cross-Connection Control AWWA M14*, (2nd ed.). Denver, CO. Author.

4. American Water Works Association. (1974) *Cross-Connection and Backflow Prevention*, (2nd ed.). New York, NY. Author. P.50.

5. TREEO-FS/AWWA Cross-Connection Control Committee. (1991, February 15) *Backflow Prevention Program Survey Results* (draft). (Available from TREEO, 3900 SW 63rd blvd, Gainesville, FL 32608-3848) R. L. Ritland.

6. Snead, M. C., V. P. Olivieri, K. Kawate, and C. W. Kruse. (1980) The effectiveness of chlorine residuals in inactivation of bacteria and viruses introduced by post-treatment contamination. *Water Research* 14, 403-408.

7. *Guidance Manual for Cross Connection Control Programs.* (1988) State of California Health and Welfare Agency, Department of Health Services, Public Water Supply Branch. p. V-16.

8. Crook, J., (1990, July) Water reuse in California. *Journal AWWA.*

RECLAIMED WATER USE AND TREATMENT

TABLE 1

TYPE OF USE	COLIFORM LIMITS	TREATMENT LEVEL
FLORIDA		
SLOW-RATE LAND APPLICATION SYSTEMS; PUBLIC ACCESS AREAS, RESIDENTIAL IRRIGATION, EDIBLE CROPS ALSO: golf courses, cemeteries, parks, fire protection, toilet flushing	75 % of the samples: no fecal coliforms 25 % of the samples: ≤ 25 fecal coliforms per 100 ml., for any one sample	Secondary Treatment High Level Disinfection Filtration ≤ 5 mg/L TSS Industrial Pretreatment
SLOW-RATE LAND APPLICATION SYSTEMS; RESTRICTED PUBLIC ACCESS Uses: Applied to vegetative land surfaces, cattle grazing areas	≤ 200 fecal coliforms per 100 ml. ≤ 10% of the samples can exceed 400 fecal coliforms No one sample can exceed 800 fecal coliforms	Secondary Treatment Basic Disinfection ≤ 10 mg/L TSS
EFFLUENT DISPOSAL; OVERLAND FLOW SYSTEMS: Uses: Applied to upper reaches of terrace sloped vegetated surfaces e.g., sod farms, forests, fodder crops, pasture lands	≤ 2400 fecal coliforms per 100 ml.	< Secondary Treatment: Low-Level Disinfection BOD: 40-60 mg/L TSS: 40-60 mg/L
CALIFORNIA		
Spray irrigation of food crops Landscape irrigation (parks, playgrounds) Non-restricted recreational impoundments	≤ 2.2 total coliforms per 100 ml. ≤ 23 total coliforms per 100 ml.	Tertiary treatment: oxidation, coagulation, clarification, filtration, & disinfection ≤ 2 ntu average 5% samples: ≤ 5 ntu/24 hours

Appendix Z

Enclosures for Freeze Protection

Enclosures for Freeze Protection

DUNCO Manufacturing, Inc.
PO Box 156
Laurel Hill, FL 32567
850/652-4705

G&C Enclosures, Inc.
60 Athens Drive
Mt. Juliet, TN 37122
888/753-6565

Hot Box - NFE, Inc.
250 N Lane Avenue
Jacksonville, FL 32254
800/736-0238

Hydrocowl
3269 Ezell Pike
Nashville, TN 37211
800/245-6333

Seafarer Boat Works
Rt 6 Box 8930
Crawfordville, FL 32327
850/926-7445

Systems Manufacturing, Inc.
PO Box 688
Laurel Hill, FL 32567
850/652-4705

Windbreaker - Birchfield Enterprises Inc.
4312 Wallace Neel Road
Charlotte, NC 28208
704/399-4450

Note: There may be many more enclosure manufacturers that are not listed on this page.

CPSIA information can be obtained
at www.ICGtesting.com
Printed in the USA
BVHW010454131219
566395BV00005BA/23/P